Peter Seyfried · Horst Kunzmann
Pat McKeown · Manfred Weck (Eds.)

Progress in Precision Engineering

Proceedings of the 6th International Precision Engineering Seminar (IPES 6)

2nd International Conference on Ultraprecision in Manufacturing Engineering (UME 2)

May, 1991 Braunschweig, Germany

Springer-Verlag
Berlin Heidelberg New York
London Paris Tokyo
Hong Kong Barcelona Budapest

Prof. Peter Seyfried
Prof. Dr. Horst Kunzmann
Physikalisch-Technische Bundesanstalt
Bundesallee 100
D-W 3000 Braunschweig, FR Germany

Prof. Dr. Pat McKeown
Cranfield Precision Engineering Ltd
Building 90, Wharley End
Cranfield, Bedford MK 43 OAL, England

Prof. Dr. Manfred Weck
Forschungsgemeinschaft Ultrapräzisionstechnik eV
Steinbachstraße 17
D-W 5100 Aachen, FR Germany

ISBN-13: 978-3-642-84496-6 e-ISBN-13: 978-3-642-84494-2
DOI: 10.1007/978-3-642-84494-2

Softcover reprint of the hardcover 1st edition

Typesetting: Camera ready by authors
Offsetprinting: Color-Druck Dorfi GmbH, Berlin; Bookbinding: Lüderitz & Bauer, Berlin
61/3020-543210 Printed on acid-free paper.

Foreword

by
Professor Pat McKeown
Cranfield Precision Engineering, UK
Member of Joint Organising Committee
IPES6/UME2

PROGRESS IN PRECISION ENGINEERING

Metal working companies in tool making, prototype manufacture and subcontract machining often use the label "precision engineering" to indicate that they are accustomed to working to finer tolerances than is normally expected in series production. But what we are concerned with in this and our preceding international conferences is much wider and deeper than this. Precision engineering is a grouping of multidisciplinary scientific and engineering skills and techniques, firmly based on dimensional metrology, by which a wide range of new advanced technology products is made possible. In the last 5 - 10 years we have witnessed dramatic progress in precision engineering, particularly by the rapid development of its important sub-sets, micro-engineering and nanotechnology.

It is a particular pleasure for me and my colleagues on the Organising Committee to welcome you to Braunschweig on the occasion of this the first joint international meeting in high precision manufacturing/precision engineering to be held in Germany. Our aim is to bring together the world's leading precision engineering practitioners from areas of application as diverse as optics for astronomy, micro and nano machining process research, design and development of ultra precision machine tools and metrology equipment, advanced materials, bio-medical research and new sensor/transducer systems.

We are fortunate in having an exciting, comprehensive programm which will prove beyond doubt that real progress is being made in precision engineering, the boundaries of which are being extended rapidly both in accuracy capability and in application.

We start with papers on high precision sensor systems and displacement metrology including X-ray interferometry with capability of resolution in the picometre range, together with developments in interferometry and grating interferometry etc., for ultra precision measurement of displacement and motion control of machines and instruments.

A subject which is the cornerstone of precision engineering is the design, development, calibration and performance testing of high precision machines and control systems. This topic is well covered in no less than five sessions.

Nanotechnology instruments are described for characterising surface/subsurface properties of materials. They include new electron tunnelling sensors and the exciting photon scanning tunnelling microscope. An important new development is the application of such instruments and related nanotechnology processes to medicine and the bio-sciences. It has been said that whole new industries will grow over the next 10 - 20 years from the exciting nanotechnology research programmes now established in several countries. At the leading edge of these new industries I am sure there will be revolutionary advances in health care and the bio-sciences. That is why this important new area is included.

Again, welcome to Braunschweig where we will all see continuing progress, internationally, in precision engineering.

Organizing Committee:
Prof. Dr. Horst Kunzmann and Prof. Peter Seyfried,
Physikalisch-Technische Bundesanstalt, Braunschweig, FRG;
Peter Waterhouse and Marija Vukovojac,
Butterworth-Heinemann Ltd, Guildford, UK;
Prof. Dr. Manfred Weck, Torsten Bispink and Ulrich Böttger,
Forschungsgemeinschaft Ultrapräzisionstechnik e.V., Aachen, FRG;
Prof. Dr. Pat McKeown and Prof. Dr. Robert Spragg,
Cranfield Precision Engineering, Cranfield Institute of Technology, UK

Table of Contents

5 TECHNIQUES AND EQUIPMENT FOR MACHINE PERFORMANCE TESTING

6 HIGH-PRECISION MACHINING PROCESSES AND MATERIALS

7 MACHINE AND PROCESS CONTROL

8 NON-CONVENTIONAL PROCESSES FOR INDUSTRIAL APPLICATIONS

Investigation and Compensation of the Non-linearity of Heterodyne Interferometers

Hou, Wenmei*; Wilkening, Günter
Physikalisch-Technische Bundesanstalt, Braunschweig, F.R.G.
*supported by the Alexander von Humboldt-Stiftung, Bonn, F.R.G.;
permanent adress: Beijing Institute of Machine Building Industry, Beijing, P.R. China

Summary

The non-linearity of heterodyne interferometers is mainly influenced by the non-orthogonality and ellipticity of the linearly polarized partial beams of the laser and by the limited extinction capability of the polarizing beam splitters. The phase errors at the interferometer output can be detected by measuring the phase difference between the two orthogonal directions in the output beam. This arrangement can be used to check an interferometer set-up for causes of non-linearity and to eliminate them by adjustment, replacement etc. When the phases of the two orthogonal output signals are measured separately and compared with the reference signal, and when the arithmetical mean value is formed, except for a small residual error the phase value obtained is free from non-linearities.

1. Heterodyne interferometer

The laser interferometer has become firmly established in length measuring technique. As the wavelength of the light is used as a measure, a relatively simple, direct link-up with the definition of the length is possible. In the acceptance and calibration of machine tools and coordinate measuring machines, laser interferometers are used as a universal means for length, angle and planeness measurements. Laser interferometers are used as measuring systems integrated into highly precise machines for microfinishing and for the manufacture of integrated circuits. The manufacture of ever smaller structures implies that the uncertainty requirements to be met by length measurements in the submicrometer range become increasingly higher. The light wavelength by no means suffices as an incremental unit, and subdivisions down to the nanometer region are therefore necessary. The two-frequency interferometer has special advantages for this purpose, as it allows a very stable subdivision of the light wavelength with high resolution on the basis of a simple phase measurement.

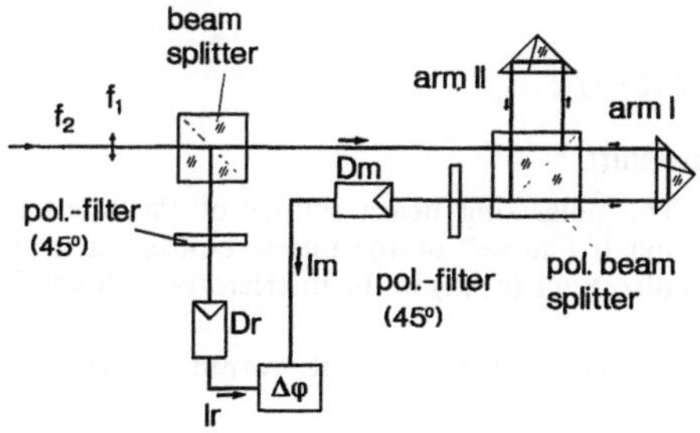

Fig. 1a Heterodyne interferometer; principle

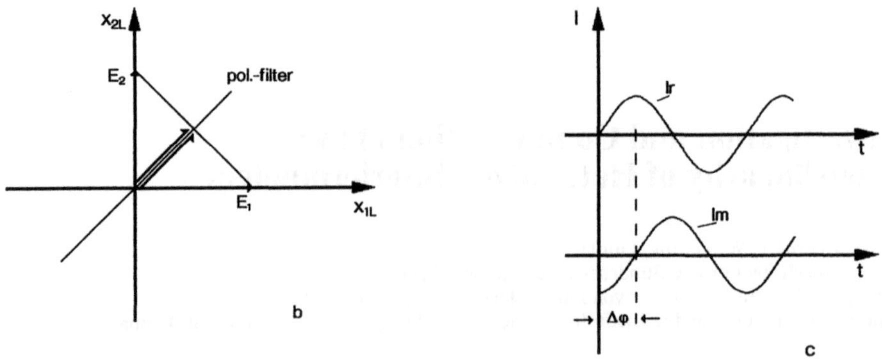

Fig. 1b, c Heterodyne interferometer

b.Components of the orthogonally linear polarized beam at the photodiode

c.Phase difference $\Delta\varphi$ between measurement and reference signal

The basic set-up of a two-frequency interferometer is shown in Fig.1. In a beam of Gaussian profile, the laser (so far, He-Ne lasers are exclusively used) emits two partial beams of different frequencies (f_1, f_2) polarized orthogonally to each other (Fig.1b). These two partial beams are described as plane waves[1]:

$$\begin{aligned} E_1 &= E_o \sin(2\pi f_1 t + \varphi_{r1}) \\ E_2 &= E_o \sin(2\pi f_2 t + \varphi_{r2}) \end{aligned} \tag{1}$$

φ_{r1}, φ_{r2} are arbitrary phases.

One part of the laser beam is split off in front of the interferometer, passes through a polarizing filter (at 45° to the polarizing directions of partial beams x_{1L}, x_{2L}), and falls on diode D_r. The partial beams are superimposed and generate a photoelectric current I_r in the diode:

$$\begin{aligned} I_r' &\sim E_r^2 \\ &\sim \tfrac{1}{2} E_o^2 \, [\sin(2\pi f_1 t + \varphi_{r1}) + \sin(2\pi f_2 t + \varphi_{r2})]^2 \\ &\sim \tfrac{1}{2} E_o^2 \, \{\sin^2(2\pi f_1 t + \varphi_{r1}) + \sin^2(2\pi f_2 t + \varphi_{r2}) \\ &\qquad + \cos[2\pi(f_1 - f_2)t + (\varphi_{r1} - \varphi_{r2})] - \cos[2\pi(f_1 + f_2)t + (\varphi_{r1} + \varphi_{r2})]\} \\ &= I(f_1) + I(f_2) + I(f_1 - f_2) + I(f_1 + f_2) \end{aligned} \tag{2}$$

The photodiodes can transform only the difference term $I(f_1-f_2)$ into an a.c. current signal. The other terms lead to d.c. current contributions without information and are suppressed along the signal path.

$$\begin{aligned} I_r &= I(f_1 - f_2) \\ &\sim I_o \cos[2\pi(f_1 - f_2)t + \varphi_o] \end{aligned} \tag{3}$$

Here φ_0 is an arbitrary, constant phase shift.

This signal serves as a reference. The polarizing beam splitter of the interferometer splits the incident beam up into directions I and II ("arms" of the interferometer), in compliance with the polarization. It fixes the polarization directions (x_1, x_2) in the interferometer. It will be assumed in the

[1]For the sake of clarity the amplitude unit Eo is assumed to be the same for the two partial beams.

following that these directions are orthogonal. They are used as a reference. If the directions of the laser system x_{1L}, x_{2L} agree with those of the interferometer (given by the polarizing splitter), the frequencies are completely separated. The partial beams cover paths of different length l_1 and l_2, are recombined in the beam splitter and detected by diode D_m. The photocurrent is obtained:

$$\begin{aligned} I_m' &\sim E_m^2 \\ &\sim \tfrac{1}{2}E_0^2\,[\,\sin(2\pi f_1 t+\varphi_{01}+\varphi_1)+\sin(2\pi f_2 t+\varphi_{02}+\varphi_2)]^2 \\ &= I(f_1)+I(f_2)+I(f_1-f_2)+I(f_1+f_2) \end{aligned} \tag{4}$$

$$\varphi_1 = \frac{4\pi n l_1}{\lambda_1} \quad ; \quad \varphi_2 = \frac{4\pi n l_2}{\lambda_2}$$

where n is the refractive index; λ_1 and λ_2 are the wavelengths belonging to f_1 and f_2 ; φ_{01} and φ_{02} are arbitrary initial phases.

Here, too, the difference term is retained:

$$\begin{aligned} I_m &= I(f_1-f_2) \\ &\sim I_0\cos[\,2\pi(f_1-f_2)t + \varphi_{0m} + \Delta\varphi\,] \\ \Delta\varphi &= \varphi_1 - \varphi_2 = 4\pi(nl_1 - nl_2)\,/\lambda_m \end{aligned} \tag{5}$$

λ_m is the mean wavelength

Compared with the (constant) signal I_r , I_m is affected by a phase shift $\Delta\varphi$, which varies with the path difference in the interferometer and which can be obtained by measuring the phase difference between I_r and I_m (cf. Fig.1c). (If, for example, the resolution of the phase measurement is 1°, in the interferometer shown in Fig.1 a length resolution of 0.9 nm can be attained for a displacement of the reflector.)

Several solutions for the phase measurement have been suggested and put into practice /1,2,3,4/ some of which allow resolutions down to the range of 0.1 nm /3,4/.

2. Non-linearities in the heterodyne interferometer

Eq.(5) is exactly valid when only one frequency occurs in each interferometer arm. This ideal condition is frequently not met in practice. Due to various influences, mixed frequencies are found in both interferometer arms. This results in a non-linear relation between the measured phase difference and the respective displacement /2,4,5,6,7,8,/. Four of the most frequent causes are investigated in the following:

- mixing due to non-orthogonality of the polarization directions of the incident radiation;
- mixing due to elliptic polarization of the incident radiation;
- mixing due to incomplete separation of the frequency components in a non-ideal polarizing splitter;
- amplitude variations in one interferometer arm.

Imperfect or incorrectly adjusted polarizing optics behind the beam splitter (e.g. retardation plates) or optics which show polarizing behaviour (e.g. reflektors) do not cause first order non-linearity.

For the discussion below it is assumed that the polarization axes in the interferometer are orthogonal to each other.

2.1 Mixing due to non-orthogonality of the polarization directions of the incident radiation

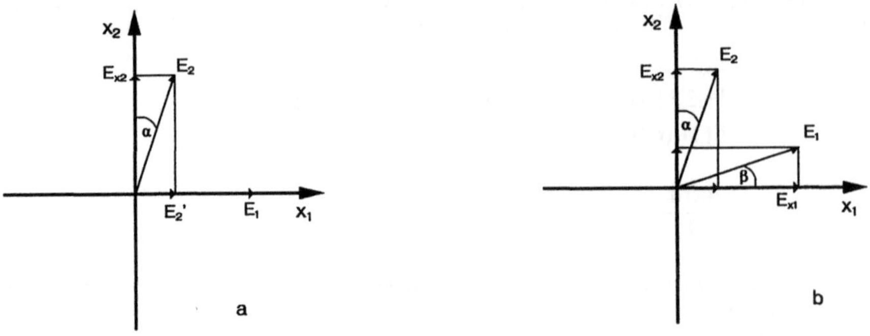

Fig. 2 Beam components in the orthogonal reference system of the interferometer (x_1, x_2).
a. Deviation from orthogonality in one direction
b. Deviation from orthogonality in both directions

The deviation of the two polarizations from orthogonality be α (cf. Fig.2a). As a result, in addition to frequency f_1, a part of frequency f_2 is also found in arm I of the interferometer. The following relationships hold for the partial beams in the two interferometer arms (i.e. for the two polarization directions x_1, x_2):

$$E_{x1} = E_0 \sin(2\pi f_1 t + \varphi_{01} + \varphi_1) + E_0 \sin\alpha \sin(2\pi f_2 t + \varphi_{02} + \varphi_1)$$
$$E_{x2} = E_0 \cos\alpha \sin(2\pi f_2 t + \varphi_{02} + \varphi_2) \qquad (6)$$

The following relation then results for the a.c. signal of diode D_m:

$$I_m \sim I_0 A^* \cos[\, 2\pi(f_1 - f_2)t + \varphi_{0m} + \Delta\varphi - \gamma] \qquad (7)$$

$$\gamma = tg^{-1} \frac{\sin\alpha \sin\Delta\varphi}{\cos\alpha + \sin\alpha \cos\Delta\varphi} \qquad (8)$$

$$A^* = \sqrt{1 + \sin 2\alpha \cos\Delta\varphi} \qquad (9)$$

To simplify matters, the following is assumed:

$\alpha \ll 1$ $(\alpha < 5°)$

The following is then valid in first approximation:

$$\gamma \approx \alpha \sin\Delta\varphi \qquad (10)$$

A comparison between eq.(5) and eq.(7) shows that the phase measurement is affected by a systematic error. γ is a periodic error which varies with Δφ. The maximum error γ_{max} is about 2 α. (If α is 5°, for example, the maximum error of phase measurement will be 10°. This corresponds to an error of 9 nm in the length measurement.)

It should also be noted that the amplitude of the measured signal I_m is no longer constant (cf. eq.(7)) but represents a periodic signal modulated by α and $\Delta\varphi$. When there is frequency mixing in both arms (cf. Fig.2b) the phase errors behave additively. Therefore, the simple rotation of the polarization directions of the laser beam relativ to those of the interferometer (i.e. $\alpha = -\beta$) theoretically do not cause a first order phase error (cf. /8/).

2.2 Mixing due to elliptic polarization of the incident radiation

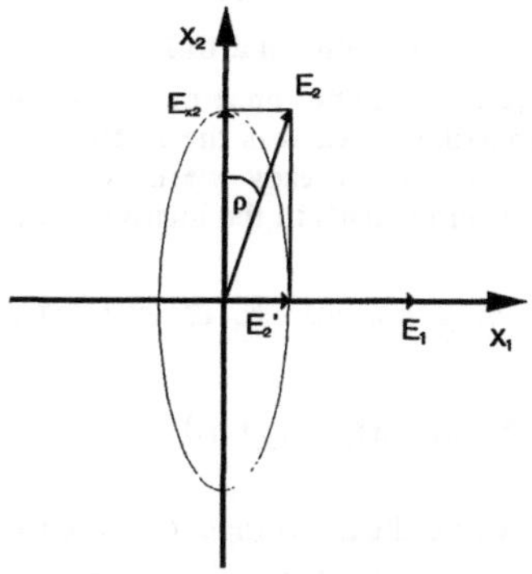

Fig. 3 Beam components in the orthogonal reference system of the interferometer (x_1, x_2); ellipticity in one direction

Frequency mixing effects also occur when the incident radiation is not ideally linear but elliptically polarized.

On the assumption that component E_2 is elliptically polarized and the ratio of the ellipse axes described by the angle ρ (cf. Fig.3), the following relations are valid for the partial beams in both interferometer arms:

$$E_{x1} = E_0 \sin(2\pi f_1 t+\varphi_{01}+\varphi_1) + E_0 \sin\rho \cos(2\pi f_2 t+\varphi_{02}+\varphi_1)$$
$$E_{x2} = E_0 \cos\rho \sin(2\pi f_2 t+\varphi_{02}+\varphi_2) \qquad (11)$$

The a.c. component of the signal formed in photodiode D_m then is as follows:

$$I_m \sim I_0 U^* \cos[2\pi(f_1-f_2)t+(\varphi_{01}-\varphi_{02})+(\varphi_1-\varphi_2)-\eta] \qquad (12)$$

$$\eta = tg^{-1} \frac{\sin\rho \cos\Delta\varphi}{\cos\rho-\sin\rho \sin\Delta\varphi} \qquad (13)$$

$$U^* = \sqrt{1-\sin2\rho \sin\Delta\varphi} \qquad (14)$$

η and U^* correspond to γ and A^* for $\rho = \alpha$ in eq.(7). In case both partial beams are elliptically polarized but orthogonal (i.e. opposite ellipticity and equal amplitudes), no first order phase error occurs (c.f. /8/).

2.3 Mixing due to incomplete separation of the frequency components in a non-ideal polarizing splitter

On the assumption that the fraction a of component E_2 reaches arm I, the following is valid for the splitting of the components:

$$\text{Arm I:}\quad E_{x1} + E_2' = E_0 \sin(2\pi f_1 t + \varphi_{01} + \varphi_1) + E_0\, a \sin(2\pi f_2 t + \varphi_{02} + \varphi_1)$$

$$\text{Arm II:}\quad E_{x2} = E_0\,(1-a) \sin(2\pi f_2 t + \varphi_{02} + \varphi_2) \tag{15}$$

In this case, both states of polarization are found in arm I.

After being reflected the beams pass the splitter once more and only the fraction a of the "wrong" component finally reaches the photodiode, whereas the fraction (1-a) leaves the splitter at the side. Similarly, only the fraction (1-a) of the already weakened component E_2 finally reaches the photodiode. Therefore the effective components in the interferometerarms are described as follows:

$$\text{Arm I}^*:\quad E_{x1} + a\,E_2' = E_0 \sin(2\pi f_1 t + \varphi_{01} + \varphi_1) + E_0\, a^2 \sin(2\pi f_2 t + \varphi_{02} + \varphi_1)$$

$$\text{Arm II}^*:\quad E_{x2} = E_0\,(1-a)^2 \sin(2\pi f_2 t + \varphi_{02} + \varphi_2) \tag{16}$$

The following expression is obtained for the a.c. component produced in diode D_m:

$$I_m \sim I_0\, K^* \cos[2\pi(f_1 - f_2)t + \varphi_{0m} + \Delta\varphi - \zeta] \tag{17}$$

$$\zeta = tg^{-1} \frac{a^2 \sin\Delta\varphi}{(1-a)^2 + a^2 \cos\Delta\varphi} \tag{18}$$

$$K^* = \sqrt{1 + 2a^2\,(1-a)^2 \cos\Delta\varphi} \tag{19}$$

a^2 corresponds to the intensity extinction ratio κ of polarizing beam splitters.

For $\kappa \ll 1$ eq.(19) can be simplified:

$$\zeta \approx K \sin\Delta\varphi \tag{20}$$

(Polarizing beam splitters usually have extinction ratios of › 300:1 ($\kappa < 3\text{x}10^{-3}$) for the inferiour polarization direction, whereas the other one often is nearly perfect. The resulting phase error leads to an error in the length measurement of 0.17 nm.)

2.4 Influence of amplitude variations

The relative amplitude variation in an interferometer arm due to the displacement of the measuring reflector may be another important cause of phase errors when there is already incorrect frequency mixing (c.f. /8/).

In the following it is assumed that in addition to an existing orthogonality error α the amplitude in one interferometer arm is attenuated. This attenuation is described by the amplitude factor k:

$$E_{x1} = E_0 \sin(2\pi f_1 t + \varphi_{01} + \varphi_1) + E_0 \sin\alpha \sin(2\pi f_2 t + \varphi_{02} + \varphi_1)$$

$$E_{x2} = k\, E_0 \cos\alpha \sin(2\pi f_2 t + \varphi_{02} + \varphi_2) \tag{21}$$

For the a.c. component of the signal formed in diode D_m, the following expression is obtained:

$$I_m \sim I_0 A^{**} \cos[2\pi(f_1 - f_2)t + \varphi_{0m} + \Delta\varphi - \gamma^*] \tag{22}$$

$$\gamma^* = tg^{-1} \frac{\sin\alpha \sin\Delta\varphi}{k \cos\alpha + \sin\alpha \cos\Delta\varphi} \tag{23}$$

$$A^{**} = \sqrt{k^2 \cos^2\alpha + \sin^2\alpha + k \sin 2\alpha \cos\Delta\varphi} \tag{24}$$

For the sake of simplicity, the following is assumed:

$$\alpha \ll 1 \; (\alpha < 5^\circ), \; 1 > k > 0.5$$

The following is then valid in first approximation:

$$\gamma^* \approx \frac{\alpha \sin\Delta\varphi}{k} \tag{25}$$

An unequal amplitude distribution in the arms of the interferometer increases the influence of non-orthogonality on non-linearity. The effect of an amplitude variation will be similar also for the other errors.

In case the amplitude attenuation varies with the displacement of the reflector (e.g. due to varying onerlapping of the measurement and reference beam) the non-linearity depends on the actual displacement.

3. Detection of phase errors

To investigate the non-linearity of interferometers, it is desirable to obtain signals which allow conclusions to be drawn concerning the prevailing phase errors.

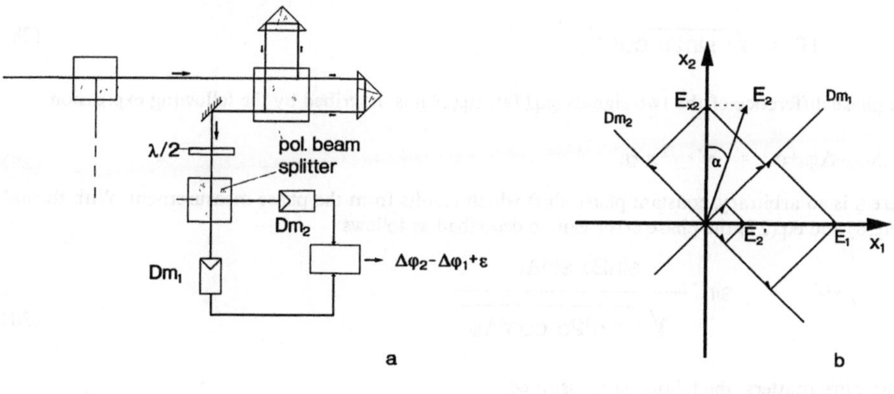

Fig. 4 Heterodyne interferometer with detection of non-linearity

a. Principle

b. Beam components at the photodiode (at the presence of non-orthogonality)

The heterodyne interferometer shown in Fig.4a has been extended to comprise two signal outputs: Superposition of the individual components from the two interferometer arms is not only detected in one direction (as is usually the case behind the polarizing filter), but also in the direction orthogonal to it. The conditions on the diodes are represented in Fig.4b for the case of non-orthogonality (cf. Fig.2a). This description also holds for the case where the incident beam is elliptic.

The necessary splitting was achieved by using a $\lambda/2$ plate (rotation of the polarization direction by 45°) in combination with another polarizing splitter. One of the two output signals (here assigned to diode D_{m1}) corresponds to that described by eq.(7):

$$I_{m1} \sim I_0 A^* \cos[2\pi(f_1-f_2)t+\underbrace{\varphi_{0m}+\Delta\varphi-\gamma}_{\Delta\varphi_1}] \tag{7a}$$

$$\gamma = tg^{-1} \frac{\sin\alpha \sin\Delta\varphi}{\cos\alpha+\sin\alpha \cos\Delta\varphi} \tag{8}$$

$$A^* = \sqrt{1+\sin2\alpha \cos\Delta\varphi} \tag{9}$$

The following relation is obtained for the other signal (I_{m2}):

$$I_{m2} \sim I_0 B^* \cos[2\pi(f_1-f_2)t+\underbrace{\varphi_{0m}+\Delta\varphi+\gamma'+\pi}_{\Delta\varphi_2}] \tag{26}$$

$$\gamma' = tg^{-1} \frac{\sin\alpha \sin\Delta\varphi}{\cos\alpha-\sin\alpha \cos\Delta\varphi} \tag{27}$$

$$B^* = \sqrt{1-\sin2\alpha \cos\Delta\varphi} \tag{28}$$

The phase difference of the two signals (eq.(7a), eq.(26)) is described by the following expression:

$$\Delta\varphi_2-\Delta\varphi_1+\varepsilon = \gamma'+\gamma+\pi+\varepsilon \tag{29}$$

where ε is an arbitrary, constant phase shift which results from the phase measurement. With the aid of eq.(8) and eq.(27) the phase error can be described as follows:

$$\gamma+\gamma' = \sin^{-1} \frac{\sin2\alpha \sin\Delta\varphi}{\sqrt{1-\sin^2 2\alpha \cos^2\Delta\varphi}} \tag{30}$$

To simplify matters, the following is assumed:

$$\alpha \ll 1 \ (\alpha < 5°)$$

The following is then valid in first approximation:

$$2\Delta = \gamma+\gamma' \approx 2\alpha \sin\Delta\varphi \tag{31}$$

Δ corresponds to the phase error due to non-orthogonality. The maximum error is approximately 2α. It is a direct measure of the interferometer's non-linearity.

The measurement of the phase difference of the two output signals may serve to detect the causes of non-linearity and eliminate them if possible. Phase errors due to incomplete separation of the polarization components cannot be detected with this arrangement. This contribution can be determined by separately measuring the extinction ratio and by using eq.(17). This amount is usually small. If the cause is known (non-orthogonality of the polarization components of the laser beam, ellipticity, etc.), quantitative statements about its parameters can also be made.

(The test of an interferometer with a two-mode laser, for example, furnished a maximum value of the phase difference (Δ_{max}) of about 15°. Additional investigations revealed elliptic polarization which was eliminated by using a $\lambda/4$ and a $\lambda/2$ plate. The adjustment could at once be checked on the basis of the result of the phase measurement. The remaining phase error was 0.5°, the corresponding non-linearity in the displacement measurement 0.4 nm.)

A means is thus available which allows essential components of the phase error - and consequently the non-linearity of an interferometer - to be measured without comparison with a reference interferometer, the particular advantage being that a phase measurement forms the basis and not amplitude measurements which are much more inaccurate.

4. Compensation of phase errors

Elimination of phase errors by means of adjustment is not always possible. Examples of this are the non-orthogonality of the polarization direction of laser radiation or the influence of a glass-fibre connection between laser and interferometer (mainly, a mixture of ellipticity and non-orthogonality). These errors can be compensated by using the measurement and evaluation method suggested in the following.

For the sake of simplicity the phase error due to non-orthogonality is chosen to be the only one. However, the method works equally in the case of ellipicity, amplitude variation, and mixtures of these errors. The phase error due to incomplete separation in the polarizing beam splitter is not compensated.

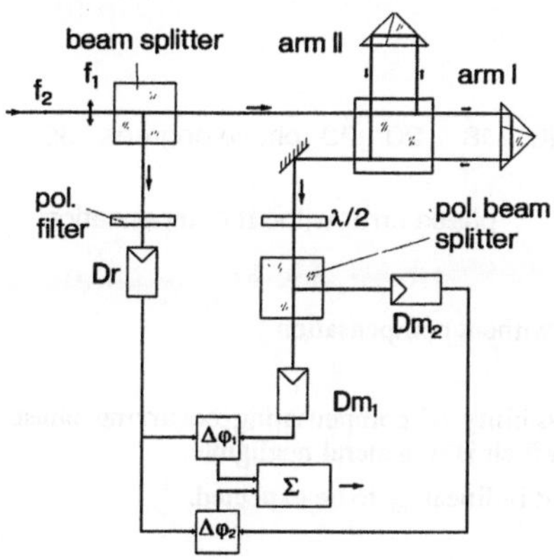

Fig. 5 Heterodyne interferometer with compensation of non-linearity

As shown in Fig.5, the differences $\Delta\varphi_1$ and $\Delta\varphi_2$ of the reference signal and the phases of the two output signals are measured. The arithmetic mean value of these two phase values is described by the following:

$$\Sigma = \tfrac{1}{2}(\Delta\varphi_1 + \Delta\varphi_2 + \vartheta) = \tfrac{1}{2}(\Delta\varphi - \gamma + \Delta\varphi + \gamma' + \vartheta) = \Delta\varphi + \tfrac{1}{2}(\Gamma + \vartheta) \quad (32)$$

ϑ is an arbitrary, constant phase shift which results from the phase measurement.

Γ results from eqs.(8) and (27):

$$\Gamma = \gamma' - \gamma = \sin^{-1} \frac{\sin^2\alpha \sin\Delta\varphi}{\sqrt{1-\sin^2 2\alpha \cos^2\Delta\varphi}} \tag{33}$$

This phase difference is represented in Fig.6 as "Residual error". It can be seen that Γ is negligible for smaller γ.

The mean value Σ contains the phase $\Delta\varphi$ which is proportional to the true displacement in the interferometer, the residual error Γ, and a constant phase shift, ϑ. This constant shift is of no importance for a displacement measurement.

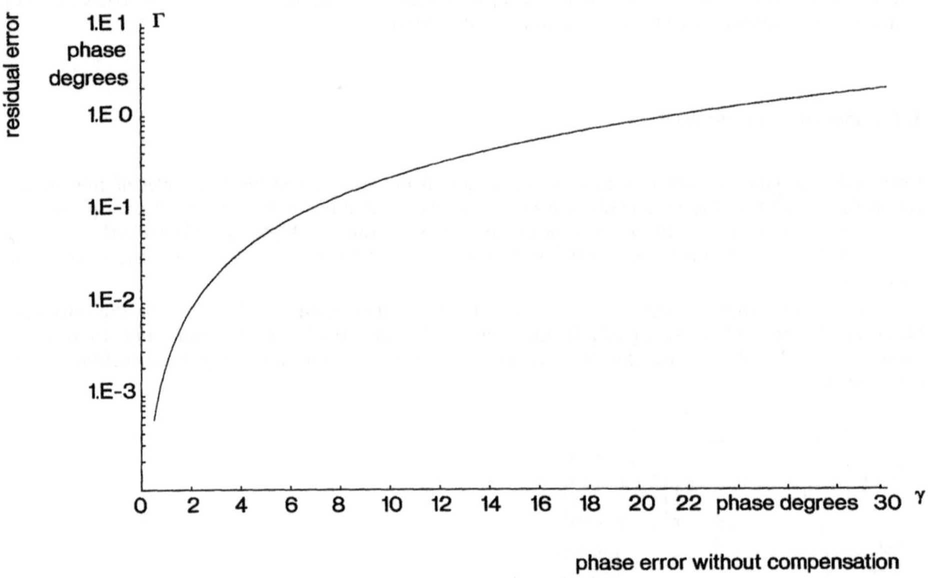

Fig. 6 Residual error after compensation vs. phase error without compensation

The measuring method described here opens up the possibility of compensating occurring phase errors "on-line", with the exception of a residual amount which is in general negligible.

This measuring method allows a considerable improvement in linearity to be expected.

5. Experiments

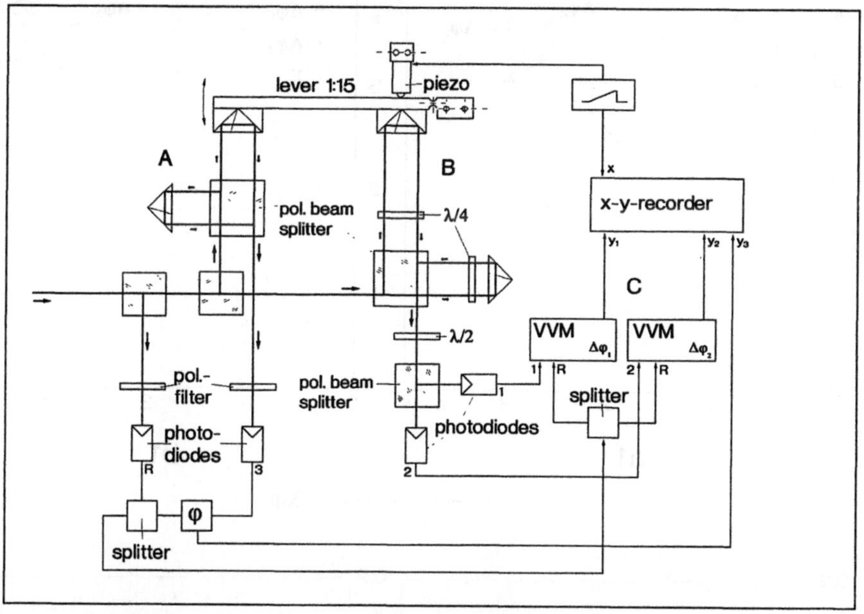

Fig. 7 Experimental set-up for the detection and compensation of non-linearities

A: Reference interferometer; zero crossings of the difference phase are taken as equidistant markings

B: Interferometer with complete phase measurement under inspection

C: Evaluation (VVM: Vector Voltmeter)

The arrangement shown in Fig.7 was set up to experimentally investigate the detection and compensation method. The cube corner of the interferometer to be examined is fastened to the short arm of a lever which is moved by a piezoelectric element. The long arm (transmission ratio appr. 1:15) carries the cube corner of another interferometer. The zero passages of the phase output of this interferometer furnish equidistant subdivisions for the displacement in the interferometer to be investigated. The zero passages are not influenced by potential non-linearities. According to eqs.(8,13,18), the phase error is periodical with $\Delta\varphi$. The same value for $\Delta\varphi$ (e.g. "zero") always furnishes the same amount of the phase error. The phase cycle of the interferometer to be investigated is provided with 15 equidistant marks, and the phase shifts $\Delta\varphi_1$ and $\Delta\varphi_2$ of the interferometer in question are measured at these marks.

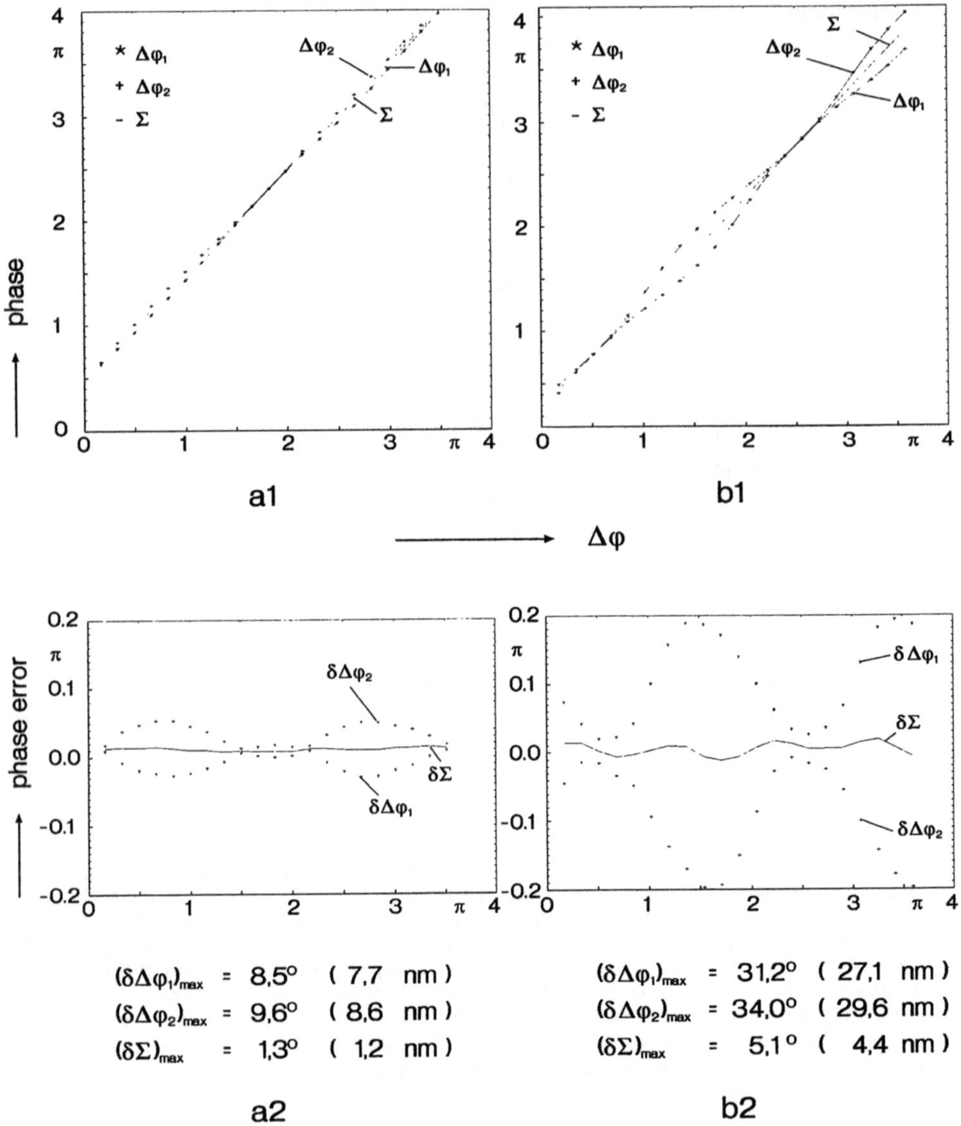

Fig. 8 The compensation of phase errors of two different interferometer set-ups (a,b)

a1, b1: phase differences $\Delta\varphi_1$, $\Delta\varphi_2$ and the arithmetic mean value Σ, vs. the true displacement $\Delta\varphi$

a2, b2: phase errors (non-linearity) $\delta\Delta\varphi_1$, $\delta\Delta\varphi_2$ and arithmetic mean value $\delta\Sigma$, vs. the true displacement $\Delta\varphi$

Fig.8 shows typical measurement results of two different interferometer set-ups. The measured phase values and the mean values have been plotted as a function of the displacement (Fig.8a1, 8b1). Figs.8a2 and 8b2 indicate the linearity errors.

The following conclusions can be drawn from the measurements:

- the traces of the two phase outputs $\Delta\varphi_1$ and $\Delta\varphi_2$ show sinusoidal deviations from linearity (according to eqs.(7a) and (26));
- the deviations from linearity are opposite in phase (according to eqs.(7a) and (26)) (i.e. the measurement of the phase difference leads to the sum of the deviations, according to eq.(30));
- the arithmetic mean value Σ indicates a clear improvement in linearity;
- the residual non-linearity $\delta\Sigma$ decreases with decreasing phase errors (according to Fig.6);
- the residual non-linearity $\delta\Sigma$ shows a doubly periodic dependence on the displacement $\Delta\varphi$ (according to eq.(33));
- constant phase components (shift into the direction of the ordinate) do not influence linearity.

The measured residual errors are greater than those calculated. A considerable amount originates from the uncertainty of the phase measurements (also in the reference interferometer!) and their processing. Another contribution is due to the incomplete separation in the polarizing splitter. Influences from the lever geometry are negligible.

6. Conclusion

In heterodyne interferometers, non-linearities in the relation between the output phase signal and the displacement result from the incorrect mixing of frequency components. The investigation of various causes of this mixing has shown that the strongest influence is exerted by the properties of the laser beams used (non-orthogonality of the linear polarization directions, ellipticity). The non-linearity is periodic with a complete phase cycle of the output signal. Zero passages of the phase ("interference fringes") are exactly equidistant. The measurement of the phase positions in two orthogonal directions of the output beam provides almost full information on linearity errors.

The measurement of the difference of these phases gives information about the quantity of the non-linearity and can be used to investigate interferometer set-ups, whereas the absolute measurement (in difference to the reference signal) may serve to compensate the non-linearity. Experiments on interferometers with different non-linearities revealed that both the check by means of the difference measurement and the compensation with the aid of the absolute measurement are effective.

There are several possible arrangements for the detection of the signals in two orthogonal planes (according to fig. 4b). Either, the polarization directions are rotated by means of a $\lambda/2$ plate (figs. 4a, 5), or polarizing beam splitter and diodes are rotated by 45°, instead, or a normal beam splitter and two polarizing filters in front of the photodiodes are taken.

For the purpose of the detection of non-linearities such arrangements can be placed in the output beam of interferometers in the form of separate instruments. Therefore the detection method even is suitable to check commercial laser interferometer systems. Determination of the occurring maximum non-linearities requires the measuring reflector to be displaced by at least one interference fringe ($\lambda/2$, in special cases $\lambda/4$). In most cases only small phase errors will occur and the phase difference measurement can be carried through in a simple way (e.g. by means of a "double balanced mixer").

The compensation method in general cannot be achieved with a separate instrument, as in this case the internal reference for absolute phase measurements must be made use of. For modular commercial multi-axes instruments, instead of the diodes shown in Fig.7, the application of receivers of two axes and the calculation of the mean value of the counts of both axes are conceivable.

Literature

/1/ Sommargren, G.E.:
A new measurement system for precision metrology
Prec. Eng. 9 (1987), 179 - 184

/2/ Quenelle, R.C.; Wuerz, L.J.:
A new micrometer-controlled laser dimensional measurement and analysis system.
Hewlett Packard Journ. 34,4 (1983), 3 - 13

/3/ Dorenwendt, K.; Probst, R.:
Hochauflösende Interferometrie mit Zweifrequenzlasern
PTB-Mitt. 90, (1980), 359 - 362

/4/ Reinboth, F.; Wilkening, G.:
Optische Phasenschieber für Zweifrequenz-Laser-Interferometrie
PTB-Mitt. 93, (1983), 168 - 174

/5/ Bobroff, N.:
Residual errors in laser interferometry from air turbulence and nonlinearity
Appl. Opt. 26, (1987), 2676 - 2682

/6/ Sutton, C.M.:
Non-linearity in length measurement using heterodyne laser Michelson interferometry
J.Phys.E 20, (1987), 1290 - 1292

/7/ Steinmetz, C.R.:
Sub-micron position measurement and control on precision machine tools with laser interferometry
Prec.Eng.12, (1990), 12 - 24

/8/ Rosenbluth, A.E.; Bobroff N.:
Optical sources of non-linearity in heterodyne interferometers
Prec.Eng.12, (1990), 7 - 11

Technology and Application of Grating Interferometers in High-precision Measurement

A. Teimel

DR. JOHANNES HEIDENHAIN GmbH, Traunreut (Germany)

Increasing requirements on the positioning accuracy of precision machine tools in recent years, e.g. for the production of optical surfaces and in semiconductor technology, have led to a stronger demand for linear encoders with resolutions in the submicrometer range.

At first glance the widely-used He-Ne laser interferometers, whose emitted wavelength of light can have a long-term stability of $2 \cdot 10^{-8}$, seem to be ideal [1]. Commercially available systems of this type offer resolutions up to 1.25 nm [2].

Optimal conditions for high measuring accuracy can be achieved with the laser operating in a vacuum. Practical considerations, however, preclude this possibility with high-precision machining tools or with most coordinate measuring devices.

The accuracy which can be attained with measurement in air is determined to a large extent by the accuracy with which the refractive index can be determined. A mathematical compensation according to the Edlén formula is often utilized [6].

Table 1 shows how accurately the parameters contained in this formula - air pressure, air temperature and relative humidity - must be measured and maintained for a relative length measurement error of 10^{-7}.

Parameter	Nominal value	Change for a relative displacement error of ± 0.1 ppm
air temperature	20° C	± 0.11 K
air pressure	1013 hPa	± 0.37 hPa
relative humidity	40%	± 10%
(workpiece temperature, steel $\alpha = 10^{-5}$ 1/K)	-	± 0.01 K

Table 1: Environmental parameters and their influence on measurement accuracy

Since this formula is referenced to standard air, deviations are to be expected in practice according to the amount of gases and vapors not considered. With direct measurement of the refractive index using refractometers, these influences are allowed for. Nevertheless, due to the thermal capacity of the surrounding material and imperfect gas exchange, lag between the measured value and the true refractive index can occur.

Further problems arise because even in rooms temperature-controlled to 1/10 K there can exist local temperature variations of 0.3 to 0.4 K. Since the refractive index of air reacts to short-term variations in air pressure and air temperature practically lag-free, the stability of the measurement signal is reduced with increasing measuring length, a situation which is particularly disadvantageous for positioning accuracy and which thus affects maximal surface quality with high-precision machine tools. In addition, it should be realized that for measurement with an accuracy of 10^{-7} of a steel workpiece with a thermal expansion coefficient of 10^{-5} 1/K, the workpiece temperature must be known to an accuracy of 10^{-2} K.

In contrast, grating interferometers offer a stable material measure with known temperature behavior and possess optical path lengths in air which have been reduced to a minimum.

Grating Interferometers

While He-Ne laser interferometers generate the measured value by interference of a working beam of variable length and a reference beam of constant length and by evaluation of the phase difference between these two beams, grating interferometers produce the measured value by the interference of two light beams, which are diffracted on the scale grating, and subsequent evaluation of the phase difference.

By selection of the orders of diffraction which are induced to interfere or by using multiple diffraction on the scale grating, the ratio of the phase grating period to the period of the interference signal can be varied.

For a relative displacement between a scale with the grating period C and the beam diffracted, the phase shift Ω of the order of diffraction n is determined by the following equation:

$$\Omega = \frac{n \cdot 2\pi \cdot x}{C}$$

The interference signal arises by superimposing two diffracted wave trains. If the modulus of the amplitudes of the two waves are set to 1, the complex amplitude of the sum of the two waves is given by:

$$U = U_1 + U_2 = e^{i\Omega_1} + e^{i\Omega_2}$$

For $\Omega = \Omega_1 = -\Omega_2$, this results in the intensity:

$$I = U \cdot U^* = 2(1 + \cos 2\Omega)$$

A relative movement between the scale grating and the diffracted beam of one grating period and interference of the +1st with the -1st order of diffraction (fig. 1) creates two periods of the interference signal.

The generation of the phase-shifted signal necessary for the recognition of direction and for interpolation can be achieved by means of a specially-dimensioned phase grating, or - using polarized, monochrome illumination - by means of optical polarizing elements.

Grating Interferometer with Coherent Light Source

Coherent light sources allow simple optical configuration of grating interferometers. The acceptably-priced laser diodes which have recently become available on the market have given new impulses to the development of interferential linear encoders.

These semiconductor lasers can be divided into index-guided and gain-guided types, depending on the method of confining the lasing mode in the lateral direction along the active medium which limits the lasing area. In table 2, several important characteristics are summarized:

Gain-guided	Index-guided
· low noise in case of optical feedback	· sensitive to optical feedback
· short coherence lengths	· long coherence lengths (up to several meters)
· astigmatism and thus more difficult to collimate to a plane wave front	· no or very little astigmatism
· several longitudinal modes	· at nominal output power basically single-moded longitudinally

Table 2: Characteristics of laser diodes

The temperature behavior will be discussed later in detail. For applications in a grating interferometer the emitted beam must be collimated to a parallel bundle of rays, which means a plane wave front.

The quality of the wave front influences the ratio of the modulated to the unmodulated component of the interference signal as well as the mounting tolerances.

Fig. 1 shows a simple arrangement for generating modulation signals with a period of ½ C (where C = grating period).

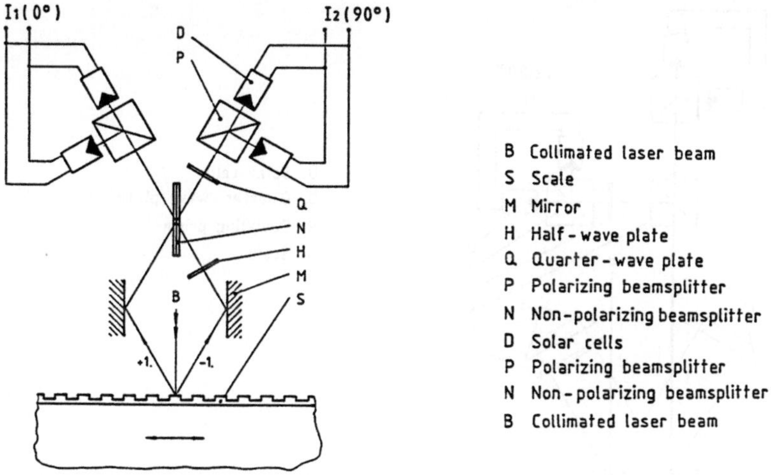

Fig. 1: Grating interferometer with optical polarizing elements

The measurement signal is produced through superimposition of the diffracted and, corresponding to the relative movement between scale and scanning unit, phase-shifted waves.

Through optical polarizing elements, and by using methods known from Michelson interferometers, four 90° phase-shifted signals result, from which two sinusoidal 90° phase-shifted signals can be generated.

Through multiple diffraction on the scale grating, the ratio of the scale grating period to the period of the interference signal can be increased. Fig. 2 shows an example in which, by means of double diffraction and the imposition of 1st-order diffracted beams, interference signals with a period of ¼ C are produced. If the amplitudes of the waves are again set to 1 and the effects of the optical polarization elements are neglected, then the amplitude of the left beam path with a displacement of the grating by the amount x with $\Omega = \frac{2\pi x}{C}$ is:

$$U_L = e^{i\Omega} \cdot e^{i\Omega} = e^{i2\Omega}$$

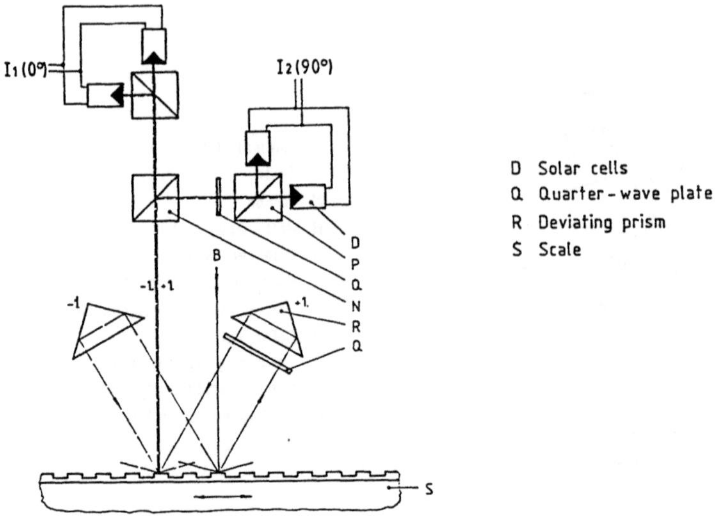

Fig. 2: Grating interferometer with double diffraction on the scale grating

The amplitude of the right beam path is:

$$U_R = e^{-i\Omega} \cdot e^{-i\Omega} = e^{-i2\Omega}$$

The superimposition of the waves after the 2nd diffraction on the scale grating results in:

$$U = U_L + U_R = e^{i2\Omega} + e^{-i2\Omega}$$

The intensity is thus:

$$I = U \cdot U^* = (e^{i2\Omega} + e^{-i2\Omega}) \cdot (e^{-i2\Omega} + e^{i2\Omega}) = 2 + 2 \cos 4\Omega$$

A grating period of the scale of 1.6 µm therefore results in a period of intensity modulation of 0.4 µm.

With such a small grating period, the order of diffraction n, with a typical wavelength λ of 780 nm, can only assume the values 0, 1 or 2 due to the condition for the diffraction angle

$$\sin \alpha = \frac{n \cdot \lambda}{C} \leq 1.$$

The angles in which the various orders are diffracted differ from each other considerably. Isolation of the desired order of diffraction is thus easily achieved, so that only the selected order of diffraction contributes to the interference signal.

The modulation signals are thus free of higher-order harmonics and can easily be utilized for high interpolation. Resolutions of up to 1 nm are possible.

On the one hand, a long coherence length of the emitted beam allows large mounting tolerances. On the other hand, with all design principles using a laser diode as light source it should be noted that the change of length of one beam path relative to the other can also lead to modulation of the intensity of the interference signal. In fact these designs are always a combination of grating interferometer and Michelson interferometer. Beam paths of different lengths also cause temperature-dependent modulation, mainly due to the dependence of the emitted wavelength of light of the laser diode on temperature and operating current.

In grating interferometers available today utilizing laser diodes as light source, the light output is usually stabilized. This control loop can easily be realized via the current of the monitor diode which is integrated into the diode housing. This will also compensate for the efficiency reduction due to aging as well as for the reduction in light output at constant current with rising temperature.

The center wavelength of the emission spectrum of laser diodes is temperature-dependent. The mean temperature coefficient of the wavelength $\frac{d\lambda}{dT}$ in the temperature range of 10 - 40° C is approx. 0.25 nm/K. With an optical path length difference Δl of e.g. 0.25 mm and a reference wavelength λ of 780 nm, a mean phase drift of

$$\frac{d\Phi}{dT} = 2\pi \cdot \frac{\Delta l}{\lambda^2} \cdot \frac{d\lambda}{dT} = \frac{2\pi}{10}\ K^{-1}$$

results, which corresponds to 1/10 of the period of the interference signal. Despite the fact that the mean temperature coefficients for gain-guided and index-guided laser diodes are practically identical, differences in the wavelength-temperature characteristics do exist.

Index-guided laser diodes often emit, at nominal power, only light of a certain longitudinal mode. This is principally determined by the chip dimensions and the doping of the active layer. With a cavity length L of 250 µm, an emission wavelength of 780 nm and a refractive index of n = 3.5, the mode number can be approximated with

$$m = \frac{2nL}{\lambda} = 2244.$$

With warming of the laser the band gap energy in the active layer is reduced. With a single-mode laser, the oscillation jumps from one resonance to the next-lower one, leading to a discrete wavelength jump in the emission spectrum. Solving the equation of the mode number for λ and differentiating by m,

$$\frac{d\lambda}{dm} = -\frac{2nL}{m^2}.$$

is obtained. For a change in mode number of -1 (dm = -1) when the laser temperature rises a change in the wavelength of light of $\Delta\lambda_m = -\frac{2nL}{m^2} \cdot dm = 0.34$ nm results.

The same result occurs when, with increasing injection current and higher output power, the temperature rises. If one assumes again an optical path length difference of 0.25 mm, with mode hopping a change in phase of

$$\Delta\Phi_m = 2\pi \frac{\Delta l}{\lambda^2} \cdot \Delta\lambda_m = \frac{2\pi}{7}$$

results. This causes an abrupt change in the measured value. Often the laser oscillates at a high frequency during transition from one mode to the next until it restabilizes, causing high demands on the processing speed of the subsequent electronics.

Between these jumps in wavelength there are intervals in which the wavelength-temperature diagram shows a continuous rise of approx. 0.06 nm/K, which is determined by the temperature dependency of the diffraction index of the active layer. The width of these intervals can amount to several K.

Fig. 3 shows a typical wavelength-temperature diagram of a single-mode laser diode. Unfortunately the position and width of the intervals vary from diode to diode, even among diodes of the same type. In addition, the jump positions change during the aging process.

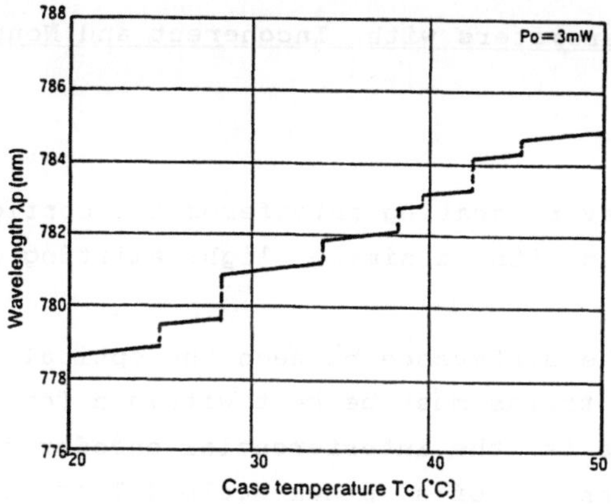

Fig. 3: Wavelength-temperature diagram of a single-mode laser diode

Depending on the laser diode utilized, the wavelength-temperature diagram also shows a certain hysteresis, especially at the mode transition points, between heating and cooling of the heat sink. A suitable laser diode should have either a very wide temperature range without mode hopping (> 10 K) or many small instances of mode hopping and little hysteresis.

In order to linearize the temperature behavior, multi-mode laser diodes can be used. These emit several longitudinal modes of various intensities simultaneously. Effects on the phase position of the interference signal originate practically only from the shift of the emission spectrum. The short coherence length of typically 100 - 300 µm limits the possible optical path length difference, and therefore the mounting tolerances, but there is an advantage in temperature behavior in comparison to single-mode laser diodes.

The thermal characteristics of a grating interferometer with a laser diode as light source are thus principally determined by the beam path length difference and by the laser diodes used.

Interferential encoders using laser diodes as light source and scales having very small grating periods (1 - 2 µm) are coming on the market. With these systems, resolutions of 1 nm over measuring lengths of 100 - 200 mm can be expected.

Grating Interferometers with Incoherent and Nonpolarized Light Sources

There are, however, grating interferometer configurations which can be realized with a simple light-emitting diode as light source.

In this case the difference between the optical paths of the interfering wave trains must be kept within a few micrometers. An example of this is the interferential encoder described below. The encoder consists of a steel scale (glass or ZERODUR scales are also possible), with a reflection-type phase grating vapor-deposited on its surface.

Scales with 8 µm and 4 µm grating periods are produced. The grating consists of rectangular strips with a height of approx. 0.2 µm, alternating with gaps of the same width. The entire grating structure is made of chrome and is thus very hard and wear-resistant.

The scanning signals are produced in a scanning head which contains a transmitted-light phase grating (index grating) with the same grating period as the scale, an LED as light source, a condenser and three solar cells (fig. 4, left).

The optical configuration of the grating interferometer is depicted in fig. 4, right, greatly simplified but with the beam paths shown "unfolded". The light source, the condenser and the solar cells are not shown.

The plane wave produced by the LED and the collimator as well as the refracted waves are represented by light beams perpendicular to the wave front. A plane wave of wavelength λ is assumed, which falls perpendicularly onto the index grating.

The interaction of the illustrated components will now be explained. While passing through index grating G_1, the wave is mainly diffracted in the three directions -1, 0 and +1. The index grating is designed such that the phase of the beam of the

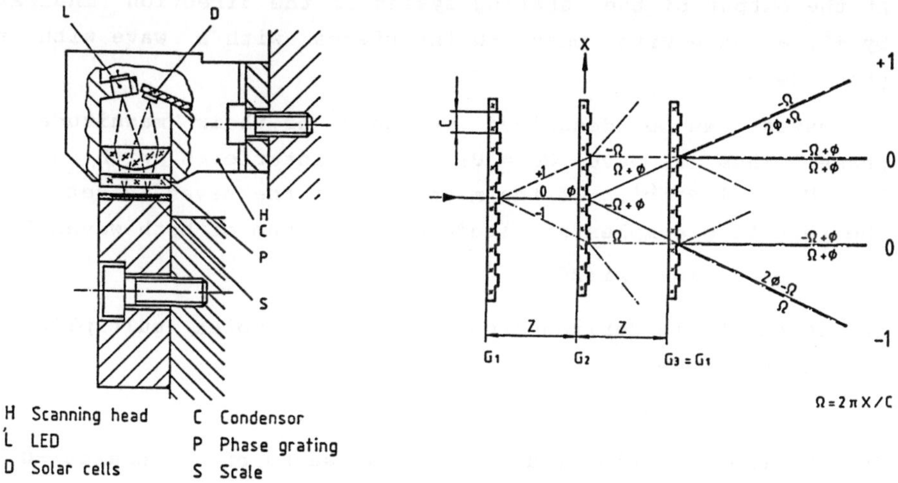

Fig. 4: Linear Encoder LIP 101 (HEIDENHAIN) - Left: design scheme; right: operating principle

0th order lag the beams of the ± 1st order by the amount Φ. As the beams strike the scale G_2, they are each diffracted in two directions of the order ±1. The scale is designed such that no 0th order diffraction exists.

When the scale is moved relative to grating 1 (= grating 3), the beams of the +1st (n = +1) order diffracted on the scale are phase-shifted - proportional to the displacement x - by the amount Ω, while the beams of the -1st order (n = -1) are phase-shifted by the amount $-\Omega$, whereby $\Omega = \frac{2\pi x}{C}$.

While passing through the grating G_3, the light waves are diffracted and phase-shifted again. Waves of the same direction and the same optical path length interfere. The phases of the interfering waves result from the sum of the phase shifts of the individual beams. With illumination by non-monochromatic and spatially incoherent light sources, only the interfering waves in the direction +1, 0, -1 contribute to formation of the signal.

At the output of the grating system in the direction indicated by +1, a wave with phase $-\Omega$ interferes with a wave with the phase $2\Phi + \Omega$.

The waves can be described by means of their magnitude and phase. Thus a wave $U_1 = U_0 \cdot e^{-i\Omega}$ interferes with a wave $U_2 = U_0 \cdot e^{i(2\Phi+\Omega)}$. If the magnitude of the waves is set to 1, the resulting complex amplitude U of the sum of both waves is

$$U = e^{-i\Omega} + e^{i(2\Phi+\Omega)}.$$

The intensity is derived from conjugated complex multiplication of the amplitude:

$$J_1 = U \cdot U^* = 2\ [1 + \cos(2\Omega + 2\Phi)].$$

For the the direction indicated by 0, waves of the phases $-\Omega + \Phi$ and $\Omega + \Phi$ interfere. The following equation describes the amplitude of the sum:

$$U = e^{i(\Phi - \Omega)} + e^{i(\Phi + \Omega)} = e^{i\Phi} \cdot 2\cos\Omega.$$

The resulting intensity is

$$J_0 = 4\cos^2\Omega = 2\ (1 + \cos 2\Omega).$$

Similar considerations lead to the intensity for the direction -1:

$$J_{-1} = 2\ [1 + \cos\ (2\Omega - 2\Phi)].$$

The intensities of these three interference signals are in each case detected in the focal plane of the collimator by the corresponding solar cell. Movement of the scale (G_2) causes formation of signals with a period of C/2 which are phase-shifted to each other by 2Φ. A phase shift 2Φ of 120° has proven suitable. This is attained through the geometric configuration of the index grating G_1 and G_3 [3]. The three solar cells transform the phase and modulation of the interference signal into equivalent currents from which two sinusoidal output signals phaseshifted to each other by 90° are derived by an interface electronics. By individual adjustment of the scale/scanning head system in the interface electronics, the amplitudes and the 90° phase shift can also be adjusted.

The strict symmetry of the optical path lengths prevents aging or temperature-dependent shifts of the LED emission spectrum from influencing measurement. Thus the encoder is ready for use immediately after switch-on.

The length of the scanning field which contributes to the measurement signal of approx. 1500 grating periods assures reliable operation of the encoder, even with local contamination of the scale. Since the phase detected by the solar cells corresponds to the mean value of the phase of the waves which are diffracted over the entire length of the scanning field, minor fluctuations of the scale grating period have little influence on the measurement.

The integrating effect and the short paths in air, from the splitting into various orders of diffraction to interference at the index grating, which amounts only to 1 - 2 mm, are important factors in the stability of the interference signal. Resolutions of 10 nm with an encoder with 8 µm grating period or 5 nm with a 4 µm grating period are thus entirely realistic.

The arrangement of the system as reflected light grating encoder allows relatively large mounting tolerances since the distances $G_2 - G_1$ and $G_2 - G_3$ are always equal (fig. 4, right).

The scales are presently produced with an 8 µm grating period in lengths up to approx. 1000 mm and with a 4 µm grating period up to approx. 300 mm.

Fig. 5 shows as an example the calibration chart of a scale with an 8 µm grating period. The total error lies within ± 0.2 µm for a measuring length of 420 mm.

The compact dimensions of an encoder system having a 4 µm period phase grating scale can be seen in fig. 6.

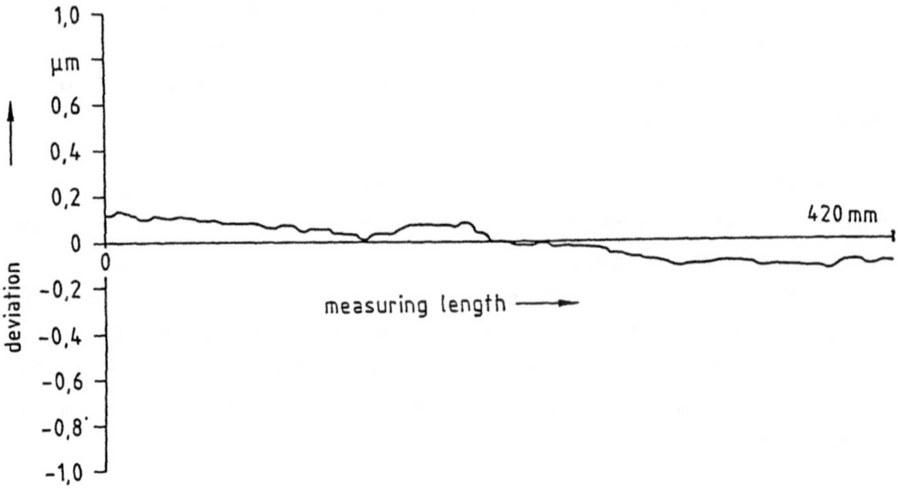

Fig. 5: Calibration chart of an interferential linear encoder, measuring length 420 mm, reference temperature 20.0° C. (Reference: laser interferometer in vacuum)

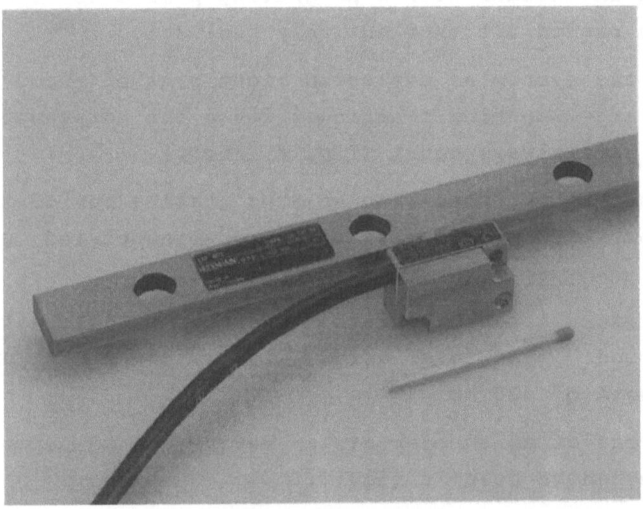

Fig. 6: Linear encoder using grating interferometry, with steel scale, phase grating period 4 µm.

Comparison of the Characteristics of Grating Interferometers and He-Ne Laser Interferometers

The following describes and compares the important characteristics of each type of interferometer in tabular form

	Grating Interferometers	He-Ne Laser Interferometers
atmospheric influences	• high accuracy in air • no influence of pressure, humidity or air composition	• can be used as length reference with laser beam in vacuum • the refractive index of air must be determined accurately by precision sensors
temperature behavior	• high heat capacity • scale carrier material can be chosen as appropriate for individual application	• spontaneous local air temperature fluctuations • temperature profile of the air path and of the workpiece must be measured and compensated by high-precision temperature sensors
measuring lengths	• Measuring length at present < 1000 mm	• Measuring lengths up to approx. 30 m
measuring accuracy	• accuracy approx. 10^{-6}, can be increased by calibration	• with measurements in air, accuracy depends on the ambient conditions and on the number and precision of the sensors
reproducibility	• reproducibility limited mainly by the thermal characteristics of the scale	• the measuring uncertainty grows with the measuring length; dead path increases the measuring uncertainty

Literature:

[1] C.R. Steinmetz: Sub-micron position measurement and control on precision machine tools with laser interferometry. Precision Engineering January 1990.

[2] G.E. Sommargren: A New Laser Measurement System. Zygo Corporation.

[3] Alfons Spies: Längen in der Ultrapräzisionstechnik messen. Feinwerktechnik & Meßtechnik 98 (1990).

[4] F. Hock, Dissertation Universität Stuttgart 1975.

[5] W. Tyler Estler: High-accuracy displacement interferometry in air. Applied Optics Vol. 24, No. 6 (March 1985)

[6] Dokumentation Laserinterferometer in der Längenmeßtechnik. VDI-Berichte 548 (1985)

Silicon as a Length- and Angle Ruler in Nanotechnology

J. Stümpel and P. Becker

Physikalisch-Technische Bundesanstalt

D-3300 Braunschweig, Federal Republic of Germany

Abstract

Highly pure silicon crystals with an almost perfect lattice structure constitute a powerful metrological tool. In real Si-crystals lattice spacings with an accuracy better than 1 part in 10^{7} are realized. These have been applied as length standards in the sub-nanometer region using the scanning X-ray interferometer, and as angle standards in the nanorad region with angle X-ray interferometers. Crystallographic angles between different sets of lattice planes within the same crystal differ only slightly (1×10^{-9} rad) from values calculated for the perfect crystal. In a double-crystal arrangement Si-crystals has been used as a single-crystal polygon in order to calibrate optical angle encoders.

1. Introduction

The controlled production of technical surfaces with the highest degree of optical quality or of highly integrated electronic devices requires the development of novel metrological techniques in order to determine the minute geometrical aberrations between the real and the ideal shape of the workpiece with a sufficient measurement uncertainty. Defects in semiconductor slices, in the X-ray and vacuum ultra-violet (VUV) mirrors of telescopes and synchrotron radiation optical devices examplify this need. For this reason it is necessary to establish linear and angular measurements in the sub-nanometer and nanorad region. A number of metrological instruments with the appropriate sensitivity are available, but they all need additional calibration procedures to reduce errors. Work on X-ray interferometry has demonstrated that absolute metrological length measurements down to 10 pm are feasible [1]. The techniques of X-ray diffraction allow the use of a crystal of sufficient quality (silicon or germanium) as an incremental encoder for high precision angular measurements.

The main idea is to apply a suitable sequence of consecutive symmetric or asymmetric reflections, either all from parts of one single crystal or from different crystals, and then to tilt or rotate these against each other to reach specified positions. The choice of the sequence used depends on the angular and lateral range to be measured and is determined by the characteristics of the single-crystal reflection curves, by the reflection order and by the relative orientations of the surface and the lattice planes.

2. X-ray Diffraction by Crystals

X-ray metrology is based on X-ray diffraction procedures using almost perfect silicon crystals. Physically, the diffraction of X-rays is an elastic scattering process in which a large number of atoms cooperate. Because of the periodic arrangement of the atoms in the lattice, the scattered waves are related in phase. As is known from optical diffraction theory, constructive interference occurs if the path difference between successive wave fronts is $n\lambda$, n being an integer and λ the wavelength of the incident X-rays. For X-rays this is valid if the Bragg law

$$n\lambda = 2d \sin\Theta_B \quad (1)$$

is fulfilled, with d being the distance between the adjacent lattice planes in the crystal and Θ_B is the angle between the incident wave and the reflecting plane. Several features not apparent in (1) are important in the application of X-ray diffraction in X-ray metrological devices:

i) If inside the crystal there are two waves whose difference in wave vector precisely equals the reciprocal lattice vector corresponding to the Bragg reflecting plane a standing wavefield will be set up whose spacing and orientation are exactly those of the diffracting planes. This standing wavefield diffracts the X-rays so making interferometry available for small length and angle measurements (see Fig. 1).

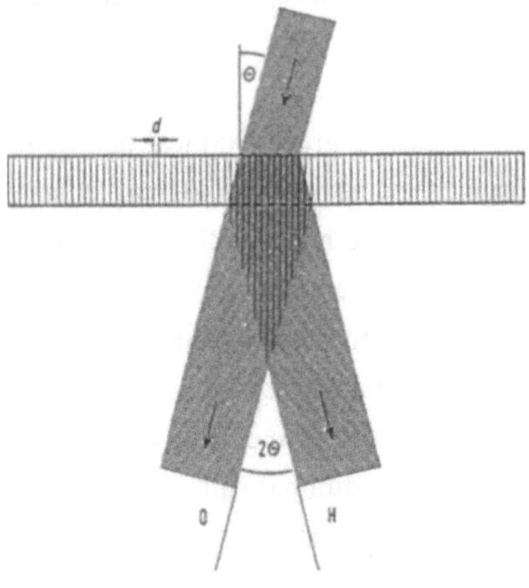

Fig. 1 X-ray diffraction in a single crystal. O original beam; H diffracted beam.

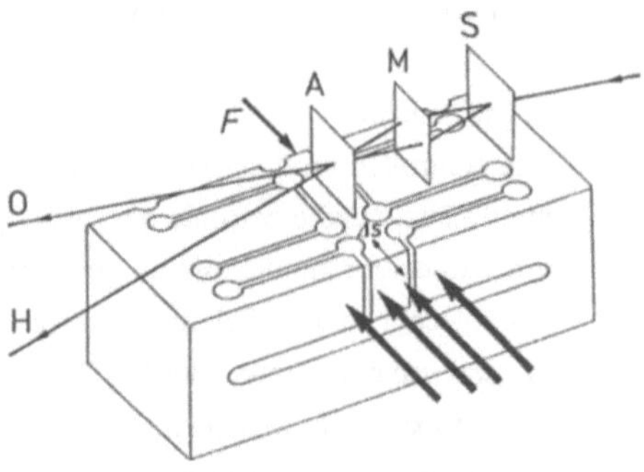

Fig. 2 Integrated X-ray scanning interferometer and translation stage. Arrows indicate the impinging points of laser beams from optical interferometers. F driving force for movement Δs of the analyzer A.

ii) Bragg reflection curves have a finite angular width of about one second of arc and show an additional fine structure. The narrowest intensity profiles may be obtained with high order Bragg reflections and short X-ray wavelengths. In double-crystal arrangements, X-ray reflection curves (rocking curves) are powerful tools for angle measurements of an extremely high resolution, e.g. in Bragg angle determination of wavelengths and in energy calibration for the X-ray or γ-ray region.

iii) In pure high quality single-crystals, the angles between lattice planes are very close to those of a perfect crystal. In real silicon, impurity atoms may replace the regular Si atoms or enter the interstices between the host atoms, and the interatomical distances are therefore sligthly changed. Because of the local displacements produced by impurities, each cell is distorted to some extent; its shape and size varies depending on the impurity atoms and the neighboring cells. Oxygen and carbon atoms are the main impurities in zone refined silicon crystals. Their atom number density is between 1×10^{15} cm^{-3} and 5×10^{16} cm^{-3}. The observed lattice parameter and the shape of the unit cell of high quality silicon are strongly related to the residual number density of carbon and oxygen and differ only slightly from those of an ideal crystal.

The devices that make the application of X-ray optics possible to metrological measurements are based on double-crystal spectrometers in which a photon beam from an X-ray tube is consecutively reflected from two crystals, or on interferometers consisting of three thin crystal plates acting as coherent beam splitters.

3. X-ray Interferometer

The ray tracing of an X-ray interferometer is shown in Fig. 2. Coherent splitting of the X-ray beam is achieved by a crystal plate S using the Laue geometry shown in Fig. 1. Mirrors are needed to superimpose the two coherent beams: a second crystal plate, M, of equal thickness acts as a mirror, and a third crystal, A, is used as an analyzer. A smooth displacement of the analyze perpendicular to the lattice planes results in a sinusoidal intensity modulation of the X-ray beams O and H, with a period which equals the lattice spacing.

Some years ago, two-crystal interferometers were designed and successfully used to carry out wavelength-independent precision measurements of the interplanar spacing of a perfect Si crystal [2]. The high positional sensitivity of this type of interferometer arises mainly from the fact that the two waves bridging the gap between the separated crystals are not parallel to each other. Fig. 2 shows the scheme of an integrated X-ray interferometer with a translation stage manufactured from one silicon single crystal. Almost perfect monocrystalline silicon is an excellent raw material for mechanical engineering devices, especially applicable whenever miniaturized mechanical components must be integrated or interfaced with electronics. The interferometer crystals used in the work were made by conventional sawing and grinding, and the affected surfaces and edges were afterwards etched to remove the strongly damaged surface areas.

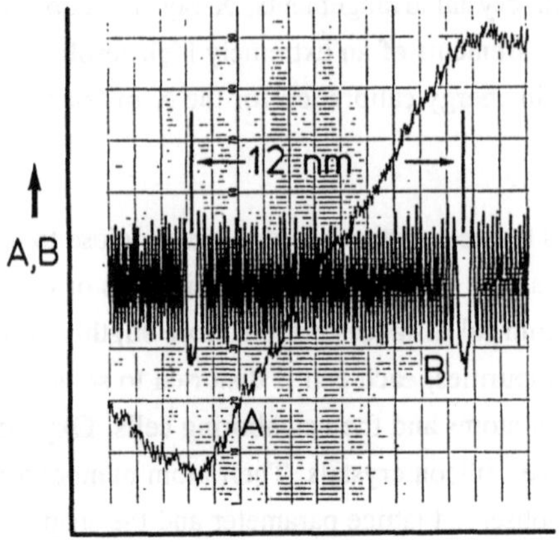

Fig. 3 Calibration of a linear voltage differential transducer (LVDT, signal A) by means of an X-ray interferometer (signal B).

The scanning X-ray interferometer has been operated successfully in the calibration of a linear-voltage differential-transducer (LVDT) (see Fig. 3) and for the interpolation of fringe fractions in optical laser interferometers.

In high resolution interferometry very precise phase measurements between the measuring and reference beam are necessary, to evaluate the fringe pattern. In experimental set ups using two-mode laser interferometry, the phase shifts are often traced back to the rotation of a polarizing filter. Such a system was combined with an integrated X-ray interferometer. The measuring beam points onto the mirror on the movable part (analyzer A), and the reference beam is reflected by a mirror on the fixed part (beam splitter and mirror) of the silicon crystal. The modulation of the X-ray signal and the azimutal angle α of the polarizing filter for zero phase difference are simultaneously recorded during a displacement of A. About 1648 X-ray fringes are observed for $\alpha = 180°$, corresponding to $\Delta s = \lambda/2$. Deviations from a linear relation between Δs and α measured for an arbitrary state of adjustment of the experimental set up are shown in Fig. 4.

In contrast to the requirements of lateral sensitivity (lattice-spacing measurement), angular sensitivity can be achieved by arranging that the gap-bridging beams lie parallel to one another. One possible realization, the skew-symmetric two-crystal interferometer was reported in 1974 [4].

Here the construction and successful operation of a monolithic version with an integrated pivot is demonstrated. The interferometer crystal is illustrated in Fig. 5.

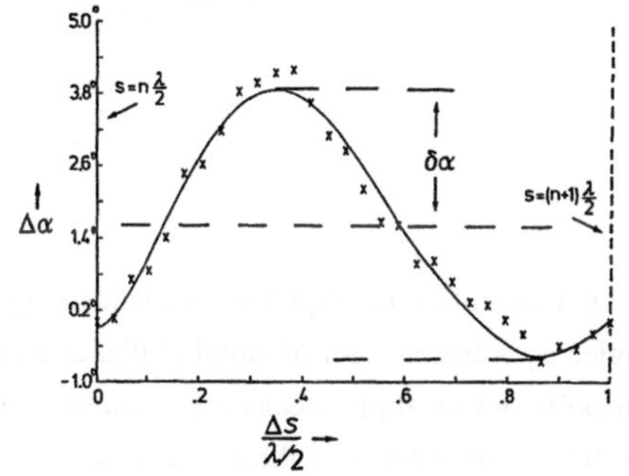

Fig. 4 Calibration of an optical length measuring system by an X-ray interferometer.

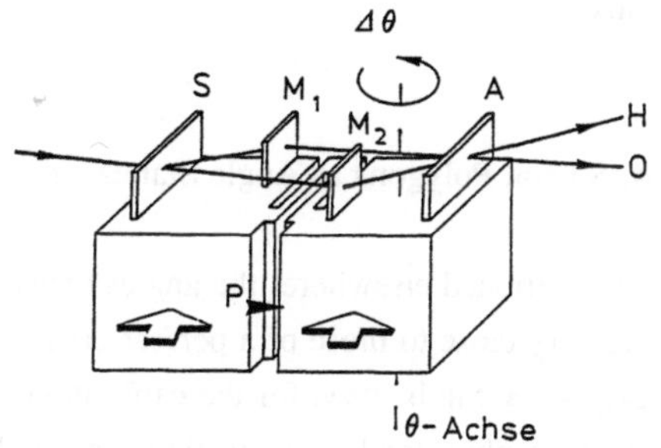

Fig. 5 Integrated X-ray angle interferometer and tilting device. P indicates the driving force.

The angular sensitivity is related to a movement analogous to the lateral shift normal to the reflecting planes used in the symmetric two-crystal interferometer and may be understood in the following way: Mirrors M_2 and A are rotated about the Θ-axis with radius vectors of different lengths: the movements Δs (normal to the net planes) of A relative to M_2 are then

$$\Delta s = \Delta\Theta\,(x + t_{M2}) \qquad (2)$$

with x the distance between mirror M_2 and analyzer A and t_{M2} the thickness of M_2.

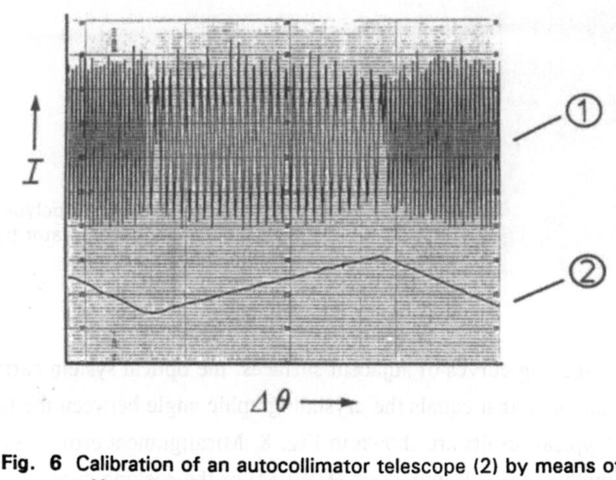

Fig. 6 Calibration of an autocollimator telescope (2) by means of an X-ray interferometer (1). One X-ray fringe corresponds to about 2×10^{-3} sec of angle.

In Fig. 6 the intensity modulations following a $\Delta\Theta$-tilt between the two interferometer parts are shown together with the output signal of an autocollimator telescope to be calibrated. The total range, of about one second of angle, is limited by the angular range of the Bragg reflection curve.

4. Silicon Polygons as Angle Standards

As mentioned elsewhere, the angles between lattice planes in almost impurity-free silicon crystals are very close to those of a perfect crystal. The angles α_i between a set of suitable planes with $\Sigma\alpha_i = 2\pi$ can be used for the calibration of optical polygons or angle-measuring devices. The stereographic standard projection for the (100) and (111) orientation of the diamond structure found by the well-known Laue method shows angles α_i between net-planes of e.g. 30°, 60°, 45° and 90°. Each value is certain to some
10^{-8}rad [5].

Fig. 7 shows a photograph of a regular form of a silicon polygon used for the calibration of an optical angle measuring system. The six faces of the crystal are cut parallel to the (220) lattice planes and can be rotated in a double crystal arrangement. With Mo-Kα_1 radiation and (880) reflection order on both crystals, the angular range in which the polygon faces reflect the X-rays makes only 1×10^{-6}rad. The position of maximum intensity can be determined with an accuracy of better than 5×10^{-8}rad.

Fig. 7 Photograph of a silicon polygon including the monochromator (left) and the signal detectors.

From the observed maxima of the rocking curves of adjacent surfaces, the optical system carrying the Si crystal has been turned by an angle that equals the crystallographic angle between the two sets of lattice planes (e.g. 60°). Typical results are shown in Fig. 8. Misalignment errors can be avoided by minimizing the half width of the full maximum (HWFM) of the rocking curves.

Because the Bragg angle is temperature dependent, it is necessary to stabilize the temperature of the crystal to ± 0.01 °C.

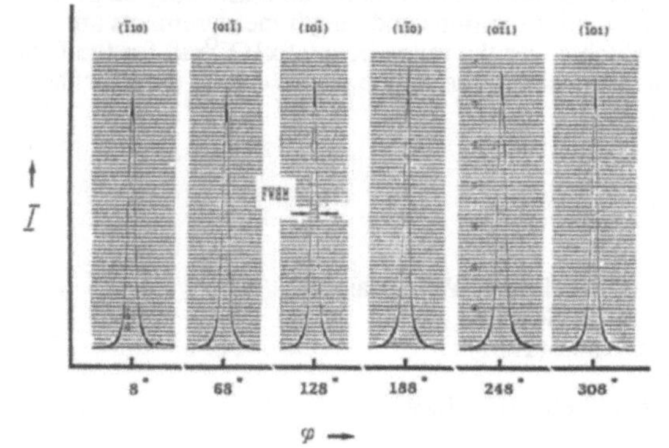

Fig. 8 Set of X-ray rocking curves. Beams are reflected by the six (220)-faces of the silicon polygon as shown in Fig. 7. The full width half maximum (FWHM) is about $9{,}5 \times 10^{-7}$ rad (Mo-Kα_1 radiation, (880) Bragg reflection).

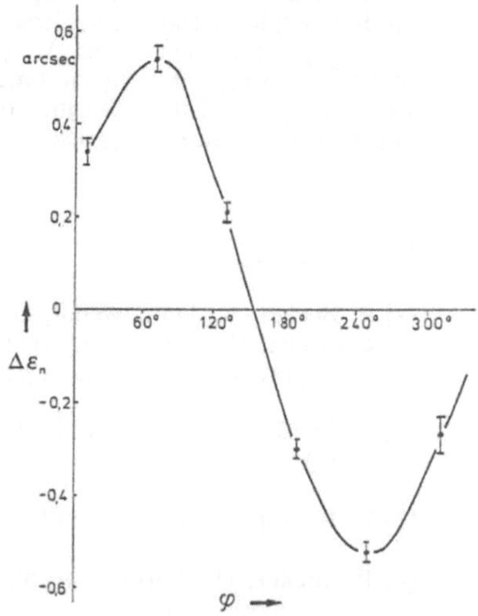

Fig. 9 Calibration curve of a precision angular encoder: The systematic deviation $\Delta\varepsilon_n$ (see text) versus angle of rotation φ.

With the six-faced silicon polygon acting as a reference angular standard, a precision angular encoder has been calibrated and the measurement uncertainty of the system investigated.
The calibration procedure was as follows: The silicon crystal was carefully mounted on a high precision commercial angle encoder functioning as a goniometer. The maximum position of the X-ray rocking curves and the corresponding angle on the encoder scale were measured by rotating the goniometer by 60° steps. The same procedure was repeated starting with a different polygon face in almost the same encoder position. A method similar to that of the so-called three-rosette method was used to evaluate the data. The systematic errors e_n for certain positions on the encoder scale, shown in Fig. 9, are close to a curve presenting one cycle of a sinusoid and can easily be corrected with software. The errors are mainly due to the mechanical connection between the encoder scale and the rotating system. The reproducibility of the silicon angle standard was also investigated and found to be better than 5×10^{-9}rad. This value is an indication of the high quality of the silicon material and corresponds to an inhomogeneity of the lattice parameter of less than 1×10^{-8} a_0. The reproducibility which can be achieved is not limited by the flatness of the reflecting surfaces, as is the case in calibration procedures using light optical methods. The overall measurement uncertainty of the determination of systematic errors is $1{,}25 \times 10^{-7}$rad at a confidence level of 95 % which is mainly caused by the error of the partial scanning pattern of the encoder.

Summary

A wide variety of methods using X-ray optics is available to determine atomic dimensions or to calibrate angle measuring devices. The possibilities reviewed here can reasonably be grouped according to their sensivities. Very small angular rotations and length measurements are achievable using X-ray interferometers which cover the range below 5×10^{-6} rad for the former and 1 nm for the latter. The macroscopic angles between lattice planes can be used for the calibration of optical polygons or turntables.

References

[1] P. Becker, K. Dorenwendt, G. Ebeling, R. Lauer, W. Lucas, R. Probst, H.-J. Rademacher, G. Reim, P. Seyfried, H. Siegert: Phys. Rev. Lett. 46, 1540-1543 (1981)

[2] R. D. Deslattes, A. Henins: Phys. Rev. Lett. 31, 972-975 (1973)

[3] P. Becker, P. Seyfried: SPIE Vol. 1015, 124-129 (1988)

[4] P. Becker, U. Bonse: J. Appl. Cryst. 7, 593-598 (1974)

[5] P. Seyfried, P. Becker, D. Windisch: Precision Engineering 10, 35-42 (1988)

Electron Tunnel Sensors

T.W. Kenny and W.J. Kaiser
Center for Space Microelectronics Technology
Jet Propulsion Laboratory
California Institute of Technology
Pasadena, CA 91109

Abstract

We have used the extreme sensitivity of electron tunneling to variations in electrode separation to construct a novel, compact displacement transducer. Electrostatic forces are used to control the separation between the tunneling electrodes, thereby eliminating the need for piezoelectric actuators. The entire structure is composed of micro-machined silicon single crystals, including a folded cantilever spring and a tip. Measurements of displacement sensitivity and noise are reported. Applications of this displacement transducer to the measurement of acceleration and infrared signals are discussed.

Introduction

Devices used in the measurements of many physical quantities (position, motion, pressure, temperature, radiation) rely on conversion of the physical quantity to a change in the relative separation between a pair of components. Changes in relative position are then detected through changes in the capacitance between the two elements,[1-3] through optical techniques,[4-6] or through flexing of piezoresistors.[7] In each case, large structures are often required to achieve sensitivities that are not limited by readout noise in the transducer. Recently, electron tunneling through a narrow vacuum barrier has been employed in Scanning Tunneling Microscopy (STM), to study the atomic scale structure of surfaces.[8] The tunneling current, I, has the following dependence on the separation between a pair of metallic electrodes, s :

$$I \propto V \exp(- \alpha \sqrt{\Phi}\, s), \qquad (1)$$

where Φ is the height of the tunneling barrier, V is the bias voltage, V is small compared to Φ, and $\alpha = 1.025$ (Å^{-1} eV$^{-1/2}$).[8] For typical values of Φ and s, the current varies by an order of magnitude for each Å change in electrode separation. This sensitivity to relative position is superior to that available in other compact transducers and creates an opportunity for miniaturization of a broad class of sensors without loss of sensitivity. A position sensor based on electron tunneling has already been incorporated into the design for an accelerometer,[9,10] a magnetometer,[11] and several other applications are being considered.

In all tunneling devices, an analog feedback circuit compares the measured tunnel current to a set-point and adjusts the separation between the tunneling electrodes with an electro-mechanical actuator. Typical tunneling devices rely upon piezoelectric actuators. These actuators are known to suffer from sensitivity to thermal drifts, hysteresis,and creep in the response of the piezoelectric materials; these effects can impose limitations on the performance of existing tunneling devices. The construction of miniature sensors which incorporate piezoelectrics is complicated by the variety of different materials required for the construction of the device.

There have been recent advances in the sophistication of micro-machining of silicon through the use of anisotropic etchants and doping to control etching.[12] These advances have led to the development of a new class of sensors composed entirely of micro-machined silicon. The advantages to the development of sensor components in silicon include the use of single crystals as raw material, use of photolithography for precision patterning, and use of batch processing techniques to reduce fabrication costs. Micromachining has been recently used in the fabrication of miniature STMs.[13]

Accelerometer

We have designed and constructed a micro-machined silicon electron tunnel sensor. Micro-machining has been used in this case to produce cm-scale components with μm-scale precision. In contrast to conventional tunneling devices, the relative position of the electrodes is controlled through use of electrostatic forces applied between the elements. The electrostatic forces induce deflection of a micro-machined silicon cantilever spring. Replacement of the piezoelectric actuator with an electrostatic actuator is important for the following reasons. First, the electrostatic actuator is insensitive to thermal drifts and immune to the problems of creep associated with piezoelectric actuators. Also the response of the electrostatic actuator is a function only of the geometry and mechanical properties of the device, whereas the response of piezoelectric actuators is also dependent on the characteristics of the material, which may not be reproducible between devices or over time. Finally, the electrostatic actuator may be miniaturized more easily because the scaling laws are known exactly, and the fabrication is less complex than for the piezoelectric actuator. The device described here can be modified for incorporation into a wide variety of sensors. Monolithic devices including sensor and control electronics are also feasible.

The tunnel sensor consists of two components which are constructed from 200 μm silicon wafers. The wafers have been polished on both surfaces and are coated with > 0.5 μm SiO_2 which is patterned by standard photo-lithographic techniques. The wafers are etched in ethylene diamine pyrocatechol (EDP), removing the parts of the silicon wafers not covered by the SiO_2 mask. After etching, the remaining oxide is removed in a buffered HF etch. A new oxide layer >1 μm thick is then grown on all surfaces of the structure. Gold electrodes are thermally evaporated onto the components of the sensor through shadow masks which have been fabricated by the same micro-machining techniques. The SiO_2 serves as a dielectric isolation layer between the metal films and the silicon substrate. Figure 1 shows a sketch of the components which result from etching and metallization. The components are approximately 4 cm^2 in area. The inner rectangular area of the folded cantilever spring can be deflected upward or downward relative to the outer segments by application of an electric potential between the large electrode and a corresponding deflection counter-electrode which is deposited on another component of the sensor. Given the mechanical properties of the silicon as well as the dimensions of the spring structure, we can calculate the properties of the spring. Because the spring constant scales rapidly with the dimensions of the legs, its characteristics can easily be tailored to meet the needs of a specific application.

Various methods for manufacturing a suitable tunneling tip are available.[8,14] We have formed silicon tips directly from the substrate by undercutting a 60 μm x 60 μm square of SiO_2 with EDP. When the undercutting is complete, the square fragment of oxide is carried away,

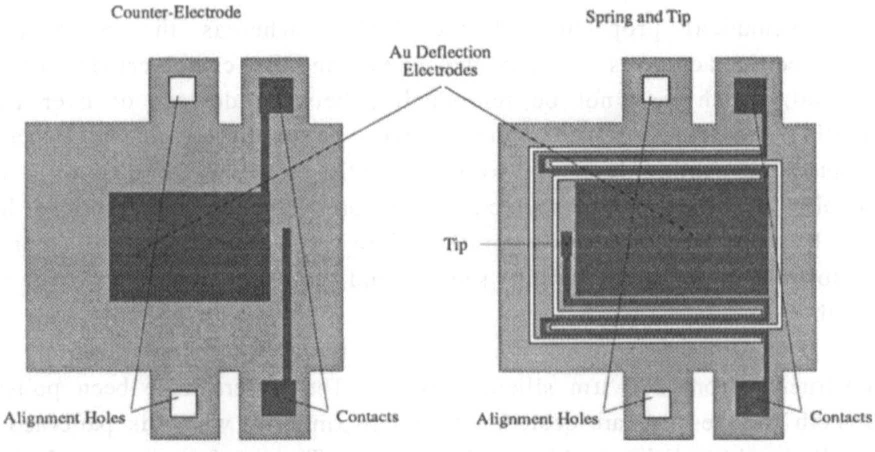

Figure 1. A drawing of the components of the tunnel sensor is shown. The folded cantilever spring with integral tip is shown on the right. The component which supports the tunneling counter-electrode and the deflection counter-electrode are shown on the left. Au films are deposited through micro-machined shadow masks to form the deflection and tunneling electrodes. The tabs with the square holes allow the components to be constrained laterally by alignment pins.

leaving a pyramid-shaped silicon tip, approximately 40 μm tall. The active surfaces of the tips are prepared by evaporation of 3000 Å thick Au films through a shadow mask.

Once assembled, a bias voltage is applied to the electrostatic deflection electrodes to close the electrodes and establish a tunnel current . Active regulation of the tip-electrode separation using feedback control of the tunneling current is carried out as for STM. The tunnel position sensor implemented in this way is generally applicable to a wide range of sensing applications.

The sources of noise observed in vacuum tunneling devices are not fully understood. At high frequency, shot noise and Johnson noise are expected to dominate. At lower frequencies, noise with a 1/f characteristic typically dominates. This noise may originate from environmental sources of noise, such as vibration, or air currents. Thermal drift, which leads to differential thermal expansion may also contribute to the low frequency noise in vacuum tunneling devices. Other mechanisms for 1/f noise have been proposed.[15-18] With the feedback loop closed and constant electrode separation being maintained, the tunnel current noise spectrum can be measured. Typical spectra for a current of 1.3 nA and bias of 100 mV are shown in Fig. 2 for different settings of the bandwidth of the feedback circuit. For these conditions, shot noise is expected to contribute 1.5×10^{-14} A/$\sqrt{Hz}$, which is below the observed noise. The current noise due to the components of the electrical feedback circuit is also well below the observed noise. The tunnel current noise power spectrum varies approximately as 1/f for frequencies below 10^3 Hz. This result is consistent with noise measurements reported elsewhere,[19] and is likely due to environmental noise sources such as vibrations or thermal effects. At frequencies greater than 10^4 Hz, an electrical resonance related to the deflection electrodes introduces additional noise.

By applying a low-frequency modulation to the tunnel current set-point in the feedback circuit, a modulation in the deflection voltage and tunnel current are produced. The barrier height, Φ, may be determined from the ratio of the amplitudes of the modulations. For the Au electrodes in this device, measurements of Φ have given a value of 0.5 eV, which corresponds to a displacement responsivity of 0.94 nA/Å.

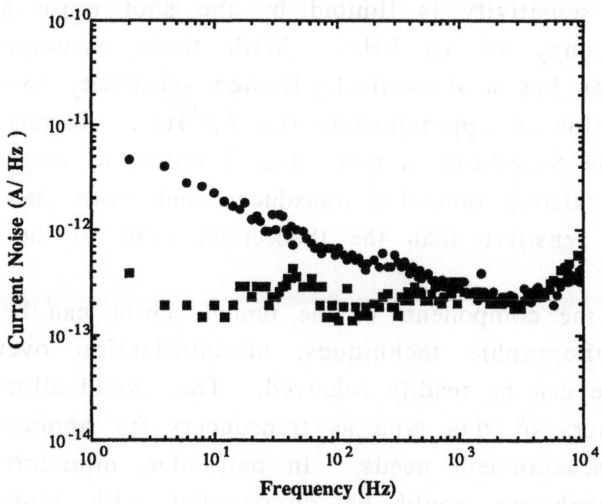

Figure 2 This figure is a plot of the spectrum of the measured noise in the tunnel current at a bias of 100 mV and current of 1.3 nA while operating in air. The bandwidth of the feedback circuit was deliberately adjusted to be less than 0.1 Hz for the data shown as circles and more than 200 Hz for the data plotted as circles.

Given the measured noise, the sensitivity to variations in the tip-sample separation is 2×10^{-4} Å/$\sqrt{Hz}$ at a frequency of 1 kHz.

For a cantilever mass of 30 mg and a measured spring constant of 60 N/m, the dynamical response to an applied acceleration may be calculated from Newtonian mechanics. Combined with the experimental measurement of the sensitivity of the displacement transducer, the responsivity of this sensor to changes in acceleration may be calculated. Below the resonant frequency of about 200 Hz, the responsivity is 3×10^{-4} A/g. With the observed noise of the sensor, the sensitivity is 1×10^{-8} g/$\sqrt{Hz}$ at 10 Hz. At 1 kHz, the sensitivity is 1×10^{-7} g/$\sqrt{Hz}$. Since the measured noise includes contributions from accelerations in the laboratory, the actual sensitivity is greater. This sensitivity is several orders of magnitude better than conventional compact accelerometers.[1,12]

For comparison, the sensitivity of a compact capacitive displacement transducer may be calculated. It is assumed that the electrode separation is no less than 1/100 of the electrode length, that the voltage is no greater than the limit for dielectric breakdown in vacuum, and that sensitivity is limited by the shot noise at the ac measurement frequency of 10 kHz. With these assumptions, the capacitive transducer has a theoretically-limited sensitivity to variations in electrode separation of approximately 0.2 Å/$\sqrt{Hz}$. In this limit, the theoretical capacitive transducer is more than 3 orders of magnitude less sensitive than the existing tunneling transducer, and more than 4 orders of magnitude less sensitive than the theoretical limit to the tunneling transducer.

Since all of the components of the tunnel sensor can be prepared using standard lithographic techniques, miniaturization over several orders of magnitude can be readily achieved. This would allow low-cost production of sensors of this type as transducers for application to a wide variety of measurement needs. In particular, miniature pressure sensors and microphones could be constructed with tunnel sensor technology.

Infrared Sensor

An important application of this transducer that is presently being investigated is the use of a tunnel sensor as the transducer in a pneumatic infrared detector. The pneumatic infrared detector was originally developed by Golay. Golay's detector consisted of a small cavity filled with gas at room temperature and separated from the surroundings by a window and a thin membrane.[20-22] The membrane was coated on one side with a thin metallic film, which has significant absorption throughout the infrared whenever the sheet resistance of the film is approximately half of the impedance of free space.[23-25] The gas was heated by contact with the membrane, and expanded thermally, which forced the membrane to expand outward. This expansion is detected through optical or capacitive techniques.[26] These detectors are quite fragile, difficult to fabricate, and expensive. Nevertheless, they have been widely used, primarily because of their improvement in sensitivity over all other room-temperature detectors in the mid to far infrared.

With the above considerations in mind, we have begun the development of an improved pneumatic infrared detector. This detector is constructed entirely from micromachined silicon, which allows the precision control of lithographic techniques to be applied to the production of components from single crystals.[12] In this device, a free-standing silicon oxynitride (SiO_xN_y) membrane is deflected by the trapped gas. To detect the motion of the membrane, we use a vacuum tunnel sensor.[27]

A cavity of area (0.1 cm)^2 and thickness 0.015 cm, filled with air at atmospheric pressure is trapped between a pair of silicon wafers, one of which has been etched through to a 0.5 μm thick SiO_xN_y membrane as shown in Fig 3. The outer surface of the membrane is coated with ~100 Å of gold to serve as an electrode for tunneling as well as an efficient absorber of infrared radiation. The heat capacity of this structure is dominated by the membrane, while the thermal conductance is dominated by conduction through the trapped gas to the window. A calculation of all the thermodynamic properties of the structure as well as the responsivity and contribution from various noise sources has been carried out. This calculation is analogous to that performed for the original Golay cell, except that an electron tunneling transducer is incorporated to detect the expansion of the trapped gas. Noise sources that play an important role in this device include thermodynamic

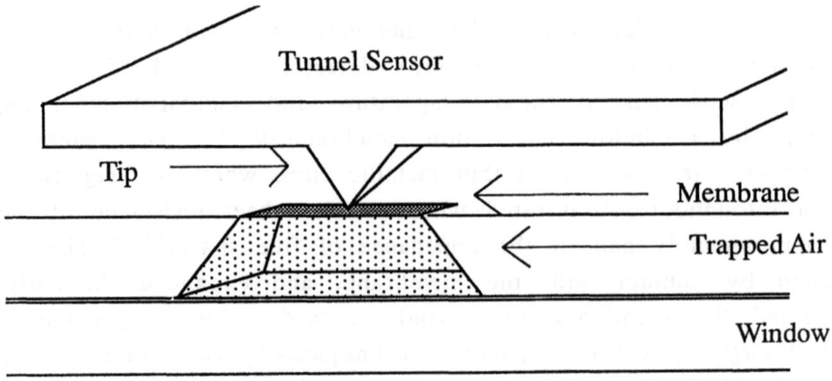

Figure 3 A drawing of the prototype detector is shown. A silicon wafer with a free-standing SiO_xN_y membrane is pressed against a second silicon wafer to trap a small volume of gas at room temperature. A 100 Å Au film on one side of the membrane acts at the IR absorber. The tunnel sensor detects the motion of the membrane which is driven by the thermal expansion of the trapped gas.

fluctuations in the temperature of the cell, shot noise in the transducer and fluctuations in background radiation.

If we sum the contributions to the noise, and consider the 50% efficiency of the absorber, the predicted noise equivalent power (NEP) of the prototype is 6×10^{-11} W/$\sqrt{\text{Hz}}$ at low chopping frequencies. At frequencies below 10 kHz, the NEP is dominated by thermal fluctuations. At frequencies above 10 kHz, the contribution from shot noise dominates the other sources of noise and degrades the NEP of the detector. The contributions to the NEP of the prototype are plotted as a function of frequency in Fig. 4.

The original pneumatic infrared detectors featured NEPs as low as 7×10^{-11} W/$\sqrt{\text{Hz}}$, but were not suitable for miniaturization. The need for focal-plane arrays of room temperature detectors stimulated the development of pyroelectric detector arrays, the best of which feature NEPs of about 10^{-9} W/$\sqrt{\text{Hz}}$.

The NEP of this detector may be further improved by reducing the thermal conductance (G). Techniques for reducing G include using Xenon, reducing the area of the membrane and increasing the thickness of the cavity to reduce the surface area-to-volume ratio. These

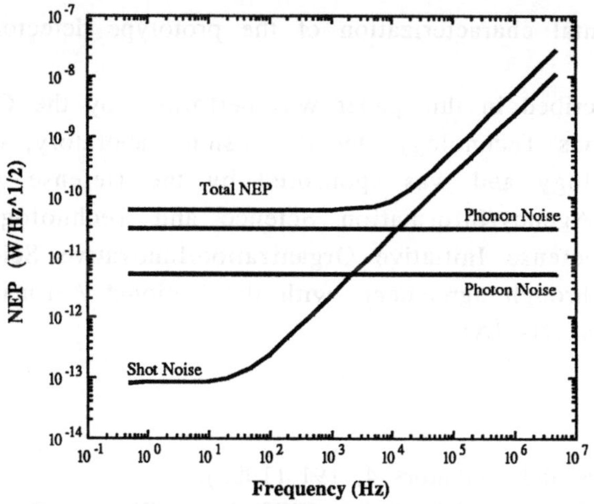

Figure 4 The contributions to the NEP of this detector are plotted as a function of the modulation frequency. Temperature fluctuations, or phonon noise limit the performance at modulation frequencies below 10^4 Hz, and shot noise in the transducer limits the performance above 10^4 Hz. Noise due to statistical fluctuations in a 300 K background is also plotted.

modifications allow reductions in the NEP of more than a factor of 5. This detector can operate at lower temperatures through the use of a more ideal gas, such as helium. The contribution to the NEP from thermal fluctuations should decrease linearly with temperature. Below 100 K, other technologies, particularly HgCdTe and InSb photoconductors, should still be superior to this detector.

The tunneling infrared sensor, being assembled entirely from micromachined silicon, can be miniaturized and batch-fabricated or integrated into arrays. Individual detectors of diameter less than 100 µm can be assembled without development of additional technology. Each element would require a tunneling transducer with its own feedback circuit; therefore, some integration of sensor and control electronics would be important.

There are many potential applications of this detector since it is sensitive to radiation throughout the infrared. However, it appears uniquely suited to applications that benefit from or require the use of an uncooled detector in measurement of radiation in the mid to far infrared. Experimental characterization of the prototype detector is now underway.

The work described in this paper was performed by the Center for Space Microelectronics Technology, Jet Propulsion Laboratory, California Institute of Technology and was sponsored by the Defense Advanced Research Projects Agency/Information Science and Technology Office and the Strategic Defense Initiative Organization/Innovative Science and Technology Office through agreements with the National Aeronautics and Space Administration (NASA).

References

1 F. Rudolf, Sensors and Actuators 4, 191 (1983).

2 K.E. Petersen, A. Shartel and N. Raley, IEEE Trans. Electron Devices ED-29, 23 (1982).

3 C.S. Sander, J.W. Knutti, and J.D. Meindl, IEEE Trans. Electron Devices ED-27, 927 (1980).

4 E. Stemme and G. Stemme, IEEE Trans. Elect. Dev. 37, 648 (1990).

5 D.L. Gardner, T. Hofler, S.R. Baker, R.K. Yarber and S.L. Garrett, J. Lightwave Tech. LT-5, 953 (1987).

6 D. Rugar, Rev. Sci. Instrum. 59, 11 (1988).

7 H.V. Allen, S.C Terry and D.W. De Bruin, Sensors and Actuators 20, 153 (1989).

8 G. Binnig and H. Rohrer, IBM J. Res. Develop. 30, 355 (1986).

9 S.B. Waltman, W.J. Kaiser, Sensors and Actuators 19, 201 (1989).

10 A.A. Baski, T.R. Albrecht and C.F. Quate, J. of Microscopy 152, 73 (1988).

11 J.H. Wandass, J.S. Murday, and R.J. Colton, Sensors and Actuators 19, 211 (1989).

12 K.E. Petersen, Proc. IEEE 70, 420 (1982).

13 S. Akamine, T.R. Albrecht, M.J. Zdeblick, and C.F. Quate, IEEE Electron Device Lett. 10, 490 (1989).

14 R.B. Marcus, T.S. Ravi, T. Gmitter, K. Chin, D. Liu, W.J. Orvis, D.R. Ciarlo, C.E. Hunt and J. Trujillo,. Appl. Phys. Lett. 56, 236 (1990).

15 M.E. Welland and R.H. Koch. Appl. Phys. Lett. 48, 724 (1986).

16 J.B. Pendry, P.D. Kirkman and E. Castano, Phys. Rev. Lett. 57, 2983 (1986).

[17] M.F. Bocko, K.A. Stephensen, and R.H. Koch, Phys. Rev. Lett. 61, 726 (1988).

[18] B. Yurke and G.P. Kochanski, Phys. Rev. B 41, 8184 (1990).

[19] D.W. Abraham, C.C. Williams, and H.K. Wickramasinghe, Appl. Phys. Lett. 53, 1503 (1988).

[20] M.J.E. Golay, Rev. Sci. Inst. 18, 347 (1947).

[21] M.J.E. Golay, Rev. Sci. Inst. 20, 816 (1949).

[22] A. Hadni, Essentials of Modern Physics Applied to the Study of the Infrared (Pergamon, Oxford, 1967) pp 269-283.

[23] L.N. Hadley, D.M. Dennison, J. Opt. Sci. Am. 37, 451 (1947).

[24] C. Hilsum, J. Opt. Sci. Am. 44, 188 (1954).

[25] A. Hadni, Essentials of Modern Physics Applied to the Study of the Infrared (Pergamon, Oxford, 1967) pp. 252-7.

[26] M.J.S. Smith, L. Bowman, J.D. Meindel, IEEE Trans. Biomed. Eng. BME-33, 163 (1986).

[27] T.W. Kenny, S.B. Waltman, J.K. Reynolds, and W.J. Kaiser, Appl. Phys. Lett. 58, 100 (1991).

Diode Lasers for Industrial Length Measurements

A. ABOU-ZEID

Physikalisch-Technische Bundesanstalt
Budesallee 100, D - 3300 Braunschweig

Due to the attractive features of diode lasers such as compact size, low cost, fast tunability of their wavelength, high emission power etc., they are practical sources for a wide variety of potential applications in dimensional metrology. The diode laser is already used today as a distance sensor according to different methods such as triangulation, time of flight, auto focussing, heterodyne interferometry etc. For all these methods the relative length uncertainty amounts to 10^{-2} up to 10^{-4} and therefore a great effort for the frequency stabilization of the diodé laser is not necessary.

Despite the many advantages, length measuring systems with a relative length uncertainty better than 10^{-6} which use a diode laser as a light source have only had a limited success. The limitations are based on:

-- the relatively high frequency sensitivty to both diode parameters temperature and current [1] ,
-- the significant differences in absolute frequency and characteristics which exist even in the case of diodes of the same type [2] and
-- the spectral ageing of the diode laser [3].

These disadvantages can be compensated by suitable frequency-stabilizing methods so that the large number of advantages offered by the diode laser become effective in industrial length measurements.

Frequency stabilization of diode lasers is achieved either by parameter or reference methods, which require different efforts and have consequently various frequency stabilities. This means that the required length measurement uncertainty and the value of the measured length are determind by the choice of a suitable stabilization method.

In this work we present the important properties of diode lasers for interferometric applications and the technologies concerning the frequency stabilization of diode lasers. More emphasis was made on the own experimental results of parameter, line scale and absorption line stabilization methods. Furthermore recent developments in the field of the diode laser technology which make them more attractive for dimensional metrology are reported.

The parameter method, in which the diode's parameters are kept constant, does not require much effort and cost, but due to the spectral ageing of the diode laser, its emitted frequency must be calibrated over a period of some months to maintain a reproducible relative frequency stability of $\approx 10^{-6}$. Modular interferometers with a parameter stabilized diode laser has been developed at PTB and have proved their efficiency in combined length and angle measurements [4].

For interferometric displacements of a higher precision ($\leq 10^{-6}$), the frequency of the diode laser must be stabilized direct to a reference, e.g. a Fabry-Perot etalon [5], a line scale or an atomic or molecular absorption line [6]. In this way the necessity of considering of the spectral ageing of the diode laser is avoided, so far the reference is temporally sufficient stable.

For the stabilization of the wavelength of a diode laser to a line scale as a reference, a measuring arrangment has been set up and tested (s. fig. 1) which consists of a precision probe with an incremental linear encoder (Heidenhain: length 60mm, resolution ±50nm and accuracy 100nm) and an interferometer with a parameter stabilized diode laser. The measuring reflector of the interferometer is fixed to the head of the probe. The wavelength stabilization is done by the comparison of the displacement of the measuring reflector for both length measuring systems. With the aid of digital regulation, a deviation of the wavelength from a pre-determined value of the wavelength is fed back to the current of the diode to compensate any wavelength fluctuations due to variations of the refractive index of the air and the diode's parameters, as well as due to the spectral ageing of the diode laser. The digital regulation was tested [7] by changing of the diode's temperature (±0.2K) or the refractive index ($\pm 1*10^{-5}$). The results show that the wavelength in air can be stabilized up to a relative uncertainty of $\approx 8*10^{-7}$ (s. fig. 2).

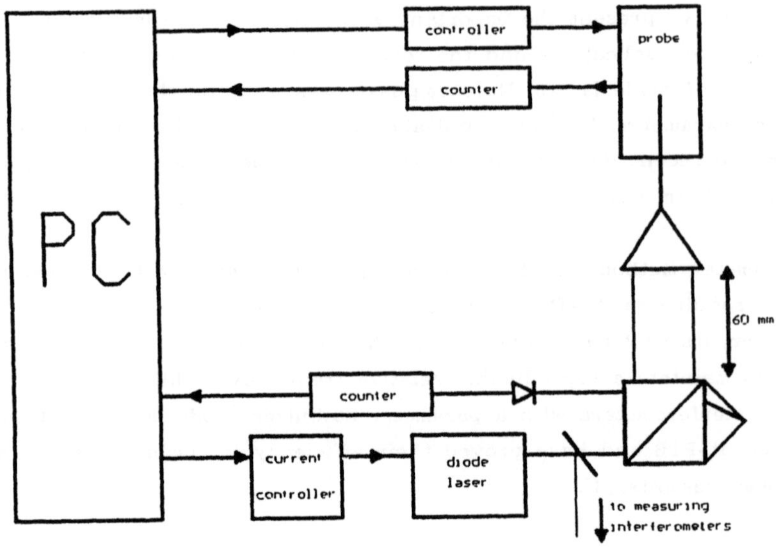

Fig. 1 Block diagram of the experimental set up for the stabilization of the wavelength of a diode laser in air to a line scale as a reference

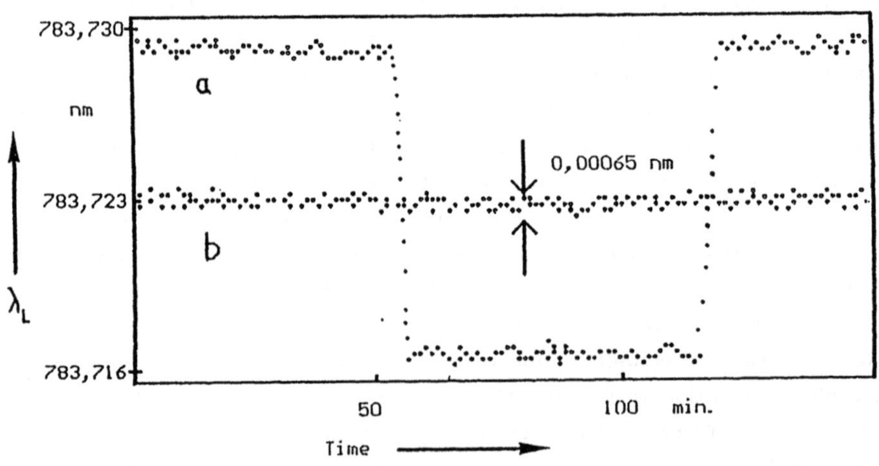

Fig. 2 Wavelength of a diode laser in air, by simulation of a refractive index change of ± 10^{-5}, a and b: without and with digital regulation respectively

This length uncertainty is sufficiently small for several applications in the dimensional metrology, e.g. guide errors of machine tools and measuring machines, leveling rods, measuring tapes, surface roughness etc. Due to the high radiant power emitted (≤100mW), multi-axis interferometers with a diode laser as a common light source are of increasing interest for measuring systems of large coordinate measuring machines.

The frequency of a diode laser has been locked to the D_2 absorption line of Rb (780.24nm). The principle of the frequency locking scheme is illustrated in fig. 3. The diode's current has been modulated to 50kHz with an amplitude of ≈10μA to gain a control signal. The collimated laser beam was directed through an 45mm long Rb cell. The absorption was detected as a difference in the laser beam intensity before and after the cell. The difference of both detected signals was sent to a lock-in amplifier. Its output signal (error signal) was amplified by a PID circuit and then fed back into the current supply of the laser diode to lock the frequency of the diode laser to the center of a Rb atom resonance line.

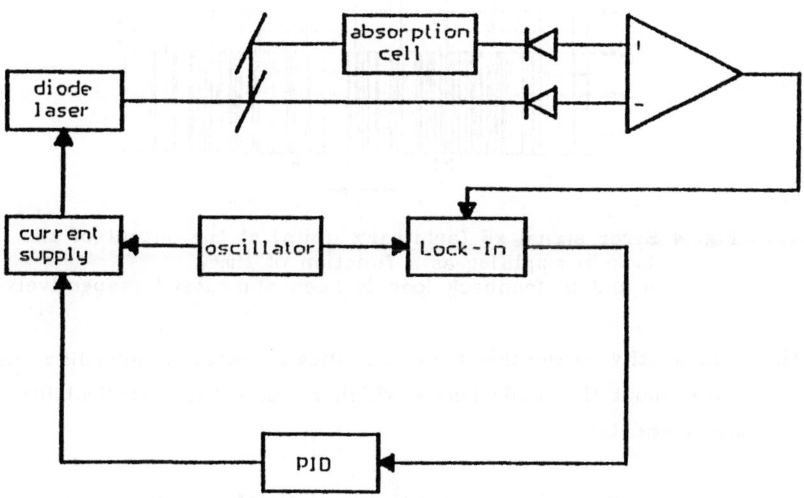

Fig. 3 Block diagram of the experimental set up for the stabilization of the frequency of a diode laser to an absorption line

The effect of the radiant power of the beam passing through the absorption cell and its temperature on the measured frequency of the stabilized diode laser has been studied.

An analysis of the relative stability of the frequency was based on measuring the size of the error signal. Preliminary studies [8] on the frequency stability of Rb-stabilized diode lasers show (s. fig.4) a short term stability (some hours) of better than 10^{-9} which promise the diode laser as an excellent candidate for precision length measurments.

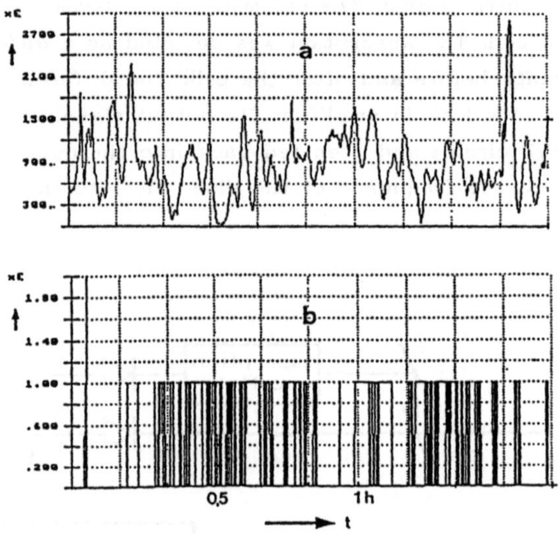

Fig. 4 Error signal xE (arbitrary units) at the output of the lock-in amplifier as a function of time,
a and b: feedback loop is open and closed respectively

Finally it is worthy to mention three significant features regarding the recent developments about the diode lasers which promote their applicability for length measurements:

-- Diode lasers emitting in the visible (Toschiba 670nm, Philips 630nm) spectrum range [9] are today commercially available

-- Diode lasers with a large hopping-free range (30K) has been developed and are also commercially available (Sharp 780nm).

-- For diode lasers as semiconductor devices, the possibility presents itself for bringing up of the diode laser together with the interferometer components on a Si chip with the aid of the integrated optics technology. In this way we have the considerable advantages opposite to conventional gas laser interferometers:

-- an enormous reduction of volume, weight and cost of the laser interferometer and

-- simultaneously increasing of mechanical and thermal stability of its components.

Such an integrated diode laser interferometer has found its first application as a displacement sensor [10].

References

1. Abou-Zeid, A.; PTB-Mitt., 94 (1984) 163
2. Abou-Zeid, A.; Wolf, M.; Technica, 20 (1988) 25
3. Abou-Zeid, A.; Proceeding of the 9th International Congress: Laser Optoelectronic Microwaves, Munich, 5.-9. june (1989)
4. Abou-Zeid, A.; Seminar Lasermeßtechnik, Haus der Technik, Essen, 25.-26. sept. 1990
5. Tsuchida, H.; Mitsuhashi, Y.; Appl. Opt., 27, 2 (1988) 322
6. Ohtsu, M.; Opt.and Quantum Elect., 20 (1988) 283
7. Abou-Zeid, A.; Erdtmann, B.;Kunzmann, H.; Prellinger, G.; to be published
8. Abou-Zeid, A.; Bader, N.; Prellinger, G.; to be published
9. Kumagai, O.; Ikeda, M.; Nakano, K.; Toda, A.; Ohatas, T.; Sato, H.; Kojima, C.; Proc. SPIE, 898 (1988) 80
10. Ülbers, G.; Springer Proc. in Physics, vol. 41 (1989) 240

Paving the Way to Thermally Stable Coordinate Measuring Machines

K.H. Breyer and H.G. Pressel

1. Introduction

The manufacturers of high and ultra-precision coordinate measuring machines (CMMs) go to great lengths when it comes to temperature control of locations for CMM assembly, adjustment and acceptance testing (1).
Ruling of the scales, fine finishing of the guideway elements, precision adjustment of the machines and the metrological determination of corrective parameters are all performed under virtually ideal thermal conditions.
The main reason for this is the demand for a metrological safety margin when working with measuring and checking devices which are themselves subject to thermal influences, such as laser interferometers, test configurations and gage blocks, for example.
The geometry of conventional CMMs, on the other hand, particularly where straightness of guideways, perpendicularity of axes and elongation of measuring standards are concerned, can only be clearly defined if the appropriate thermal conditions are maintained.

Air-conditioned assembly halls with a thermal stability of ± 0.5 K/day and precision-climatized cabins of ± 0.1 K/day guarantee that a stable condition is documented during acceptance testing at the manufacturing plant. This condition is again retrieved after installation on the user's premises if similar thermal conditions are provided.
In actual fact, however, it is quite a problem to obtain these specified thermal conditions at the site of installation:

the CMM which operates in the manufacturing area and is frequently integrated into material flow systems is inevitably subject to prevailing thermal variations (2).
A technically proven solution to this problem is protective enclosures (Fig. 1), but these are not always accepted due to the costs they involve and to the complications in CMM loading.

Fig. 1: Coordinate measuring machine inside climatized cabin

The air-conditioning of cleanroom areas for conventionally designed high-precision CMMs may frequently swallow additional costs to the tune of DM 100,000.-- to 200,000.-- and is therefore often not implemented as comprehensively as required.

The demand made on the manufacturers of CMMs therefore is: "reducing the thermal sensitivity of the CMM while maintaining or even improving its accuracy".

2. The main causes for thermal trouble in a CMM

Deviations from the reference temperature of 20 °C as defined by DIN 102 cause changes in the lengths of the measuring scales of the CMM and in those of the workpiece. Their reliable software correction is possible only to the extent to which the actual temperature can be determined and to which the expansion coefficients are known.
In addition, the temperature distribution inside the scale and the workpiece must be homogeneous.

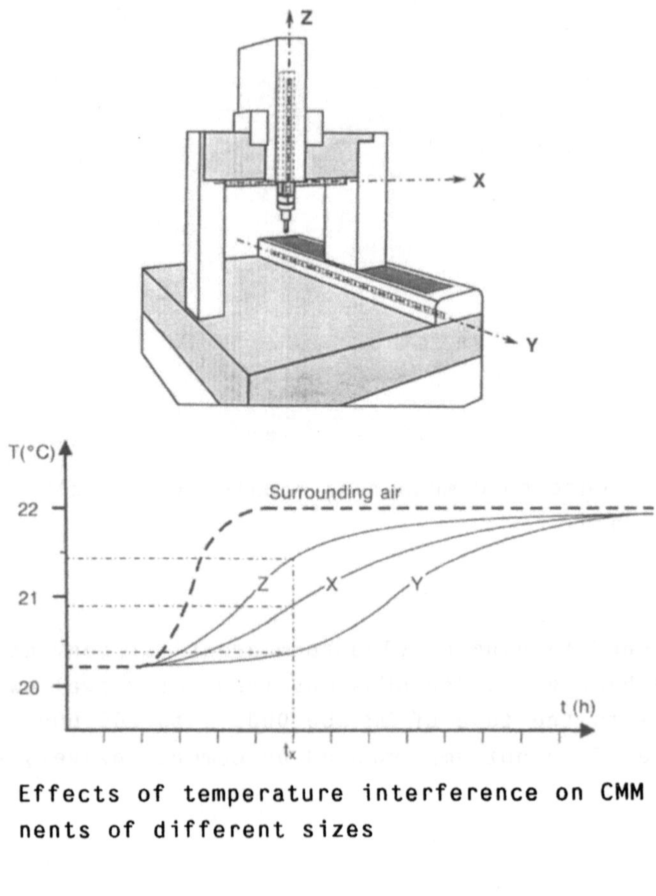

Fig. 2: Effects of temperature interference on CMM components of different sizes

Spatial thermal gradients in the air volume of the CMM and radiation from adjacent machines, walls, windows or lamps and draught impair, above all, the straightness of the guideways and their perpendicularity in relation to each other.

In the event of temporal temperature variations, the different thermal inertiae of the individual CMM components mean that different times are required to adapt to the new temperature level. As a result, the Z-scale mounted on the spindle of a bridge-type CMM may have a different temperature from the X-scale on the crossbeam and the Y-scale on the granite table (Fig. 2).

3. Fundamental possibilities for reducing the thermal sensitivity of CMMs

The methods for reducing thermal influences on the measuring result known to date can be categorized into three groups:

- Measures to keep thermal interference away from the CMM.
- Compensation of the effects of thermal interference which reach the CMM and impair its performance.
- Use of materials and design principles which display no or only minor reactions to thermal interference.

In all these options, the main objective must be to minimize the quantities to be corrected. After all, any correction can only be performed after the interference factor has previously been metrologically determined, and this is possible only within limited accuracy.

4. Details for optimum thermal stability in CMMs

4.1 Reducing the effects of internal heat sources

Drives, especially in high-speed coordinate measuring machines, constitute heat sources which may cause thermomechanical deviations.
It is therefore vital that the machines be designed in such a way that heat exerts as little influence on the scales or guideway components as possible.
For this purpose, the drives are encapsulated and air-cooled. Even better protection against internal heat sources is provided if scales and guideway components are made of thermally stable materials.

4.2 Insulation

One of the most effective measures for protecting the CMM against thermal influences is to place it in an insulated cabin (Fig. 1). This cabin also fulfills the function of a safety barrier and keeps out dirt. Fig. 3 shows to what degree variations of the external temperature are excluded.

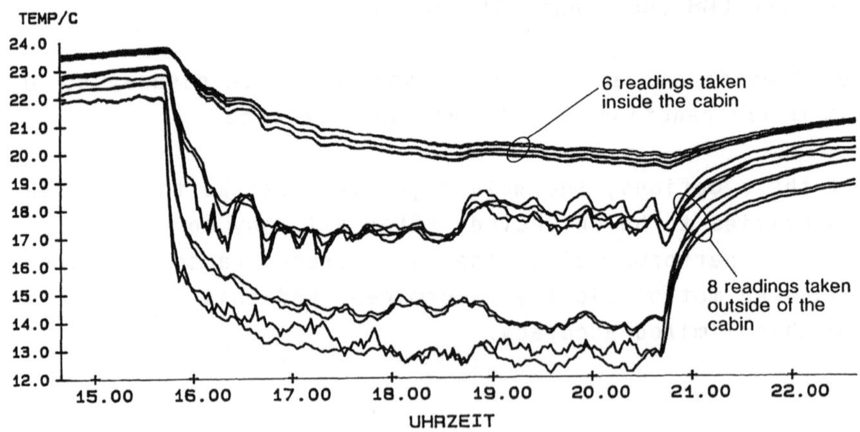

Fig. 3: Reduction of temperature interference using a protective cabin

The damping effect of a well-insulated cabin is considerable. If this is not sufficient, the cabin can be equipped with its own air-conditioning system and an air shower for the workpiece.
The previously listed disadvantages, such as cost, space requirements and complex loading, are the reasons why cabins are seldom used.

Complete machine enclosures - a standard feature in machine tool building - are a good, space-saving alternative. The FMC production measuring center shown in Fig. 4 is effectively protected against thermal influences by such an enclosure, which is a perfect example of the principle that "it is better to exclude interference altogehter than to compensate for its effect."

Fig. 4: Completely enclosed FMC CMM

The thermal protection of individual components is a step in this direction. The time-tried casing of the CMM bridge with PU foam elements has been awarded top marks in a comparison by the Physikalisch Technische Bundesanstalt in Brunswick. The tested UMC-type CMM still conformed to accuracy specifications even after the temperatures of the climatic chamber walls rose to the level shown in Fig. 5.

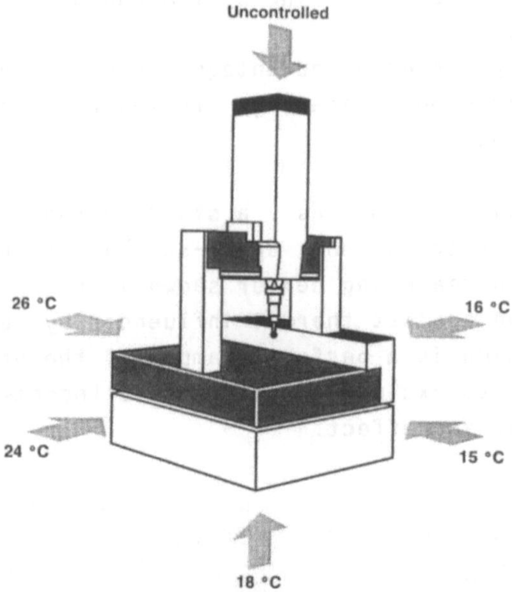

Fig. 5: Conditions for testing the thermal stability of a partially insulated CMM using different radiant temperatures

Another proven method is to insulate the granite table of those CMMs whose other guideway components are made of materials insensitive to thermal bending (Fig. 6).
But the top surface of such insulated granite tables can not be protected in the same way, as the workpiece has to be clamped on it. Thermal differences entering from above may therefore result in bending of the granite plate (see chapter 4.5).
The effects of such bending can be reduced by measuring the temperature differences and compensating them by software correction (see 4.6).

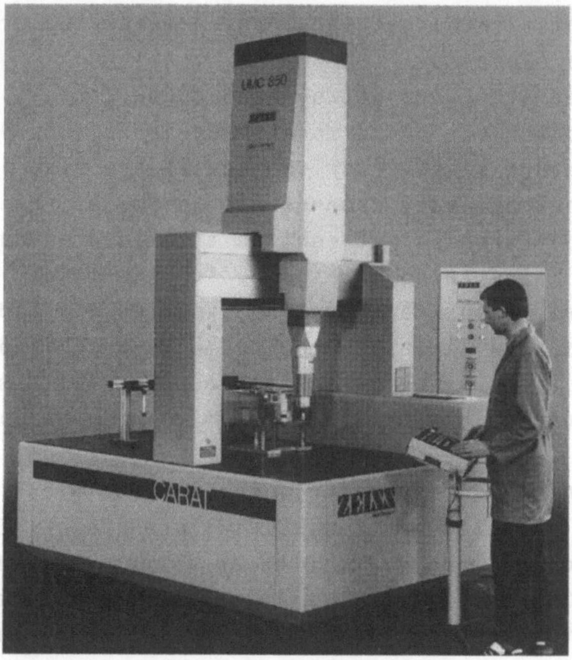

Fig. 6: Thermally insulated granite table

4.3 Thermally insensitive length-measuring systems

Coordinate measuring machines are frequently equipped with glass scales. Their expansion coefficient normally lies around 5 .. $10 \cdot 10^{-6}$ 1/K, depending on the type of glass. In higher-accuracy CMMs the glass scales are therefore provided with temperature sensors for the linear correction of thermally induced expansion.
This correction has its limits, as the actual expansion coefficient of the respective glass charge is rarely known, with variations of $0.5 \cdot 10^{-6}$ 1/K being realistic. The accuracy of measuring the scale temperature is also restricted, and is additionally aggravated by the inhomogeneous distribution of temperature in the scale. Even with

the most sophisticated technical setup measuring errors of 0.5 K must be expected.
Scales made of steel are better heat conductors, i.e. the temperature distribution is more homogeneous than in glass. Here, temperature sensors for the scales are also indispensable if higher accuracy demands are to be met, due to the temperature differences between the three scales described under 2. and illustrated in Fig. 2. An uncertainty of 0.1 K in temperature measurement must be expected. In addition, the differences in expansion coefficients described for gage blocks under 4.4 must be taken into account. With steel scales, an uncertainty of $0.5 \cdot 10^{-6}$ 1/K must be expected.

The uncertainties in expansion specified for glass and steel scales are eliminated if the material used is of ZERODUR ™ glass ceramic. ZERODUR ™ features an expansion coefficient of $0 \pm 0.05 \cdot 10^{-6}$ 1/K and unparalleled ageing resistance.
In a temperature change from 20 °C to 30 °C a Zerodur scale of 500 mm length expands by 0.25 µm at the most; in a steel scale of the same length this would be as much as 68 µm.
An essential aspect to consider when installing thermally stable scales is the way they are mounted on their respective guideway component, whose expansion characteristic is, after all, temperature-dependent. Therefore, a special mounting technique (Fig. 7) was developed for connecting the Zerodur scale to the CARAT guideways consisting of aluminum alloy (see 4.5). By fixing them in one point only and "float-supporting" the rest of the scale, it is guaranteed that no forces are transmitted to the Zerodur scale if the carrier material is subject to changes in length.

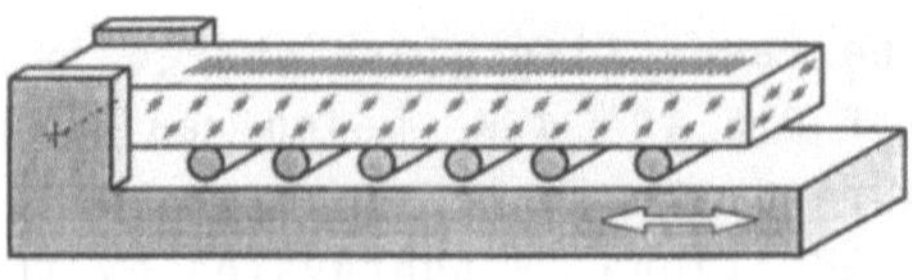

Fig. 7: Principle of mounting a Zerodur scale (compensation of expansion differences)

4.4 Interaction of thermal expansion in the scale and the workpiece

If the scales and the workpiece have different expansion coefficients, correct results can only be obtained without correction if the measurement is performed at exactly 20 °C. If the temperature deviates from 20 °C, correct results can only be obtained without correction if all three CMM scales and the workpiece have

- exactly the same temperature and
- exactly the same expansion coefficient.

This is not feasible in practical work. Even the shortest thermal variations entail temperature differences at the scales, as shown in chapter 2.
Identical expansion coefficients are also unrealistic, due to the diversity of the parts to be measured. Even if only steel parts were to be tested on a CMM equipped with steel scales, the expansion characteristics would still be different.
Two sets of gage blocks measured at Carl Zeiss showed completely different results in the expansion coefficients α of the individual blocks:

Set no. 1: α = 10.1 to $12.3 \cdot 10^{-6}$ 1/K
Set no. 2: α = 10.3 to $12.9 \cdot 10^{-6}$ 1/K

If the correction value $11.5 \cdot 10^{-6}$ 1/K specified by the manufacturer is used, the maximum correction error which may have been made is 1.4 µm/m·K. Even if the temperatures of the scale and the workpiece are absolutely identical, this would result in an error of 2.8 µm per meter at a measurement temperature of 22 °C. It must be expected that steel scales do not always have the same expansion coefficient either, and correction is subject to error when performed in temperatures other than 20 °C.
The expansion of the workpiece must also be corrected if the measurement temperature is not 20 °C. For this, calibrated sensors are needed on the workpiece and the actual expansion coefficient α must be known. In standard AlMgSi alloys, the

variations in Si content may cause differences in α of up to $2 \cdot 10^{-6}$ 1/K. If an aluminum part of 500 mm length were to be measured at 25 °C, this would result in a false correction value of 5 µm!

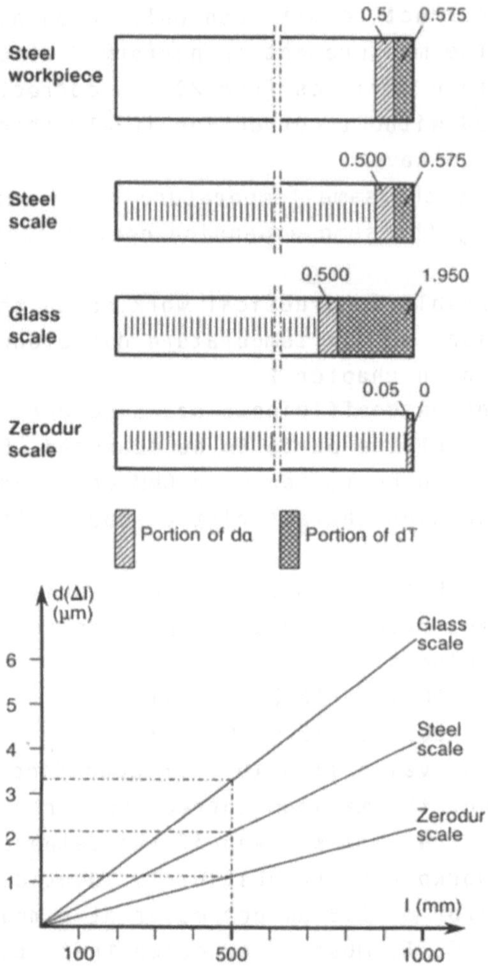

Fig. 8: Error d(Δl) when using measuring scales of different materials with T = 22 °C, workpiece = steel (L = 500 mm)

Fig. 8 shows the correction errors when measuring a steel workpiece using scales made of steel, glass and ZERODUR ™, at a temperature of 22 °C and the following uncertainties supposed:

Error of temperature measurement:	of steel workpiece	0.1 K
	of steel scale	0.1 K
	of glass scale	0.5 K
	of Zerodur scale	no influence
Uncertainty α:	steel workpiece	$0.5 \cdot 10^{-6}$ 1/K
	steel scale	$0.5 \cdot 10^{-6}$ 1/K
	glass scale	$0.5 \cdot 10^{-6}$ 1/K
	Zerodur scale	$0.05 \cdot 10^{-6}$ 1/K

Basis of the formulae

Expansion correction Δl

$$\Delta l = l_0 \cdot \alpha_M (T_M - 20) - l_0 \cdot \alpha_W (T_W - 20)$$

l_0 = measured length in mm
α_M = expansion coefficients of scale
α_W = expansion coefficients of workpiece
T_M = temperature of scale
T_W = temperature of workpiece

Uncertainty of expansion correction $d(\Delta l)$

$$d(\Delta l) = \underbrace{|l_0 \cdot dT_M \cdot \alpha_M|}_{(A)} + \underbrace{|l_0 \cdot dT_W \cdot \alpha_W|}_{(B)} + \underbrace{|l_0 (T_M - 20) d\alpha_M|}_{(C)} + \underbrace{|l_0 (T_W - 20) d\alpha_W|}_{(D)}$$

dT_M = uncertainty of temp. measurement of scale
dT_W = uncertainty of temp. measurement of workpiece
$d\alpha_M$ = uncertainty of α of scale
$d\alpha_W$ = uncertainty of α of workpiece

(A) → error portion from dT_M
(B) → error portion from dT_W
(C) → error portion from $d\alpha_M$
(D) → error portion from $d\alpha_W$

Scale material	l_0 (mm)	T_M=T_W (°C)	A (µm)	B (µm)	C (µm)	D (µm)	d(Δl) (µm)
		21	0.575	0.575	0.250	0.500	1.65
Steel	500	22	0.575	0.575	0.500	1.000	2.15
		23	0.575	0.575	0.750	1.500	2.65
		21	1.950	0.575	0.250	0.500	3.03
Glass	500	22	1.950	0.575	0.500	1.000	4.53
		23	1.950	0.575	0.750	1.500	4.03
		21	0.000	0.575	0.250	0.500	0.85
Zerodur	500	22	0.000	0.575	0.500	1.000	1.13
		23	0.000	0.575	0.750	1.500	1.40

This table clearly shows the unparalleled superiority of ZERODUR ™ as a scale material. The fact that the high-precision sensors must be monitored and possibly exchanged must be taken into account in the temperature measurement.

4.5 Thermally insensitive guideway components

Different temperature layers - caused by external heat flux - building up inside a square guideway component will result in bending (Fig. 9).
With a constant thermal gradient inside the guideway component, the amount of this bending is a function of the expansion coefficient α of the material.
This means that a component made of ZERODUR ™ glass ceramic is not subject to any bending, as α = 0. With an aluminum guideway, on the other hand, thermal bending might be expected due to aluminium's relatively high α value.
However, when comparing different materials, it cannot be taken for granted that the internal thermal gradient is the same in all guideways when a specific external thermal interference is applied. With a constant external heat flux Q,

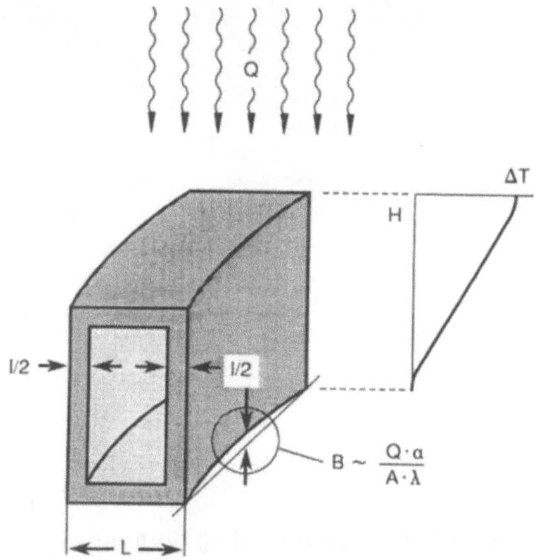

Fig. 9: Bending B of a guideway component caused by a given heat flux

thermal distribution depends on two parameters instead: on the thermal conductivity λ (material feature) and on the cross-over section A in the guideway component. The following holds for the bending of a guideway component:

$$B \sim \frac{Q \cdot \alpha}{\lambda \cdot A} \qquad \text{(see Fig. 9).}$$

This means that a guideway component with a large cross-over section (solid material) and good thermal conductivity (e.g. aluminum) builds up only a minor internal gradient and is therefore only subject to minimum bending (despite a high α value) in the case of one-sided temperature changes.

For the materials granite, ceramic, aluminum and steel currently being used, the relations for thermally induced bending are as shown in Fig. 10 if the influencing parameters are set at 1 for granite.

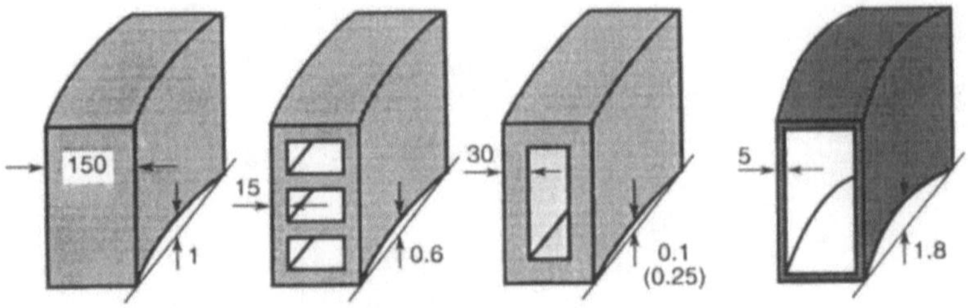

	Granite	Ceramic (Wilbanks AD 96)	Al alloy Solid profile	Al alloy Hollow profile	Steel
Thermal conductivity λ	1	10	40	40	20
Cross-over section I/L	1	0.2	1	0.4	0.067
Cross-over resistance and thermal grandient ΔT	1	0.5	0.025	0.0625	0.75
Expansion coefficient α	1	1.2	4.0	4.0	2.4
Resulting bending B	100%	60%	< 10%	25%	180%

Fig. 10: Bending of guideways at given heat flux through the material (standardized to granite)

It becomes obvious that a solid aluminum profile displays the highest stability with respect to thermal bending. It is 10-times better than granite, 6-times better than ceramic and 18-times better than steel. The table also shows, however, that the time-tried granite guideways display a more favourable thermal behaviour than steel guideways with a wall thickness of 5 mm.
The cross sections were selected as defined by the respective manufacturing technology.

A special finishing, ageing and coating procedure was therefore developed for the guideway components of coordinate measuring machines and has since been launched under the name "CARAT". CARAT stands for Coated Ageing Resistant Aluminum Technology and designates non-ageing aluminum alloy components provided with a ceramic surface coating.

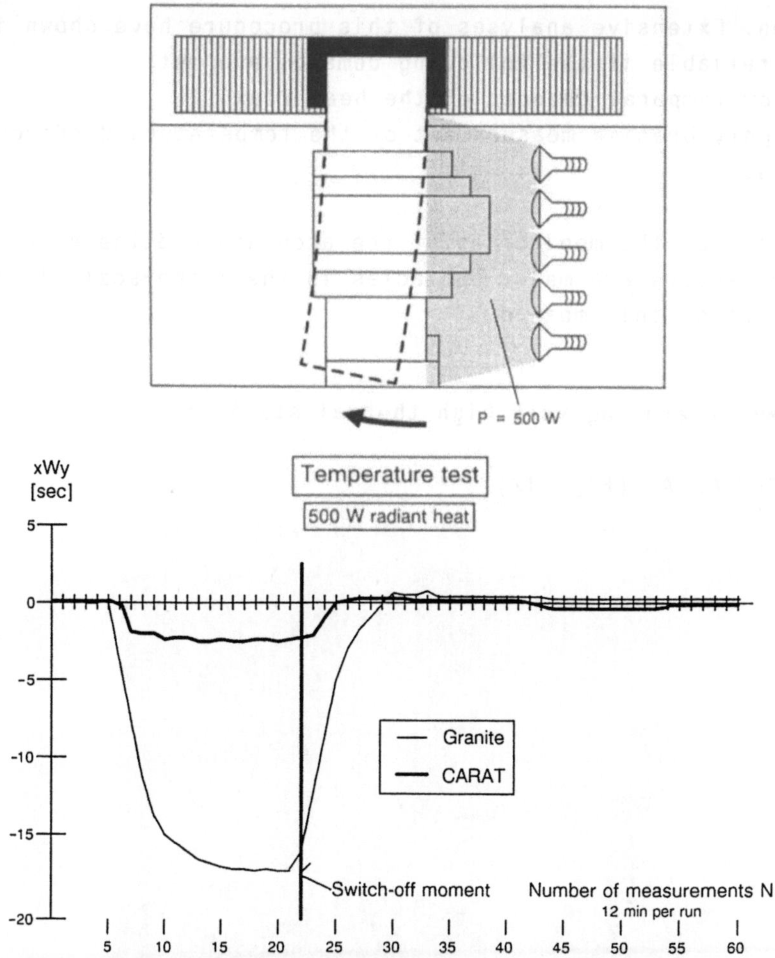

Fig. 11: Bending test on CARAT and granite crossbeams

4.6 Software-compensation of thermomechanical deviations

Thermal bending of guideway components can be software-corrected to a limited extent.
First, the thermal gradient inside the guideway component is measured by two temperature sensors. The bending of the component is calculated and software corrected on the assumption that the thermal gradient is constant across the cross

section. Extensive analyses of this procedure have shown it to be reliable if the following demands are met:

- slow temporal changes of the heat flux
- highly precise measurement of the temperature differences.

In practice, the monitoring of the accuracy and the exchange of the sensors are major obstacles in the large-scale implementation of this method.

5. CMMs featuring very high thermal stability

5.1 UPMC CARAT (Fig. 12)

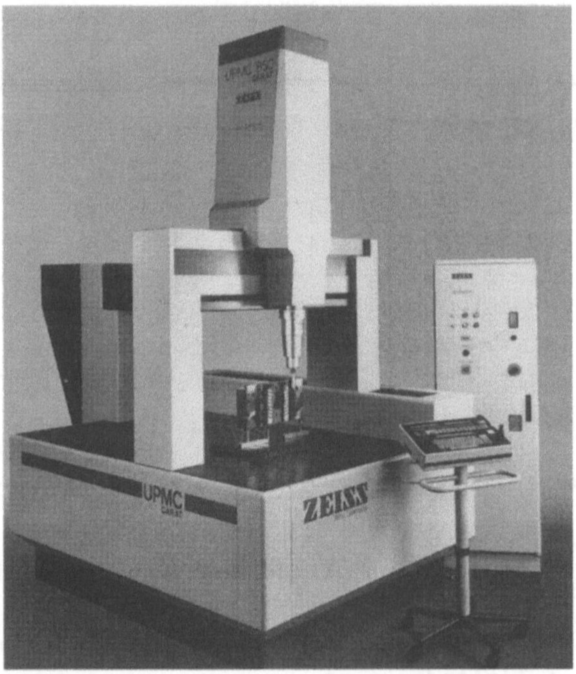

Fig. 12: UPMC CARAT Universal Precision Measuring Center

An ultra-accurate bridge CMM:
$u_1 = (0.5 + L/900)$ µm and high thermal stability:

temperature range : 20 °C ± 2K
thermal variations: 0.5 K/hour; 1.0 K/day
spatial gradient : 1.0 K/m

This was achieved by:

- thermal insulation of the bridge columns and the granite table
- temperature sensors in the granite table for the software compensation of thermally induced residual bending
- crossbeam and spindle in the CARAT technology
- thermally stable scales made of ZERODUR ™ glass ceramic
- measurement of the workpiece temperature

5.2 PMC CARAT (Fig. 13)

Fig. 13: PMC CARAT Precision Measuring Center

Thermally stable bridge CMM for which two different specifications for accuracy and temperature conditions are given.

Use in the metrology room

In the existing temperature conditions which can be achieved with an acceptable outlay:

temperature range :	20 °C ± 2K
thermal variations:	1.0 K/hour; 1.5 K/day
spatial gradient :	0.3 K/m horizontally
	1.5 K/m vertically

a measuring uncertainty of $u_1 = (2.0 + L/400)$ µm is guaranteed.

Use in the manufacturing area

The measuring uncertainty of the same CMM is still $u_1 = (2.5 + L/250)$ µm if the temperature conditions are relaxed by the following margins:

temperature range :	18 °C ... 28 °C
thermal variations:	2.5 K/hour; 8.0 K/day
spatial gradient :	0.6 K/m horizontally
	2.0 K/m vertically.

For the very first time in the accuracy specifications of a coordinate measuring machine, this "dual specification" provides quantitative information on how the measuring uncertainty changes when the temperature conditions are tightened/relaxed.

This very good thermal stability was achieved by the following measures:

- insulation of the bridge columns
- thermally stable scales made of ZERODUR ™
- crossbeam and spindle in the CARAT technology
- software compensation of thermally induced residual table bending
- measurement of the workpiece temperature

5.3 FMC Production Measuring Center (Fig. 4)

A cantilever-type CMM designed for use directly on the shop-floor. The complete enclosure of the machine provides effective protection against thermal influences, also for the probe changing magazine and the workpiece.
Due to the enclosure and the use of CARAT guideway components and thermally stable Zerodur scales, the length-measuring uncertainty of $u_1 = (2.5 + L/250)$ µm can be guaranteed under the following conditions:

temperature range: 15 °C ... 35 °C
thermal variations: 2.5 K/hour; 10 K/day
spatial gradient: 2.5 K/m

An optional temperature sensor automatically measures the workpiece temperature, for example, prior to measurement start.

6. Summary

The technical options shown herein are a major step towards achieving thermal insensitivity of the coordinate measuring machine.
New materials and the appertaining processing technologies play as great a role as software compensation procedures. As the accuracy of measuring the thermal conditions in a CMM are limited, however, these procedures can only be implemented to a limited extent. The exchangeability of high-precision temperature sensors becomes problematic, especially when there are too many of them.
The most sound approach therefore lies, as described under 3., in the screening out of thermal influences, the use of thermally stable materials and the compensation of the residual effects.

7. Literature

1) Schepperle/Zeller: "Abnahmeprüfung von Koordinatenmeßgeräten beim Hersteller", QZ 30 (1985, Heft 9)
2) Trapet, Wäldele: "Koordinatenmeßgeräte in der Fertigung ...", VDI-Berichte Nr. 751, 1989

Towards Quadrilateration using Steel Ball Retroreflectors

J.M.Burch

SIMS, Cranfield Institute of Technology, UK

Introduction

As was discussed briefly at the 3rd IPES (Interlaken 1985), studies have been made at NPL by Forno and myself on the problem of very accurate 3D calibration using multi-station close range photogrammetry. Using axicon imaging with the specially designed NPL Centrax lens, Forno has since reached close to the expected accuracy of one part in 10^6 in measuring a space-frame (with about 50 target-points defined by small polished steel balls) in a measured volume around $1m^3$. The method is "self-calibrating" by virtue of the redundant nature of the data obtained, but this accuracy may well be close to the limit of what can be achieved by angular measurement.

If 3-axis calibration with even greater accuracy is demanded, then it seems evident that angular measurement must be replaced by interferometric length measurement. For a single moveable retroreflector, this has already been achieved by incremental fringe-counting using four separate tracking interferometers which view the measured target-space from different directions.

Impressive as this is, for purposes of CMM calibration it would be useful to extend the quadrilateration method so as to measure (in rapid succession) a whole series of separate targets as in a space-frame. If three of these target-points are mounted securely near the edges of the measuring table, they can be used as "control-points" to define a stable reference triangle whose vertices are accessible to the probe of the machine being calibrated, since they are inside its working volume. Secondly, if a small triangle of targets is attached to a moving quill or end-effector, then angular as well as displacement data can be extracted.

In order to make such a system function reliably and without loss of accuracy, it seems necessary firstly to replace fringe-counting with multi-wavelength "absolute" interferometry, and secondly to replace the monocentric glass "capped sphere" by some more rugged and inexpensive form of omni-directional retroreflector. If each station incorporates equipment not only for target acquisition and tracking but also for control of beam focus, then it should be feasible to define all of the target-points (both optically and mechanically) by means of suitable polished steel balls. This contribution represents an interim report on progress in studying and verifying these two possibilities.

4-station multi-wavelength interferometry

If CMM calibration with very high accuracy is being attempted, it seems likely that the lengths being measured will not be changing rapidly, and usually they will be close to their nominal expected values within a mm or so. In that case it may be sufficient to rely on fringe fractions obtained with just three stable and non-commensurate laser wavelengths. Because of the requirement for sharp focusing the beams should be visible and should not span too large a range of the spectrum. A suitable choice might be a combination of HeNe red orange and green lasers, all coupled into the same mono-mode fibre. (Vacuum wavelengths in nm are 632.9908, 611.9708 and 543.5157). A 2-beam interferometer has been designed which will capture three pairs of signals of the form (1+Vcos(a)) and (1+Vsin(a+e)) where e represents the small departure from ideal phase quadrature which is liable to occur for some of the wavelengths. Fringe fractions obtained with the help of an ellipse learning routine may well be reliable to better than 0.01, but even so there is some risk of ambiguity arising if the distance being measured is itself uncertain. If for example the target approaches or recedes by 0.8015mm (2533 red fringes, 2620 ŏrange fringes and 2950 green) then nearly the same values for all three fringe fractions are obtained. The changes are at the 5 millifringe level, varying only slightly with the atmospheric parameters assumed.

If CMM errors are being calibrated, an error of 0.8mm may be excluded because of our previous knowledge of expected position. But such an error could be very annoying during initial setting-up operations, and it seems wise to avoid it by correct choice of the four tracker positions. If for example one of the four stations always appears <u>inside</u> a triangle formed by the other three, the measurements it gives will reject any spurious position which is in danger of being accepted by the other three. If on the other hand the four tracker stations all appear from the target to form a rectangle or a cyclic quadrilateral, then there is a degeneracy whereby a large error in a "central" direction might escape detection.

In order to achieve the best possible accuracy it might seem logical to interogate the measured volume from three orthogonal directions, but such a system is liable to pose awkward problems of obstruction. A less awkward arrangement is to group the four trackers so that they subtend an angle of about 30 degrees. Occasional interruption of some or all of the beams is unimportant, but measurement of a steel ball position will be impossible unless at least three trackers are able to operate. If the three control-points are kept uncovered they can provide occasional checks that the tracker locations have not themselves drifted slowly with respect to the table.

Focusing tracker for use with polished steel balls

Detailed design of the tracking arrangements must depend on the dimensions of the region to be measured and on the space available to accommodate the four trackers. But if accuracy to 100nm or better is being sought then it may be sensible to use a single plane tracker mirror, mounted in a gimballed pair of precision air-spindles which control bearing and elevation respectively. Such a system will be slower than can be obtained by separated galvanometer movements, but it offers a wealth of "transiting procedures" which will verify mirror alignment, orthogonal intersection of the two mechanical axes, and (most importantly) accurate alignment of the tracker beam and of the linear slide used to control beam focus. (The latter needs to be straight to 100nm or so, and the required displacement for each new target acquired can be preset to the nearest 0.1mm, the total focusing range being typically 50mm.)

For operation over ranges from 600m to about 1200mm a ball diameter of 25mm should prove suitable, and the beam expander used for each tracker then needs a main collimator lens with about 120mm focal length. The largest system contemplated uses balls of 67mm diameter over ranges from 1.7m to 8m. In that case the Gaussian beam launched from each tracker mirror will have about 15mm $1/e^2$ diameter, and the system will be less compact. After leaving the tracker mirror the beam converges slowly to a focused "waist" at the ball which is 0.2mm or less in diameter. With suitable matching conditions the beam is returned with the same geometry all the way back through the interrogating tracker to the CCD array detector where it generates three pairs of interference signals of good contrast.

The precise design depends on the depth of volume to be interrogated and on the accuracy of the angle sensors used for initial target acquisition. The requirements for the surface quality of the steel ball are readily satisfied, and the optics of this type of "Fourier focusing" system have been checked experimentally.

Testing of Algorithms for Coordinate Measuring Machines

D. Palanivelu, Y. Hari, K. Chen, J. Yang, J. Raja, and R. Hocken
University of North Carolina at Charlotte
Charlotte, NC 28223 USA

Summary

In order to obtain the final results of an inspection performed on a coordinate measuring machine (CMM) it is necessary to analyze the data using mathematical algorithms. The most common type are the least-squares algorithms, but for many inspection tasks other algorithmic choices, such as mini-max, are more appropriate [1]. As part of a program to develop sampling strategies for the measurement of mechanical parts, we have been developing and testing least-squares, mini-max algorithms, and other algorithms over the past year. In this paper we will discuss our results to date.

Introduction

Coordinate measuring machines acquire coordinate data and these data must be fit to generate geometric models. This is done by algorithms whose results are dependent upon many factors, including the sampling strategy and density, random errors in the data, systematic deviations of the measured form from the ideal form, and, of course, the correctness and robustness of the algorithms. We are examining the effects of sampling on algorithmic results for a set of basic features. These features are lines, planes, circles, spheres, cylinders, and cones. Because we could not obtain tested algorithms elsewhere, we created our own set of algorithms for use in our experiments.

Algorithm Development

Developing our least-squares algorithms we followed the methods outlined by researchers at NPL [2,3]. A mathematical description of these algorithms may be found in [4]. These algorithms were tested by us against the PTB data [5] and against a data generator developed at UNCC [6]. The results were also compared to the results obtained from an American round robin on the PTB data conducted by the National Institute of Standards and Technology (NIST) [7]. In no case were deviations numerically significant. We then developed mini-max algorithms for the same features. These algorithms have been tested against the PTB data sets. We are currently confident that most of these algorithms give the correct results; however, only in very specific cases, such as the 2D circle, can correctness be determined analytically. For figures that are "intrinsically closed", that is, have an inside and an outside, the mini-max answers may be verified by examining the residuals which should be equal and opposite inside and outside of the figure. For two cases, the 3D line and the 3D circle, such a condition does not exist. We have, however, shown that the 3D line is mathematically identical to the axis of the minimum circumscribed cylinder which surrounds the data set. For the 3D circle we are currently fitting the data to a mini-max plane, projecting the residuals on the mini-max plane, and fitting the residuals to a 2D mini-max circle. This, we believe, is a correct procedure; however, mathematical proof is still lacking. We have also developed algorithms for the maximum inscribed and the minimum circumscribed cylinder and circle. All algorithms are in Quick BASIC™, and we are contemplating releasing source code to selected collaborators for preliminary testing purposes.

<u>Algorithm Testing</u>

After testing with the PTB data sets [5], we tested the algorithms against a UNCC data generator to ensure their performance. The UNCC generator functions by creating an ideal feature and then adding to that feature form error consisting of sinusoids of various amplitudes and phases (for two-dimensional features, two sinusoids are used). Next random error, drawn from a uniform distribution, with different amplitudes, is added to the data set. The data are translated and rotated in space to provide varying offsets and angular orientations. Normally 81 data sets are created for each feature, the fits performed, and the differences between the algorithm output and the "true" value plotted. Again, for some cases there is no analytical method to define the "true" answer. In such cases we use the answer obtained from the algorithm used on the data set containing only the form and systematic error when the sampling size is very large (>500 points).

These tested 3D algorithms are used to define the starting

points for the mini-max fits. Currently we are using three methods of mini-max fitting. The first we call the geometric method, which can be used on very specific types of features. The second method is the Down-hill Simplex method, and the third is the Anderson-Osburn-Watson method. On some features run time is a problem.

Conclusions

With our completed algorithm library, we are conducting a very large number of tests on a wide variety of data sets. As expected, the results appear to show that in the presence of sinusoidal-type errors the algorithmic results are variable when the sampling density is low and quickly converge towards a "correct" answer as the Niquist sampling density is reached. Stratified sampling converges more rapidly than uniform or random sampling.

Acknowledgments

This work was funded by the National Science Foundation.

References

1. ANSI/ASME Y14-1990 Ad Hoc Committee on "Mathematization", Richard Walker, Committee Chairman, Westinghouse Electric Co., Sunnyvale, CA 94022, 1990.

2. BSI, "Assessment of Position, Size, and Departure from Nominal Form of Geometric Features", British Standards Institution, BS 7172, 1989.

3. Forbes, B. A., "Least-squares best-fit geometric elements", National Physical Laboratory, UK, Report DITC 140/89, 1989.

4. Caskey, G., Chen, K., Hari, Y., Palanivelu, D., Wilson, R., Yang, J., Raja, J., and Hocken, R., "Sampling Techniques for Coordinate Measuring Machines", Proceedings of the Division of Design and Manufacturing Engineering NSF Conference, January 1991.

5. Porta, C., and Waeldele, F., "Field Testing of Three Coordinate Measuring Machine Evaluation Algorithms", Physikalish Technische Bundesanstalt, March 1986.

6. Palanivelu, D., "Algorithm Evaluation for Coordinate Measuring Machines", M.Sc. Thesis, University of North Carolina at Charlotte, 1990.

7. Algeo, M. B., NIST, private communication, January 1991.

Experiences in the Development of Ultra Stiff CNC Aspheric Generating Machine Tools for Ductile Regime Grinding of Brittle Materials

K. Carlisle and P. Shore

Cranfield Precision Engineering Ltd.
Cranfield Institute of Technology
Cranfield, Bedford, MK43 OAL, England, UK

Introduction

Ductile regime grinding of glass and other brittle materials promises a major economic breakthrough in the manufacture of complex advanced technology components such as non-conventional optics, wave guides and opto-electronic sub-systems. This paper describes recent work at Cranfield Precision Engineering Ltd. in designing a machine for the development of ductile regime grinding technology, including the essential sub-systems of CNC truing and dressing of diamond grinding wheels.

It traces the design considerations and decisions in producing a new ultra precision CNC aspheric generator in which high loop stiffness was a main driver in ensuring that ductile regime grinding of a range of brittle materials can be carried out efficiently on components up to 600 mm diameter and 250 mm axial length. Target workzone accuracies within 50 nm, considerations of optimum configuration hydrostatic or aerostatic bearings, servo drive actuators, displacement transducers, the application of state feedback to servo drive systems, thermal drift control and CNC requirements are discussed.

Machine Design

Fig. 1 illustrates the key elements of the Nanocentre - a six axis diamond turning and ductile regime grinding machine. The 'T'-bed base supports the 'Z' axis tool slide and the moving bridge 'X' axis work slide. A second stationary bridge spanning the tool slide, supports a 'Y' axis truing and dressing station for the grinding head.

Granitan S100 is used for the bed to give excellent dimensional stability, high specific stiffness and high internal damping. Seismic vibrations are reduced by supporting the base on a three point, self levelling pneumatic isolation system which stands on a large, damped, mass concrete foundation.

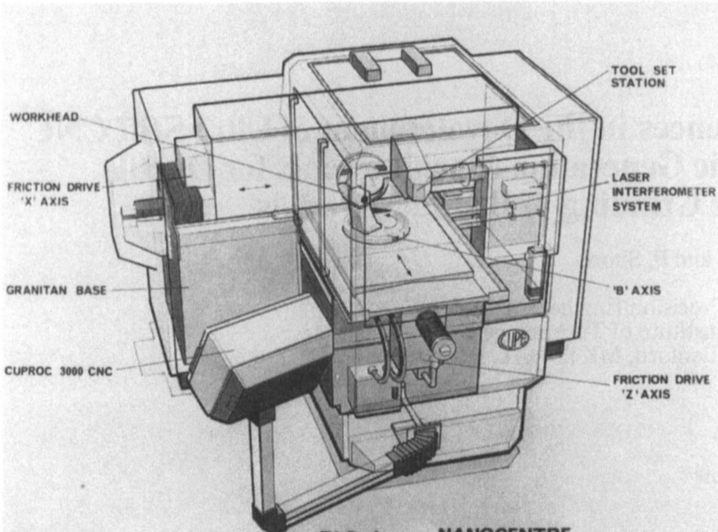

FIG.1. NANOCENTRE

Oil hydrostatic bearings were selected for their attainable high stiffness and load carrying capacity. They also provide smooth motion and good damping with freedom from slip-stick, all essential requirements when servo positioning slides down to 1.25 nm during contouring operations. Oil also has the advantage of being temperature controlled to within 0.01^{o}C or better to ensure thermal stability of the machine.

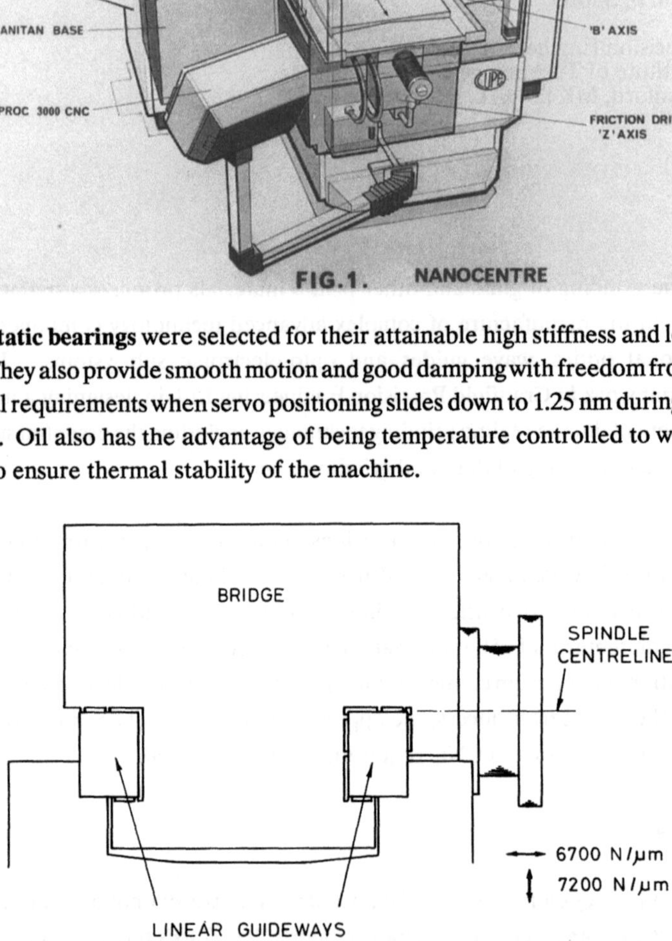

FIG. 2. CROSS SECTION OF 'X' AXIS

In order to achieve a high loop stiffness **the bridge** bearings were positioned up on the cutting plane of the machine to minimise any turning moments from the cutting forces. It also reduces the effects of pitch, yaw and roll in the straightness motion of the bridge (see Fig. 2). Another benefit of the bridge design is its long bearing engagement and the high

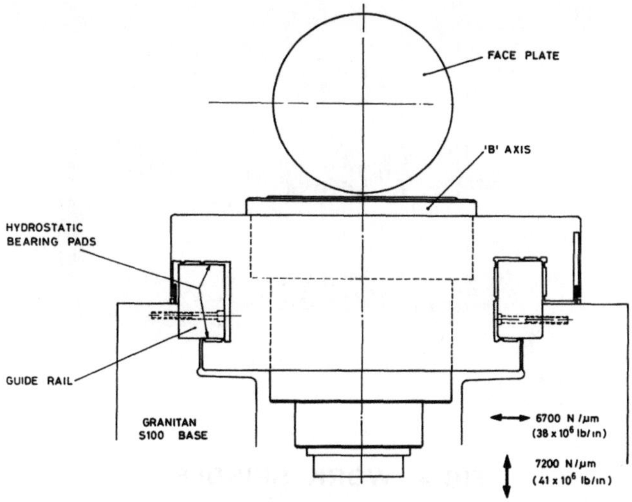

FIG. 3. CROSS SECTION OF Z AXIS

structural stiffness that could be achieved. The large surface area of the bridge also enable tuned damping devices to be applied to the surface.

The tool slide uses the same bearing design as the bridge. Original designs were developed with the bearings on the cutting plane, these were later rejected on the grounds that operational access was very limited and the 'B' axis had to be a permanent feature of the tool slide which did not meet the machine design objectives of interchangeability between fixed tool post, rotary 'B' axis and Elid (Professor Nakagawa - Riken) grinding spindle. The adopted design is illustrated in Fig. 3 where the bearings are as close to the cutting plane as possible to achieve the desired swing diameter for the workpiece. The wide spacing of the guide rails enables carriage 'roll' to be controlled more easily and provides space for the 'B' axis to be embedded within the carriage. This offered the advantage of designing a 'B' axis having bearings of sufficient size and stiffness to offset the design constraints of not having the slide bearings up at the cutting plane. Equally, with the 'B' axis removed the grinding spindle could now be flange mounted to the carriage with its drive motor assembly below the carriage and the grinding spindle above, thus providing a very stiff cantilever support to the wheel.

The workspindle (Fig. 4) is built into the bridge carriage which moves at right angles to the spindle axis. It uses oil hydrostatic bearings designed for high stiffness, high load carrying and low error motion. Thermal stability is maintained by careful selection of the material used for its construction, the bearing design and temperature control of the oil to better than $\pm$ 0.01^{o}C. All these factors have been optimised to achieve the highest standards in single point turning and enables full exploitation of ductile regime grinding

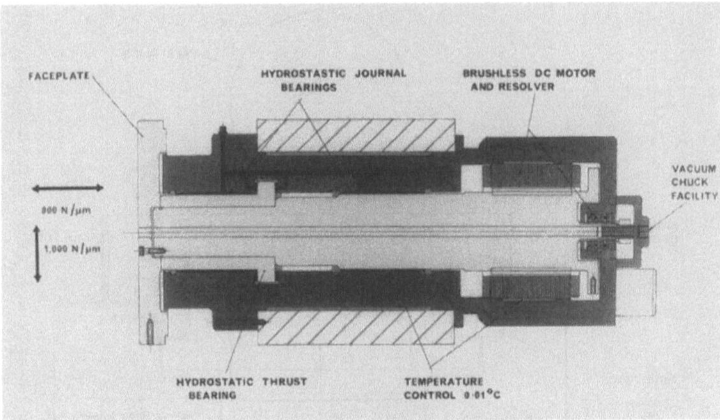

FIG. 4. WORK SPINDLE

of brittle material. The spindle is driven by a high performance brushless d.c. motor mounted directly on the spindle to eliminate the need for a coupling. This provides a torsionally stiff non-influencing drive with a built in facility to fit a tachogenerator and rotary encoder for fast tool servo applications. Motor speed is steplessly variable and bidirectional with a dynamic fail safe braking system. Facilities built into the spindle include dynamic balancing, vacuum work holding and temperature control of the drive motor.

The 'B' axis rotary table facility offers the advantage of being able to programme the tool cutting tip always at a tangent to the surface being cut, i.e. "NORMAL" to the cutting surface at all times. With the same point of the tool held constant to the cutting surface any errors due to tip irregularities are eliminated. To maximise tool life the 'B' axis can also introduce a new portion of the cutting tip by a small angular offset, it also permits versatility when machining Fresnel optics and can be used as an indexing table for fly-cutting of optical polygons. Unlike most other machines there is no loss in workpiece swing diameter when using the rotary table. This is because of the unique design of the rotary table which is embedded within the tool slide, see Fig. 3. The table is supported by hydro static bearings to ensure high stiffness and load carrying capacity. The drive system consists of a direct drive frameless d.c. torque motor/tachogenerator mounted directly onto the table and the measuring system comprises a rotary encoder. Temperature control of the drive system is maintained by cooling oil.

The tool post forms a critical element of the machine system because it is the final link in the structural loop between tool and workpiece. An objective review was made of existing tool post designs to determine their suitability for this machine. It was found that most designs incorporated 'X', 'Y' and 'Z' adjustments for the tool, especially those used with a 'B' axis where centring of the tool is necessary. These adjustments can greatly reduce the overall stiffness of the tool post due to sliding interfaces and their drive mechanisms. The overall positional performance of the tool post is limited by the precision of the individual axis adjustment which can often cause errors in the other two axis, these then require further re-adjustment. They also suffer from thermal effects due to hands on adjustment and stress relaxation due to high clamping forces. Therefore to fulfil the design goal of high loop stiffness, a radical change in tool post design and functional requirements was necessary.

The tool post adopted for the Nanocentre (Fig. 5) has only one axis of adjustment which adjusts the tool height setting. It has a range of 20 μm which requires the tool to be pre set to a nominal height and length off the machine. This requires the tool to be mounted and clamped to its own tool carrier before being transferred to the optical measuring system. Height and length adjustments are made to the carrier to bring the tool to a specified position. Once set, the tool and its carrier are transferred to the tool post on the machine. A typical tool set routine is followed, except that only tool height is adjusted and the tool co-ordinates measured. This is sufficient for a fixed tool post operation but requires additional measurement when using the rotary 'B' axis.

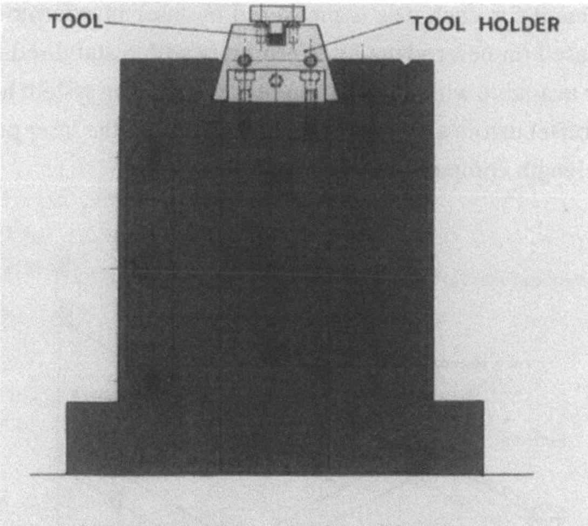

NANOCENTRE – TOOL HOLDER
FIG. 5.

Tool setting for the 'B' axis requires the tool cutting edge to be set on the axis of rotation which is normally achieved by adjustments of the tool post. The problems of achieving this have already been described. In addition to these problems there is the rotational error motion of the 'B' axis at tool height and finally the tool radius error to be considered.

Our solution is to measure all these sources of error at one tool setting by rotating the 'B' axis and taking continuous measurements of the cutting edge relative to the fixed point of the tool set station. The information is then stored in the memory of the machine CNC system for error compensation when profile machining.

Drive systems to both carriage is by the well proven CPE friction drive system consisting of a direct drive frameless d.c. torque motor/tachogenerator mounted directly onto a precision drive shaft incorporating a hardened drive roller. Motive force to the carriage is achieved through the friction drive roller driving a traction bar attached to the 'Z' axis carriage via a non influencing coupling. The 'X' axis has the traction bar secured to the base and the drive system attached to the moving bridge, a similar system of non influencing couplings is used. These couplings ensure high stiffness in the drive direction without influencing the translation accuracy of the carriage.

The friction drive system offers many advantages over the contentional ballscrew drive system, the most important for this machine being: backlash free, non-influencing, no transmitted drive vibration, good velocity bandwidth and well proven positional capability at the nanometer level.

Linear displacement for each axis is measured by laser interferometry using the Zygo Axiom system based on heterodyne interferometry with a stabilised He:Ne laser for the highest accuracy available with a resolution of 1.25 nm. The system has been designed to minimise Abbe offset errors and environmental effects on the laser path are corrected by the optical wavelength compensator built into the system.

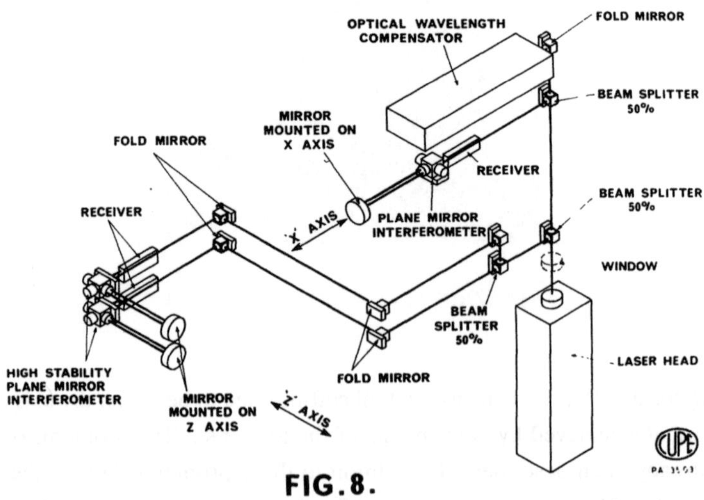

FIG.8.

Fig. 8 shows a schematic of the interferometry system. The 'X' axis mirror is mounted on the bridge adjacent to the workhead at spindle centre height. The 'Z' axis has two interferometers, one to measure displacement and the other to measure pitch of the carriage. In this way, Abbe errors can be corrected. The laser beam paths are all totally enclosed in a sealed system to resist hydro carbons from the oil etc. entering into the system.

The Nano-Grinding Facility (Fig. 7) for this machine has been designed to exploit the latest technological advances in grinding of glasses, ceramics and other hard materials. A critical factor in ductile mode machining of brittle materials is the dynamic "loop stiffness" of the machine. This has been ensured by the use of hydrostatic bearings and the unique design of the machine to achieve the very highest loop stiffness. This gives the machine the ability to take real sub-micron cuts essential to achieve good form accuracy and minimum sub surface damage.

Unlike many other machines where the grinding spindle is treated as an attachment which can be somewhat limited in dynamic performance by its supports system, our objective was to design a fully integrated machine grinding facility. The air bearing grinding spindle complete with its direct drive oil cooled brushless d.c. motor is mounted rigidly to the tool slide as can be seen in Fig. 7. Great importance was given to the method of mounting the spindle to achieve a stiff and well damped support system. This is assured by the flange mounting and ribbed column support to the wheel.

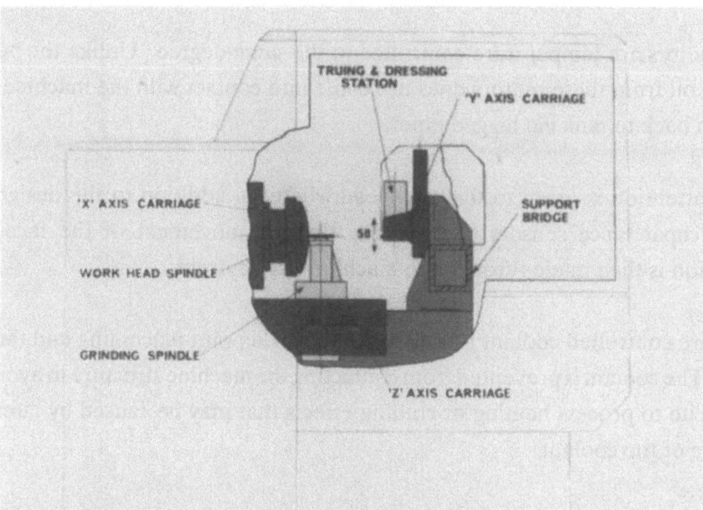

NANOCENTRE DUCTILE GRINDING FACILITY FIG.7.

Accurate positioning of the wheel in relation to the work head spindle centreline is crucial for workpiece accuracy and is obtained by an air bearing LVDT wheel setting station.

Key features of the process include on-machine precision truing and Elid dressing equipment which has to be dynamically stiff. The truing wheel is mounted on its own motorised air bearing spindle which is attached to a servo driven vertical 'Y' axis carriage to enable contouring of the grinding wheel to be carried out. This is essential for the grinding of concave shapes where wheel clearance and wheel centre height is necessary. Wheel profiles are programmable through the machine CNC system. The truing process ensures no single grit protrudes the wheels bond, subsequent use of the Elid dressing techniques provides the wheel with consistent "grit" protrusion and in a controllable manner. This Elid technique is used during all grinding operations and provides near constant grinding forces as wheel glazing (loading) is not experienced. As a result, components are produced with a defect-free surface.

Thermal Stability

Hydrostatic oil from the linear and rotary bearings are temperature controlled to within $\pm$ 0.01°C. Temperature sensors built into the machine monitor both the oil feed and exhaust temperatures to ensure thermal stability of the machine. After exhausting from the bearings, the oil is deliberately run over the internal surfaces of the machine before running back to tank.

All drive motors are temperature controlled to the same degree. Unlike the bearing oil, the cooling oil from these motors does not come into contact with the machine structure but is taken back to tank via lagged pipes.

Particular attention is given to the workhead where, in addition to the design features described, capacitance sensors monitor any thermal movement of the faceplate and compensation is then made through the machine CNC system.

Temperature controlled coolant is used for both single point machining and the grinding processes. The coolant is prevented from contacting the machine structure to avoid thermal transients due to process heating or chilling effects that may be caused by latent heat of evaporation of the coolant.

A containment protects the machine from environmental effects which is a self supporting structure having no contact with the machine, see Fig. 6. This protects the machine from

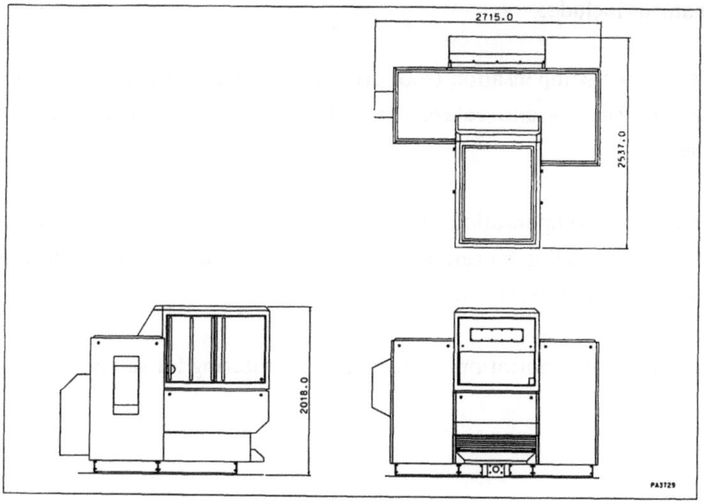

FIG 6 NANOCENTRE CONTAINMENT

inadvertent collisions or loading from the operator. Acoustic damping material is added to the containment panels to reduce air born effects that may induce structural vibration of machine elements.

The containment also supports an oil shower thermal control system for the machine structure. A laminar flow of oil, controlled to 0.01^{o}C, is directed over the structure and is gravity collected in a tray beneath the machine, thus providing insulation from any ground thermal influences.

The CUPROC CNC Series 3000 Advanced CNC System with its "Electronic Gearbox capability" which ensures zero following error performance is used on the machine.

It was specifically designed to meet the ever increasing demand for high precision generation of complex profiles. Exceptional tool path accuracy is guaranteed by unique, high resolution, fast feed forward techniques thus eliminating servo following error.

A unique feature of the controller is on-line cubic spline interpolation. Data input of the path definition can be either by the equation of shape of a given optical component surface, entered at the operator station or by a personal IBM/compatible computer, or a minimum number of datapoints; conventional post processing is eliminated.

Optional features include:-

* Software error compensation (tool error mapping). Error compensation for axis (B-axis rotary motion and scale error). Volumetric error compensation (cubic lattice, parametric).

* Dynamic error compensation. Error compensation for thermal drift (e.g. spindle growth), compensation for refractive index for laser interferometry, metrology frame methods and fast tool servos.

* Diagnostics, faulty indication, performance monitoring for safety.

Development of High-precision Aspheric Grinding/Turning Machine

T.Osaka, K.Unno, A.Tsuboi, Y.Maeda, K.Takeuchi

As industrial products acquire ever more advanced functions in the fields of optoelectronics, requirements for reduction in size and weight and for improvement in the accuracy have been increasing. This has lead to the employment of high-precision aspheric lenses in the optical systems of the products, where press-molding process has been dominant recently for its high productivity. Press-molding process requires high-precision molds which are machined contour accuracies of 0.1μm or higher.

Authors developed a CNC high-precision aspheric grinding/turning machine, model AHN10, to machine such high-precision molds. This paper describes the mechanism, performance and an example of molds machined by the aspheric grinding/turning machine.

1. Introduction

As optoelectronic products such as VTR cameras, CD players and laser disk players (LDP) come to provided with higher performance in more compact constructions, attention has been drawn, in place of conventional optical systems combining spherical lenses, to new optical systems employing aspherical lenses which have higher optical performance with simplified lens construction.

However, aspherical lenses are difficult to manufacture with high precision and higher efficiency by means of the material removal processes such as grinding and polishing of spherical lenses. For this reason, press-molding process using molds of aspherical surfaces has been introduced in electric, electronics and optics industries, for higher accuracy and lower manufacturing cost. The aspherically shaped molds used in the press-molding process must be machined to a close tolerance as high as 0.1μm (or higher). Thus, there has been awaited the development of a machine tool which is capable of such a high-precision machining with a high efficiency. AHN10 has been developed to satisfy these needs with the following two major design concepts:

1) improvement in machining accuracy through improved accuracies of the machine tool components, and

2) improvement in productivity through on-machine measurement.

2. General of AHN10

The AHN10 is a CNC high-precision grinding/turning machine which is equipped with X- and Z-axis slide tables and a B-axis rotary table and which is capable of grinding or turning various optical component parts through 3-axis simultaneous control (Fig.1). As shown in Fig.2, the bed is supported by air mounts to isolate the machine from external vibrations. The bed carries a X-axis slide table movable with a grinding wheel spindle or a tool holder being mounted thereon and a Z-axis slide table movable with a work head being mounted thereon which is capable of turning a workpiece, both tables being arranged at right angles to each other. The B-axis rotary table, placed on the X-axis slide table, is angularly controlled so that the grinding wheel or the cutting tool is oriented always in a direction perpendicular to the workpiece surface as the machined point moves.

Table 1 shows major specifications of the AHN10.

3. High-precision machine components

3-1. X- and Z-axis slide tables

The slide way of each slide table employs oil-hydrostatic bearings which have a high accuracy of linear movement, high rigidity and high vibration damping performance. The ball screw used in the feed mechanism is supported by oil-hydrostatic bearings which are conversely of a low rigidity in radial directions so that the fluctuation in the rotation of the ball screw is absorbed not to be transmitted to the tables. Fig.3 illustrates the accuracy of the slide table movement, demonstrating that a high accurate sliding movement of 0.03μm over 200mm of stroke was obtained.

A laser ranging system is used for the position feedback. Emphasis in the basic design was placed on the disposition of the optical components in such an arrangement as minimizes the Abbe Offset Error as well as on the stabilization of the optical path environment. This made the positioning control of 0.01μm/pulse possible. Fig.4 shows the measurement of the pulse response of the slide table.

3-2. B-axis rotary table

The B-axis rotary table is required to have a high rigidity, high vibration damping characteristic, stable speed in the low speed range and capability of indexing by minute angles. In order to satisfy these needs, hydrostatic bearings and direct motor drive system with encoder feedback are employed. Fig.5 shows the measurement of the pulse response of the B-axis rotary table. The figure demonstrates that stable angular positioning accuracy without backlash is obtained at the least increment movement of 1.3 sec. per 1 step.

3-3. Workhead spindle

The workhead spindle is supported by air-hydrostatic bearings of high rigidity. Direct motor drive mechanism is adopted to eliminate the sources of vibration such as belt, coupling and the like and to obtain a high output power with a compact construction. The motor stator which is a major source of heat generation is cooled by an oil jacket to prevent thermal deformation.

3-4. Wheel spindle

In order to achieve an extremely high maximum speed of 100000rpm, the through balancing of the wheel spindle was carried out to eliminate the effect of eccentric load of the spindle. The spindle is supported by air-hydrostatic bearings and is driven by an air-turbine of low vibration which does not generate heat.

4. Improvement of productivity through on-machine measurement

4-1. Contour measuring instrument

Machining accuracy of a workpiece in process is checked usually by dismounting the workpiece from the machine tool and measuring it with a stand alone measuring instrument. However, as far as the workpiece is dismounted for the measurement, runout and longitudinal positional error of the workpiece cannot be restricted within several micrometers, even if a chuck of the highest accuracy presently available is used. Thus, at least several micrometers of infeed is unavoidable when the workpiece is mounted again to resume machining. In a machining work where 0.1μm or higher contour accuracy is sought, an infeed of several micrometers is nothing other than a rough grinding, which inevitably causes contour errors due to unpredictable factors such as wheel wear. In addition to this problem, such errors accumulate every time a corrective machining is attempted, setting a limitation to the machining accuracy which can be obtained.

The AHN10 is equipped with a high-precision contour measuring instrument (Form Talysurf: Rank Taylor Hobson) for on-machine measurement, whereby a series of procedure from machining, to accuracy measurement and further to the corrective machining in accordance with the measurement result can be realized throughout a single chucking operation. This made it possible to reduce the infeed in the corrective machining to such an order that the wear of the grinding wheel can be ignored, and to obtain the desired machining accuracy reliably. Moreover, it resulted in drastic improvement of the productivity because dismounting/rechucking of the workpiece and the accompanying alignment operation are eliminated, because the machining accuracy can be checked on the machine, and because the number of machining cycles is reduced.

Fig.6 shows the on-machine measuring instrument being engaged in contour measurement of a workpiece. During measurement, the spindle is swivelled through an angle of 90° to be directed right above, thereby permitting contour measurement by means of a feeler needle.

The repetitive positioning accuracy of the spindle to the opposite end positions of the 90° swivel motion is very important, as it has influence on the machining accuracy and on the contour measurement accuracy. As shown in Fig.7, 0.03μm has been achieved for the repetitive positioning accuracy of this motion. And it was verified that reproducibility of the contour measurements made before and after the 90° swivel motion of the spindle was within 0.02μm (Fig.8).

4-2. Tool position detecting system

In machining of an aspheric surface, positional error of the tool tip greatly affects the contour accuracy. This make it necessary to position the tool to the center of B-axis rotation and to the center of the workhead spindle rotation with an accuracy of an order of a micrometer.

Conventional grinding machining uses a contact position sensor for the wheel positioning. This has disadvantages such as the wear of the sensor feeler and incapability of monitoring during wheel rotation. The tool position detecting system of the AHN10, in contrast, enlarges the rotating tool tip by means of a microscope and displays the image on a TV screen. Thus the system enables the tool to be accurately and easily positioned, without contact, to reference positions which are not visible such as the B-axis rotation center and the workhead spindle rotation center. This resulted in remarkable improvement of the productivity. It was also make possible to monitor the wheel wear of the order of a micrometer caused by machining and the runout of the wheel during high-speed revolution which have been difficult to check quantitatively. This resulted in an improvement of the machining accuracy.

5. Principle of aspheric machining

Majority of the molds for aspherical lenses ranges from several millimeters to 20mm in diameter. In a mold with a concave surface of a small diameter, radius of curvature at the apex is as small as several millimeters, and the recess of the concave surface is a relatively great. Consequently, applicable wheel shapes are limited and the machining is performed with the wheel spindle being often tilted to 45° relative to the workpiece as shown in Fig.9 (a). While in the machining process, an aspherical surface is generated through positional control of two axes, X-axis and Z-axis, tilting the wheel spindle makes it difficult to effect tool diameter compensation. Therefore B-axis rotary table must be controlled so that the wheel spindle is always oriented perpendicularly to the machined surface as the machining position changes.

Fig.9 (b) shows a plan view of the machined section. The grinding wheel is set in such an arrangement as its tip coincides with the center of the B-axis rotation.

6. Machining example of mold for aspherical lens

Fig.10 shows the AHN10 machining a mold of aspherical surface for press-molding of LDP pickup lenses. After rough-finish grinding and intermediate-finish grinding to produce a desired configuration, the mold is subject to a precision grinding process for higher surface accuracy. Because the mold is required to have a contour accuracy as high as within 0.1μm, a cycle of machining, on-machine measurement and tool position compensation is repeated several times.

Fig.11 shows the data obtained from measurement during grinding of the mold of aspherical surface to manufacture LDP pickup lenses. Contour accuracy of 0.05μm and surface roughness of 0.025μm were obtained.

7. Conclusion

As can be seen from the machining example, the development of the AHN10 had a great impact on the improvements in machining accuracy and productivity. In the machining of high-precision molds for press-molding of aspherical lenses, in particular, the AHN10 achieved a great reduction in the number of machining processes and an improvement of productivity as high as ten times that of the conventional machines.

We will make effort for practical use of the high-precision machining technology to contribute further development of high technologies.

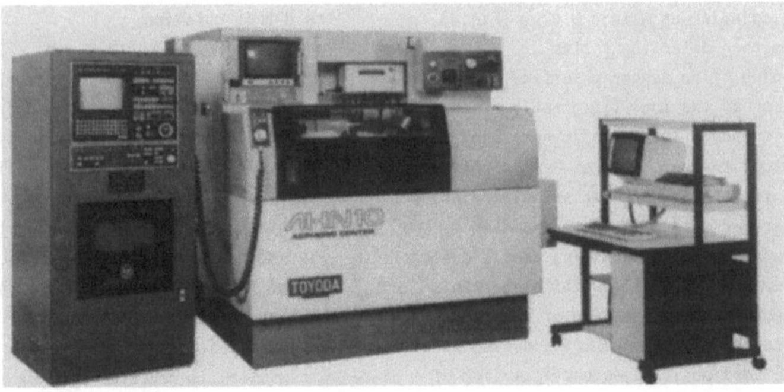

Fig.1 General view of AHN10

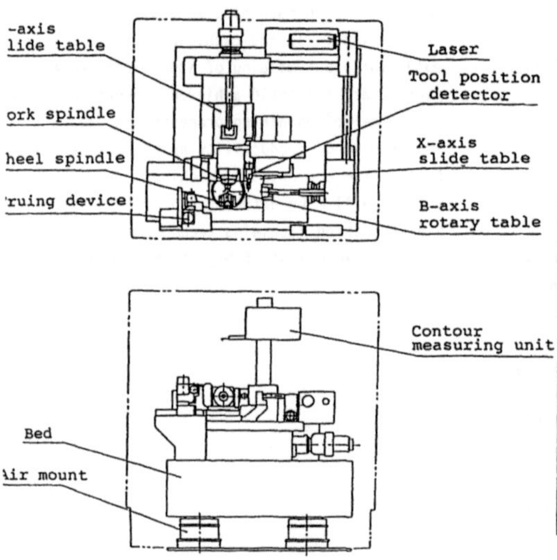

Fig.2 Mechanical construction

Table 1 Major mechanical specifications

Item	Unit	Spec.
Max. Workpiece diameter	mm	ϕ100
Work spindle revolution speed	rpm	300-3000
X-axis stroke	mm	250
Z-axis stroke	mm	200
Resolution (X,Z)	µm/pulse	0.01
Resolution (B)	sec/pulse	1.3
Wheel spindle revolution speed	rpm	Max100000

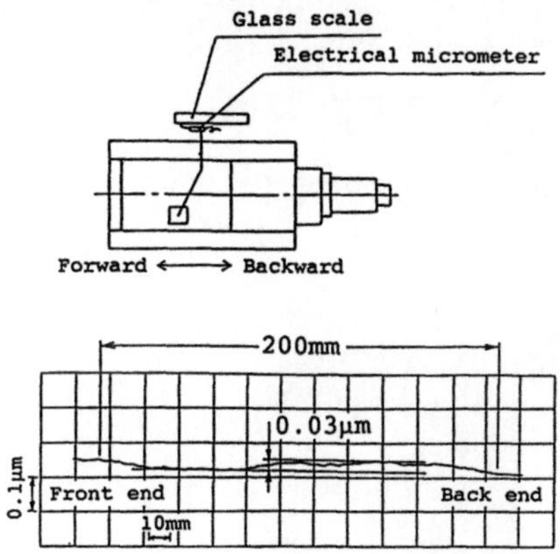

Fig.3 Straightness of slide table movement

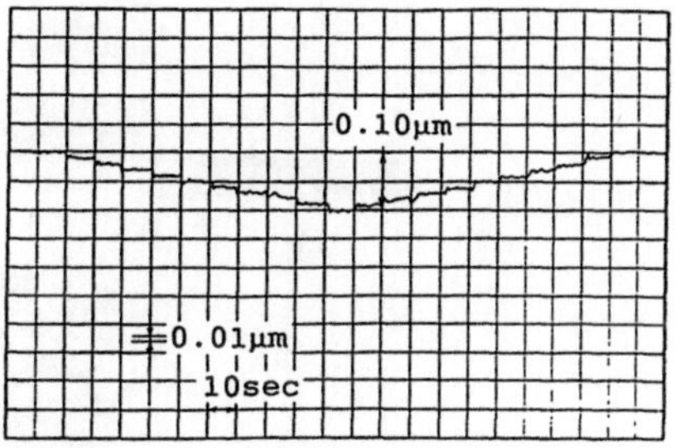

Measuring instrument: Electrical micrometer

Fig.4 One-pulse response of slide table

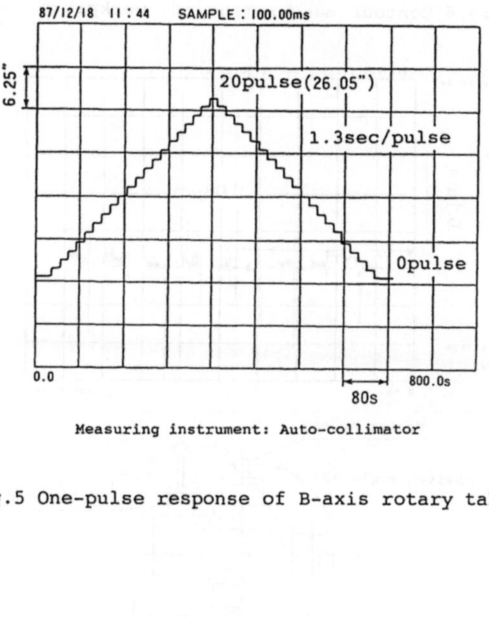

Measuring instrument: Auto-collimator

Fig.5 One-pulse response of B-axis rotary table

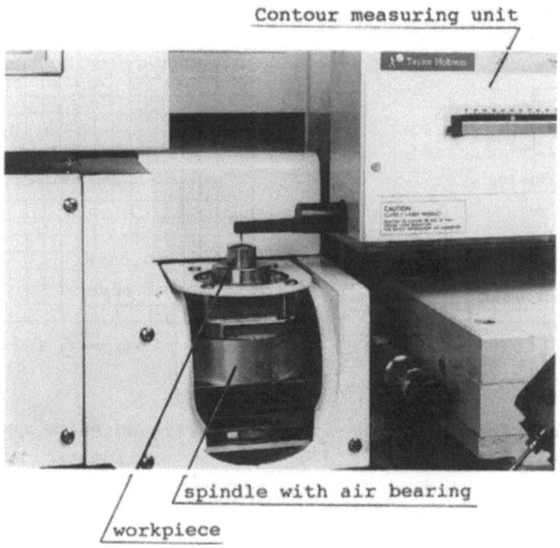

Fig.6 Contour measurement of workpiece

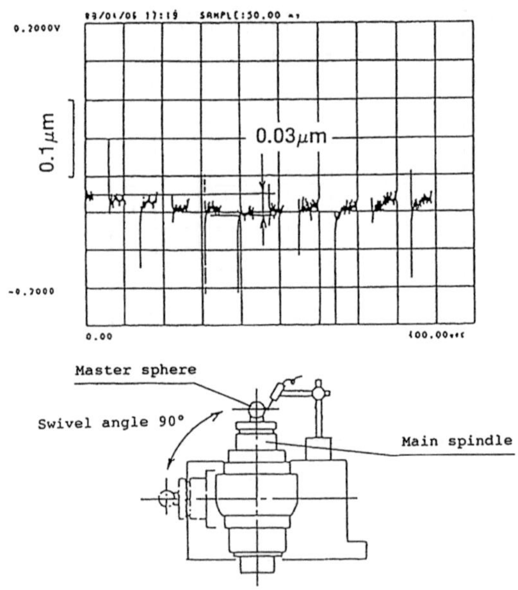

Fig.7 Repetitive positioning accuracy of spindle 90° swivel motion

Repetitive accuracy of contour measurement ±0.02μm

(a) Before 90° swivel motion

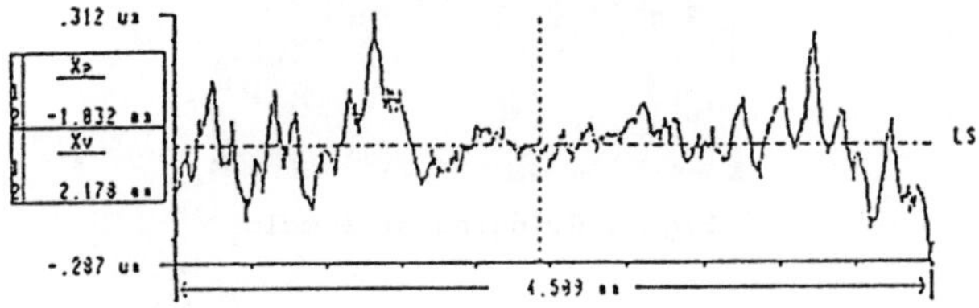

(b) After 90° swivel motion

Fig.8 Repetitive accuracy of contour measurement before and after 90° swivel motion of spindle

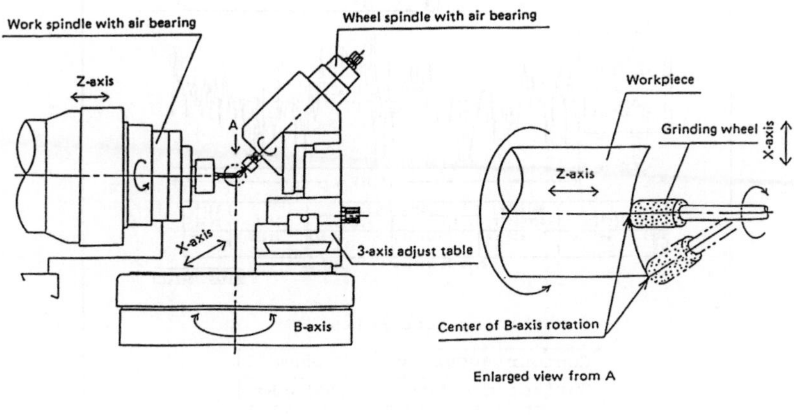

Fig.9 Schematic of aspherical surface machining

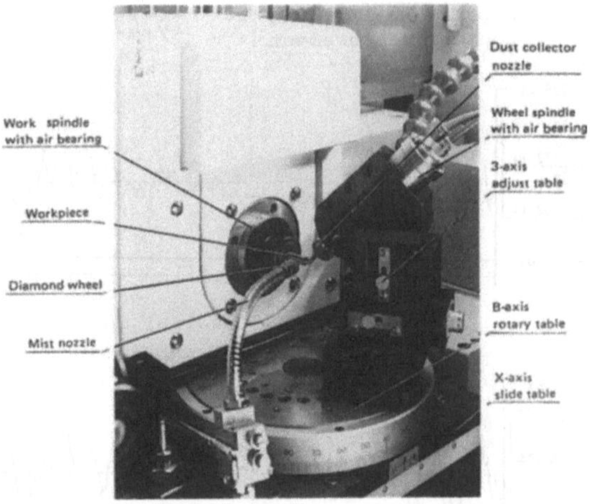

Fig.10 Grinding of a mold

Product name	Mold for pressing LDP lens
Material	Cemented carbide
Contour	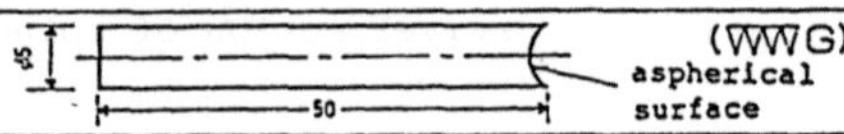

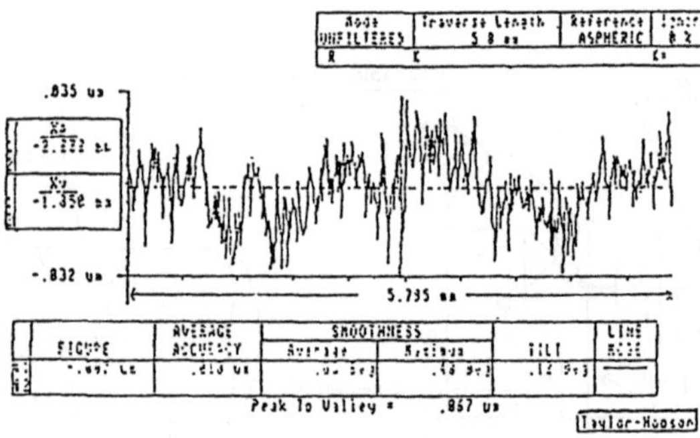

Machining accuracies

Contour accuracy	0.05μm
Surface roughness	0.025μmRz

Fig.11 Machining accuracy of aspherical mold

Design and Testing of a Self Coupling Hydrostatic Leadscrew

Alexander H. Slocum, Ph.D
P.O. Box 51, MIT Branch
Cambridge, MA 02139 USA

Abstract

This paper discusses the design and preliminary test results of a new type of hydrostatic leadscrew [1]. Like other hydrostatic leadscrews, it experiences no wear, backlash, or friction and its axial resolution is limited only by the performance of the drive motor and position sensor. In addition, the hydrostatic leadscrew discussed herein is only rigid along the axis of motion so it cannot cause errors (e.g. straightness errors) due to misalignment between the actuator and the carriage; hence the leadscrew is *self coupling* even though its axial stiffness can easily be designed to be from 10^8 to 10^9 N/m. Previous designs did not address the need for self coupling [2,3,4,5,6] and therefore carriage straightness errors often had a component with a period equal to that of the leadscrew. Tests on a full scale testbed verified that the self coupling hydrostatic leadscrew met all its design requirements, and its large self-coupling capability makes it economical to manufacture and install.

Large radial clearance is provided at the roots of the thread and hence small radial motion is allowed before the gaps closed. The amount of radial motion δ allowed is a function of the hydrostatic gap h between the thread flanks, the lead ℓ, and the inner diameter D_i:

$$\delta = \frac{h}{3\sin\left(\tan^{-1}\left(\frac{\ell}{\pi D_i}\right)\right)} \tag{1}$$

In order to prevent radial pressure differentials from occurring which would hinder the self coupling capability, the hydrostatic pocket extends uninterrupted along the full helix of the nut's thread. This also prevents the nut from having any pitch or yaw stiffness. Other designers had used many pockets along an angled thread (e.g. Acme) or specifically provided radial motion resisting pockets at the thread roots. The deep square thread used here allows for radial motion of the nut with respect to the screw. Variations in the thread geometry only cause variations in the accuracy of the lead which a modern machine's servo control and sensor system can easily correct. For high load or stiffness requirements, several nuts could be ganged together. Manufacture of the nut is made possible by replicating the thread to a thread milled screw shaft. A typical design application is shown in Figure 1.

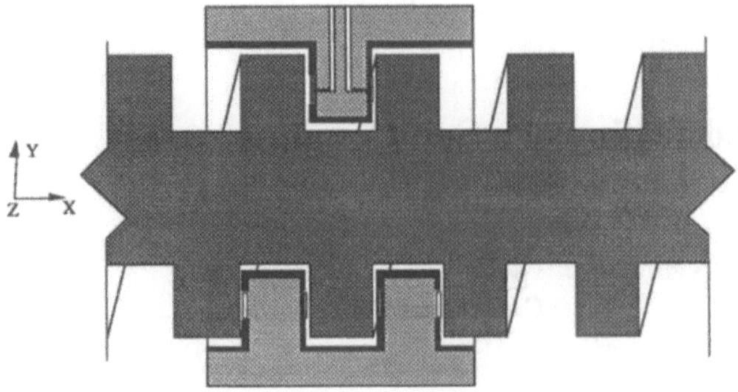

Figure 1 Cross section of self-coupling hydrostatic leadscrew and nut.

A test machine was designed to maximize manufacturability while attaining a high degree of stiffness and accuracy. It was reasoned that if a machine with sub-micron resolution could be made using parts with only ground accuracy assembled by a novice fitter, then a significant advance in the state of the art would be realized. The test machine's carriage was supported by six hydrostatic bearing pads arranged in a quasi-kinematic dovetail configuration shown in Figure 2.

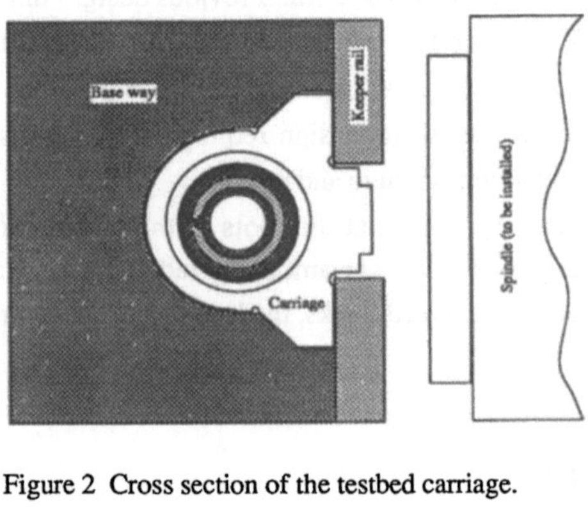

Figure 2 Cross section of the testbed carriage.

The ways were ground for economy. The design is self-checking and easier to manufacture (e.g. scraped) than a standard rectangular bearing configuration. In order to determine if self-coupling was achieved, other than by noting that there was no resistance to turning of the screw, the horizontal straightness of the hydrostatic bearing supported carriage was measured using an LVDT probe and an optical straightedge. The leadscrew was turned by hand to move the carriage over its 125 mm range of motion and the results are shown in Figure 3. The ways were only ground so the straightness measured is respectable and there is no error component with a period of 50 mm that would be caused if there was lateral coupling between the leadscrew and the carriage.

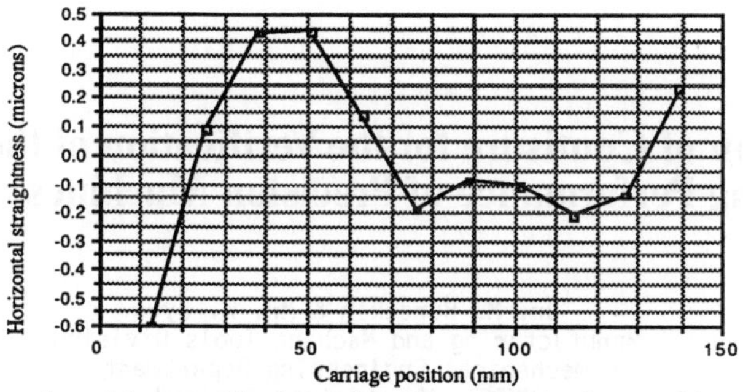

Figure 3 Horizontal straightness of hydrostatic self coupling leadscrew driven carriage. Nine measurements were averaged. Uncertainty was due to the following: Influence from hydraulic fluid supply ± 0.2 μm, LVDT lever probe ± 0.1 μm, and Straightedge calibration ± 0.04 μm.

After assembly, the position of the leadscrew with respect to the face of the carriage was measured for 0o, 90o, 180o, and 270o positions of the leadscrew and on the left and right hand sides of the 250 mm long carriage. The lateral misalignments across the 250 mm long carriage were 0.122 mm and 0.065 mm. The total radial error motion of the screwshaft at these two points was 0.008 mm and 0.005 mm respectively. These misalignment values are about an order of magnitude larger than those normally considered acceptable for a precision leadscrew. Yet despite these large manufacturing errors, the leadscrew turned without resistance over its entire range of motion and there was no correlation with straightness errors of the carriage to the period of the leadscrew. It can thus be concluded that self-coupling was achieved.

References

[1] U.S. Patent 4,836,042, June 6, 1989. Foreign patents pending.
[2] M. Weck Handbook of Machine Tools Vol. 2, pp 165-167, © 1984 Wiley Heyden Ltd.
[3] U.S. patents 2,320,353; 3,171,295; 3,183,732; 3,448,632; 3,670,583; and 4,566,345.
[4] Federal Republic of Germany patent 1550774.
[5] Japan patents 58-166161 and 58-166162.
[6] United Kingdom Patents 914165 and 1239557.

Acknowledgements

The author would like to make the following chronological acknowledgements to institutions that have helped to support this work in its various phases: The Franklin Pierce Law Center in Concord NH for performing the patent work; GCA a unit of General Signal Corp. for machining the testbed components; LeBlond Makino Corp. for providing financial assistance and moral support; Cranfield Precision Engineering Ltd. for providing shop space and personnel to complete the assembly and testing of the testbed; and financial support for the testing stage by an appointment to the University Faculty Off-Site Research Participation Program sponsored by the Strategic Defense Initiative Organization and funded through the Optics MODIL, Oak Ridge National Laboratory, which is operated by Martin Marietta Energy Systems, Inc., under contract with the U.S. Department of Energy. The program is administered by Oak Ridge Associated Universities through their management contract with the U.S. Department of Energy, Contract No. DE-AC05-76OR00033.

Application of Contisure for the Verification of the Contouring Performance of Precision Machines

Dr. M. Burdekin & Mr. W. Jywe
Manufacturing and Machine Tools Division
Mechanical Engineering Department
UMIST, Manchester, England

ABSTRACT

The methods available for assessing the contouring accuracy of precision N.C. machines are considered along with the details of the design and construction of the 'Contisure' hardware. This hardware comprises two tungsten carbide reference spheres located at the tool and workpiece positions. As the N.C. machines executes a circular path, the 'Contisure' kinematic transducer link, located between the two spheres, detects the changes in the programmed radius. The transducers in the link are connected on-line to a P.C. or compatible for analysis and plotting.

The system software which has been developed in collaboration with a number of industrial users is an essential part of the 'Contisure' system and salient features of this software are outlined.

Typical results from the application of the 'Consiture' system to machining centres and lathes are illustrated together with examples of polar and linear representation of the errors with least squares and statistical analysis. Interpretation of the error data in terms of diagnostic information such as hysteresis, servo errors etc. are considered.

Introduction

The contouring performance of a CNC machine can be established by assessing its ability to move along a specified profile by the simultaneous movement of two or more axes. The circular profile as specified in the NAS 979 cutting test for machining centres is well established and can now be emulated by instrumented techniques. Although instrumentation techniques generally check the machine in a no-load condition, they offer certain advantages over cutting conditions. In particular, tools and test specimens are not consumed and the need to metrologies test pieces is also eliminated.

In addition to the Contisure System (1) there are two basic types of circular contouring instruments available. The Cary Bi-DIM K circular test as developed by Knapp (2) uses a bi-dimensional probe in conjunction with a circular masterpiece. The deviation of the programmed circular path relative to the circular masterpiece is detected by the bi-dimensional probe and the analogue output, representing the error in the programmed path, can be displayed in polar form on an X-Y plotter. However, owing to the friction between the master disc and the probe stylus there is a relatively large angular orientation, between the clockwise and anti-clockwise data plots.

The flexible ball-bar as developed by Bryan (3) offers an alternative approach. This system was developed from the well established fixed ball bar used to assess the volumetric accuracy of coordinate measuring machines. The flexible element of the bar, in conjunction with an LVDT transducer enables changes in distance between the two balls to be determined. The balls at the end of the bar are supported in magnetic cups which in turn are mounted at the tool and workpiece positions. Thus the change in the radius between the two magnetic cups as the machine performs a circular contouring operation can be detected. This flexible ball bar has advantages over the BI DIM-K system since it can be used as a three dimensional device if required. The magnetic support cup arrangement however requires steel balls and the ball error and uncertainty of the elastic contact is considered to be a limiting factor for high precision applications.

Contisure Hardware

The Contisure hardware performs a similar task to that of the flexible ball bar but has a fundamentally different configuration. The magnetic support cups are replaced by two high precision tungsten carbide reference balls which are rigidly attached at the tool and workpiece positions; the centre of these two balls defining the two reference positions on the

Fig.1 A 100 mm kinematic link & setting block

machine. The absolute distance between these two reference balls as the machine performs a circular contouring operation is measured with the aid of two linear displacement transducers mounted in a kinematic transducer link and acting directly upon the two balls. Thus as the machine describes a circular contour, the two transducers are orientated normal to the reference surface of the balls. A Contisure link in operation on a machine tool is shown in Fig.1.

As the sum of these two transducer outputs represents the change in the separation between the two reference balls, the influence of unknown elastic deflection between the link location and the ball is not significant.

The roundness error of the balls used in the standard Contisure hardware is in the order of 0.1 microns which can be neglected when the system is applied to a standard range of CNC machines. However, for application to high precision machines, this error forms a significant portion of any uncertainty. This uncertainty associated with the ball roundness error can however be reduced by the Contisure software using an error mapping technique. For high precision applications, the contacting type of LVDT transducers are replaced by non-contacting capacitance probes. This further reduces uncertainties resulting from sliding contact with the reference ball surface.

Traceability of the absolute length of the measured contour radius is assured by setting the kinematic transducer link against a calibrated block. This block consists of two stem mounted tungsten carbide balls located in a rectangular steel block at a nominal calibrated centre distance. Under normal use this setting block can be temperature stabilised by placeing on the machine structure and thus eliminating the need for length correction due to differential temperature. Invar setting blocks are sometimes used to minimise the influence of temperature. For high precision applications, the associated Contisure software includes facilities for compensation of setting block and machine temperature.

The kinematic transducer link is manufactured from carbon fibre tube and aluminium which ensures a stiff, lightweight and thermally stable construction. Long term stability is not a problem as the link is set against a setting block each time it is used. Where improved thermal stability is required, the aluminium is replaced by a titanium alloy as this has a much lower coefficient of expansion.

For testing the machine geometry & servo drive performance different effective lengths of links are desirable. With the standard transducers mounted with internal configuration with respect to the reference distance between the two reference balls, the shortest possible link length which can be manufactured is in the order of 100 mm, whilst maintaining possible 3D orientation of the link. Where servo characteristics are to be investigated with minimum influence from the geometric errors in the machine, links with an effective length in the order of 20 mm can be manufactured by mounting the transducers in an external configuration. In this case, however, the link is restricted to 2D contouring which is generally adequate for most practical applications.

For link lengths of 200 mm and longer, a modular link design is adopted using carbon fibre tubular extension pieces. Such links have been produced up to 500 mm long but longer lengths could be produced if required. With this modular link, a composite setting block is used. In one such application a rectangular invar block supported at its airy points was used to set links of 100 mm, 200 mm, 300 mm, 400 mm and 500 mm.

The LVDT transducer probes used with the link are manufactured by Tesa of Switzerland although alternative manufacturers probes could also be used. The Tesa probes have a specified repeatability of 0.01 micron and an anlogue output facility which can be directly output to a simple strip chart recorder if required. For efficient and effective application of the Contisure system a PC compatible computer is employed for data acquisition, analysis and presentation of the results. The data acquisition is carried out with the aid of a 12 bit A/D interface card mounted in the PC and connected directly with the analogue output of the transducer probes. Peripheral equipment such as an Epson compatible dot matrix printer and a HP compatible plotter are used to produce hard copies of the results.

Contisure Software

An essential part of any modern calibration system is its software and the Contisure system is no exception. The current version (1.2) covers a range of facilities which have been developed in collaboration with both users and manufacturers of CNC machines. This software is menu driven and its salient features are outlined as follows:

a) System Set-up

This facility enables the system to be configured to any PC compatible (XT, AT, 386 etc) having a range of graphics adaptors such as CGA, EGA, VGA & Hercules.
The VDU printer aspect ratio is specified so that polar plots appear circular when displayed on the computer screen and also printed.
Hardware addresses of the A/D converter and plotter is important as these may have to be changed from the default values in order to avoid interfering with other devices in the computer. Both IEEE and RS232 H.P. compatible plotters are supported. Additional but important cosmetic features such as pen colours can be selected along with a metric or inch units for the convenience of North American users.

b) Transducer Calibration Transfer & Verification

When the A/D interface is read by the computer, the digital reading returned will be within the range 0 - 4096 (12 bit). This value then requires conversion to the equivalent micron or mill units by the application of a calibration factor. In order to establish the appropriate calibration factors and at the same time extend the traceability of the transducer unit through to the computer, a simple calibration transfer procedure has been devised. This involves setting the transducer to at least three different values over the selected range of the instrument and reading the corresponding A/D value with the computer. The results of a typical transfer process is shown in Fig.2. The two constants defining the linear regression line through the data are subsequently used by the software to convert from the A/D readings to the corresponding transducer value. Transfer calibration factors for up to six different ranges can be stored. These values can be up-dated periodically when the transducer unit has been re-calibrated. Verification of the transducer reading by the computer prior to carrying out a contouring test can be selected by the user. This facility checks for possible mismatch between the instrument range setting and the calibration factors selected by the computer.

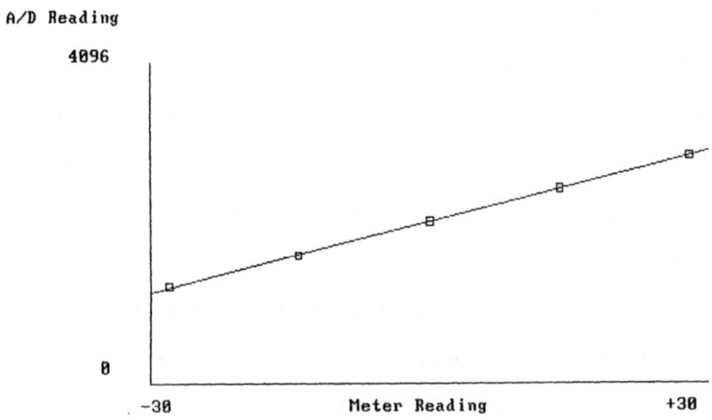

Fig.2 Typical transducer transfer calibration

c) Test Specification

An N.C. program corresponding to the desired circular path is first prepared; the programmed radius corresponding to the set length of the kinematic link. Where possible a 360° swept angle of the radius should be used but for application to lathes etc., the angle may be limited to 90° or 180° .

The start and end points of the contour are detected by a software algorithm which assumes a tangential approach and exit at the start and end points. The clockwise and anti-clockwise direction of the programmed path is also relevant and when multi-scans are used to assess repeatability, the machine is programmed alternately clockwise and anti-clockwise to ensure that the transducer cables do not tangle. Where possible, the start and exit points on the path should avoid the salient positions such as axis reversal points etc. With a 360° scan, a convenient start and exit point would be at say 22 degrees from one of the principal axes.

```
CONTISURE :                              DEFINE TEST DATA

Machine Name = Takisawa                       Serial Number= 12634
Test Date                  = 22-10-90

File Identifier            = DEMO1ST          Link Radius (mm)          =    99.987

Machine   Temp. (\C)       =  20.0            Machine   Exp.Coef.(ppm)=  11.0

Set'g Blk Temp. (xC)       =  20.0            Set'g Blk Exp.Coef.(ppm)=  11.0

Plane (e.g. XY, XZ, YZ) = XY                  No. of Clockwise Scans    = 3

No. of Anti-Clock Scans = 3                   Start Anti/Clock (A/C)    = C

Start Angle                = 22               Swept Angle               = 360

Feedrate (mm/min)          =  1000.0          Nominal Samples/360x      = 9000
                                                            Max=16000 Min=1000
```

Fig.3 Example of test data input to software

The programmed feed rate should correspond to the practical range of the process although higher feed rates can be used to highlight the servo drive errors. The machines feed-rate override can be applied, if necessary, to select different rates. An example of the input data to the software is shown in Fig.3. In addition to machine and file identifiers, the results can be corrected to $20^{o}C$ by specifying machine and setting block temperatures along with the corresponding expansion coefficients.

The nominal number of data samples read by the computer per 360° of swept profile can be selected by the user within the range 1000 - 12,000; the default being 9000 samples. In general the more samples the better, but processing time will get longer and data files bigger. After each scan of a test has been completed, the readings are stored in integer form with nano-metre resolution for subsequent analysis. For repeatability testing of the machine, up to three clockwise and three anticlockwise scans can be specified.

d) Analysing the Data

The measured data from each test are stored in files defined by the file identifier. After selecting a particular file identifier for analysis and confirming that the test specification is appropriate, coarse or fine resolution can be selected. The rėlevant files containing the measured data are read from the disc and converted to 'per degree' values for coarse resolution or 'per 0.2 degree' for fine resolution before being transferred into memory. This data averaging and reduction process is necessary prior to analysis in order to reduce computing time and to satisfy the computer RAM limitations as up to six scans may have to be stored in memory at any one time. The analysis mode is also selected. For example Bi-directional or unidirectional analysis; single or multi scans. Two basic types of data analyses are available and one must be selected at this stage.

i) Measurement Centre Analysis

If the set-up of the reference balls on the machine is not perfect, there will be an eccentricity error between the programmed circle and the actual path travelled by the kinematic link. Measurement centre analysis uses the data as measured without any mathematical compensation for the set-up error.

ii) Least Squares Centre Analysis

This analysis of the data permits the deviation in the contour radius to be specified with respect to a 'best-fit' or least squares radius thus eliminating the influence of any set-up error. The magnitude of the error band about this radius represents the error in the contour profile. The computerised eccentricity between the measured and least squares centre of the profile is specified on the output plots of the results. This information is useful in that it can be used to reduce the actual set-up error of the machine and also for subsequent analysis of thermal drift and machine geometry. When multi-scan data has been recorded, the repeatability of the machine can be assessed by processing the data in terms of mean and standard deviation. The basic analysed data can be displayed in either polar or linear form. With the linear display, salient features of the error can be accurately evaluated from display by a cursor controlled readout. Zoom facilities of selected parts of the measured profile are also available.

For accurate correlation between the Contisure results and those of a cutting test similar to NAS 929, the influence of the cutter radius should be considered. Since the basic Contisure data is representative of a cutting test with a zero radius cutter a further analysis of the results is available to show the effect of machining both external and internal circular profiles using a cutter of specified radius.

Practical Applications

Clockwise and anticlockwise scans were made on a recently refurbished machining centre with a 200 mm length link at 500 mm/min feed rate. The clockwise and anticlockwise scan results are shown in polar form in Fig. 4a and 4b using a least squares analysis.

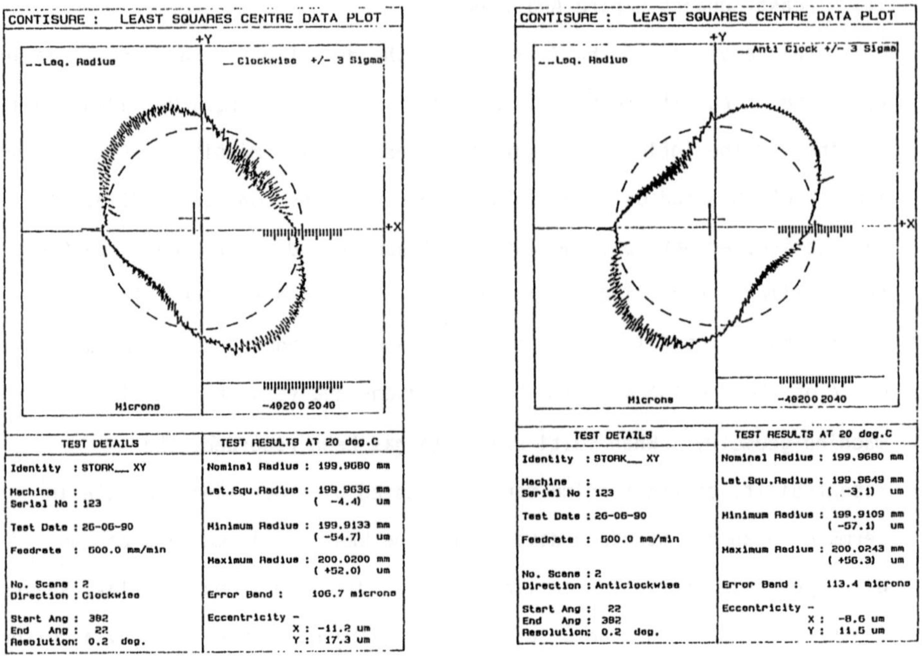

Fig.4a/b Example of servo mismatch on a machining centre

The results clearly show the classical servo mismatch problem where the eliptical profile with a principal axis at 45° moves through 90° when the scan is made in the opposite direction. With this size of link and at a feed of 500 mm/min, one would not expect such a dominant servo error and this clearly indicated that the drives require adjustment. At least squares radius of the profile is only 3 to 4 microns less than the link radius even though the error band is in the order of 107-113 microns.

CNC Milling Machine

A moving column type CNC milling machine, fitted with a 'black-box' electronic volumetric accuracy enhancement system was tested to ascertain the effectiveness of the correction system under contouring conditions. The initial test in the XZ plane (horizontal plane) without the enhancement system

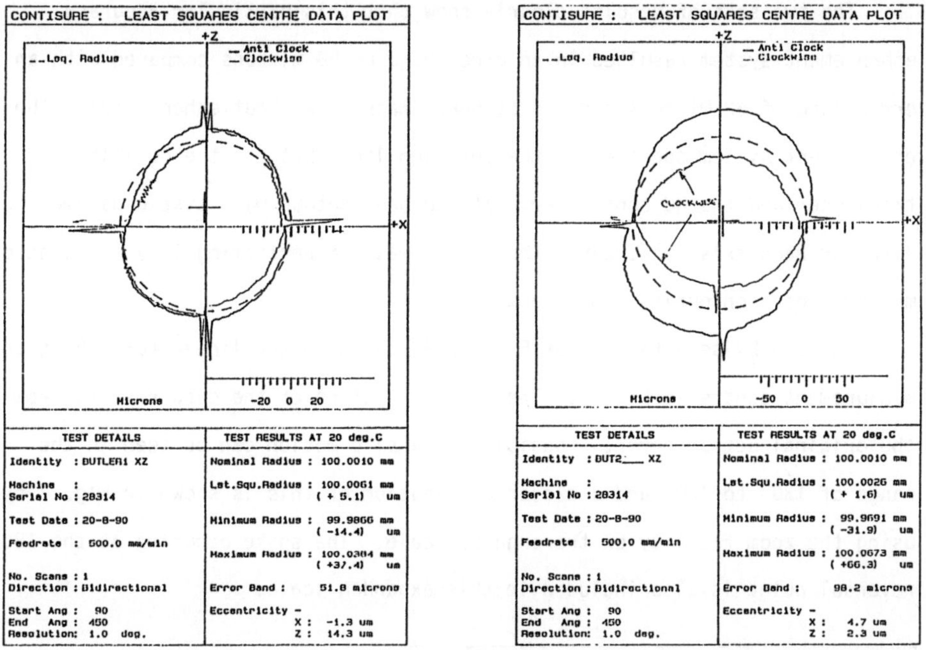

Fig.5a/b Effect of electronic accuracy enhancement on a milling machine

is shown in Fig. 5a where the clockwise and anticlockwise scans have been superimposed. At the four reversal points, lost motion is evident but in addition there are relatively large spikes particularly at the X axis reversal point. We do not have a satisfactory explanation for these spikes but such characteristics are frequently observed on machines with relatively large moving masses.

The above test when repeated with the accuracy enhancement system active is shown in Fig. 5b. The clockwise and anticlockwise scans have been superimposed but their characteristics are particularly distorted in the direction of the Z axis. With the actual colour plots this difference can be clearly seen. The inner profile is the clockwise scan and the outer profile the anticlockwise scan.

The two sets of results clearly show that this particular accuracy enhancement system resulted in an error band of 98 microns compared with an error band of 52 microns for the standard machine without enhancement. The enhancement system operated in the position loop, between the position transducer and the NC controller. It had been set-up with static parametric data for each axis but clearly the system was not performing in a satisfactory manner under contouring conditions.

The clockwise data scan of Fig. 5a is shown in the linear form about the measurement centre in Fig. 6. The periodic nature of the data results from the original set-up error. Possible stick-slip motion can be seen in the range of 130° to 160° and a closer examination of this is shown in Fig. 7 by using the zoom facility on the angular scale. The spike occuring at the 180° reversal point is also included in this expanded scale.

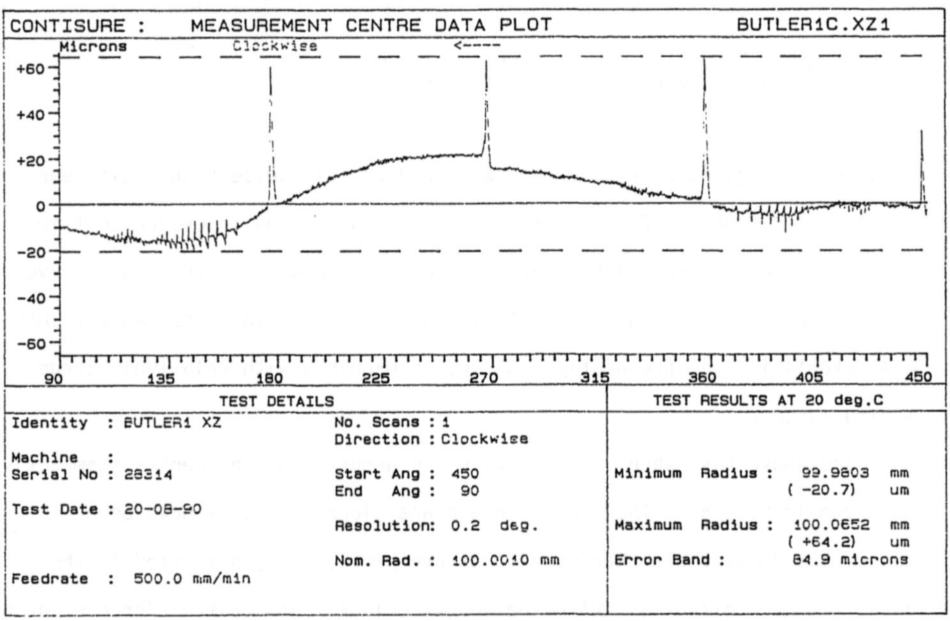

Fig.6 Linear plot representation of data from measurement centre

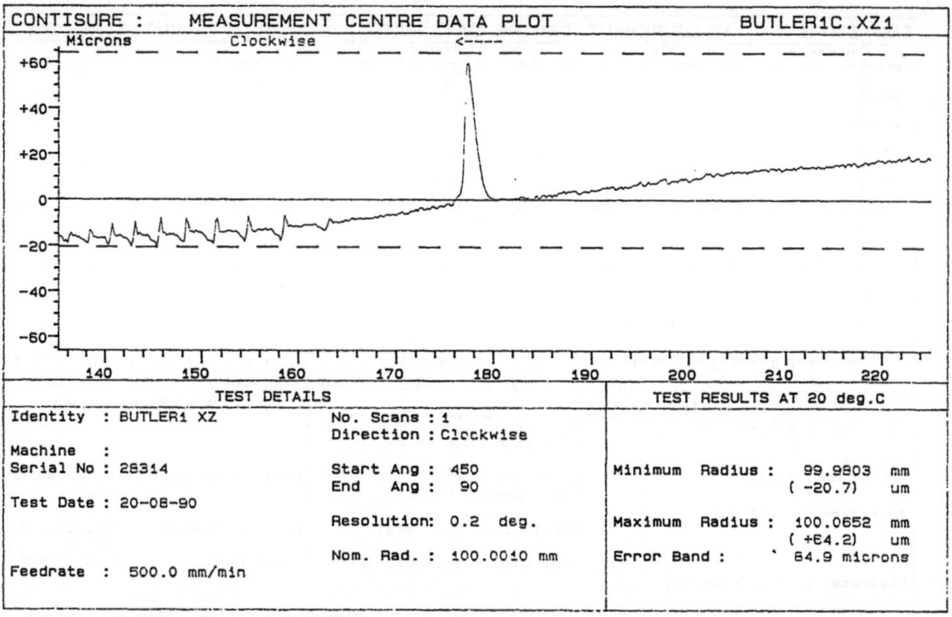

Fig.7 Expanded angular scale showing details of stick-slip and reversal spike

The influence of the cutter radius can also be considered. The basic data shown represents the profile using a zero radius cutter and with a more practical size of cutter it is not unreasonable to expect smoothing of spike like features when machining external profiles. Further analysis of the data however shows that the 'sharpness' of the spikes is greatly exaggerated due to the magnification distortion. For example, the influence of a 50 mm cutter over the region 130^{0} to 160^{0} is shown in Fig. 8 and Fig. 9 for external and internal profiles respectively. For external machining the magnitude of the 180^{0} spike is only reduced from about 60 microns to 45 microns, with additional smoothing of the peaks of the stick-slip motion.

For internal machining of the contour, the magnitude of the peaks are not affected from the basic measurements. The width of these peaks are however increased due to the cutter radius as shown in Fig. 9

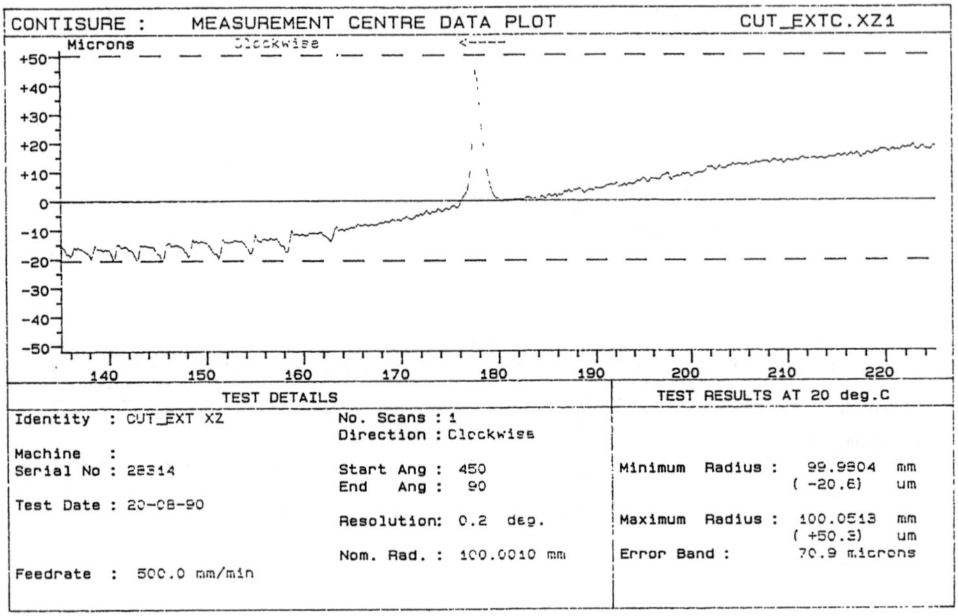

Fig.8 Influence of milling cutter for external machining

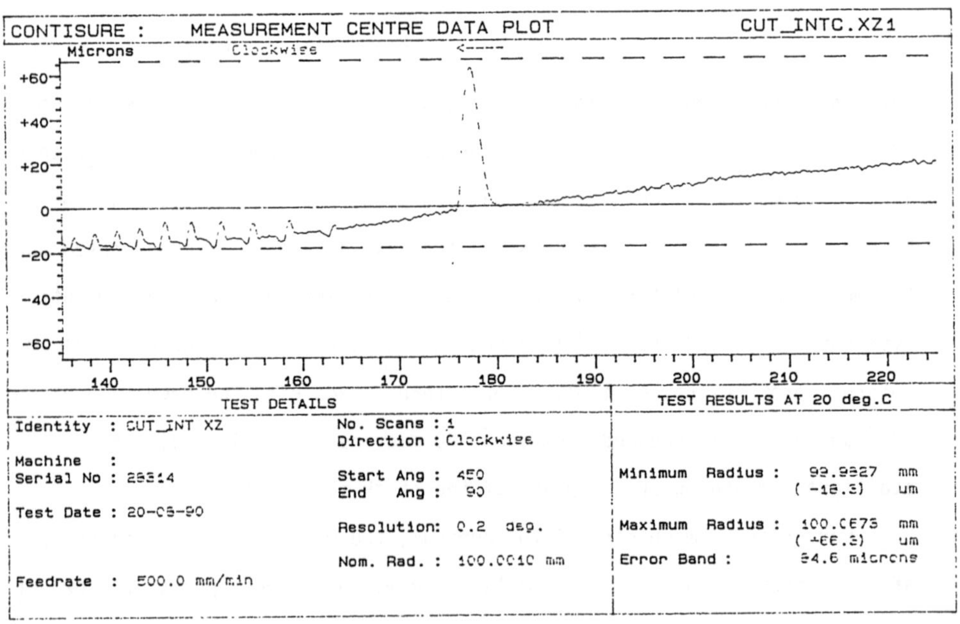

Fig.9 Influence of milling cutter for internal machining

Precision Lathes

The application of Contisure to lathes is usually confined to a swept angle of 105 degrees, nevertheless useful information can be obtained. Fig. 10 shows the rusults of a sweep from 0 to -95 degrees on a precision lathe. A relatively large negative spike was observed at the pole position in the Z axis direction which upon further investigation was found to be caused by a sticking guideway guard. After adjustment this characteristic was eliminated.

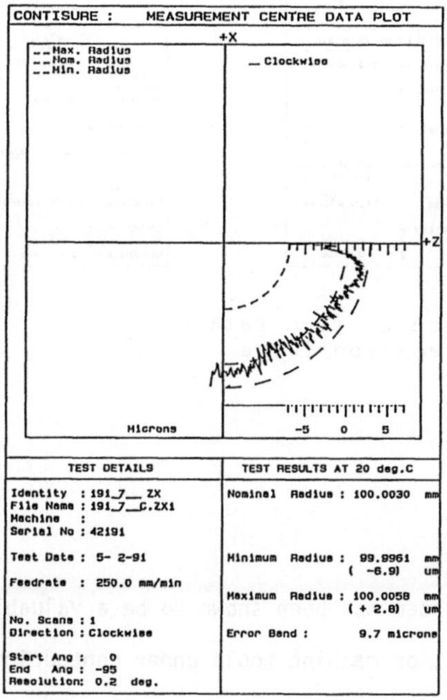

Fig.10 Influence of sticking guideway guard on lathe

Results from a high accuracy turning centre are shown in Figs. 11 & 12 for a 95 degrees scan. The difference in the two test conditions is the feed rate which was 25 mm and 125 mm/min respectively. The resulting profile shape is remarkably similar in every respect with a resulting error band about a least squares radius of 1.5 micron.

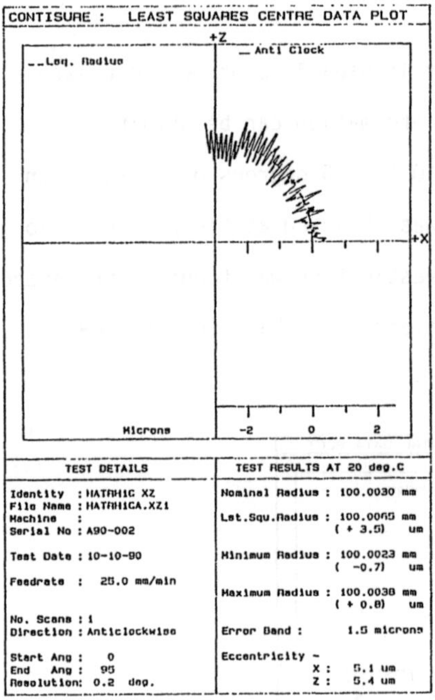

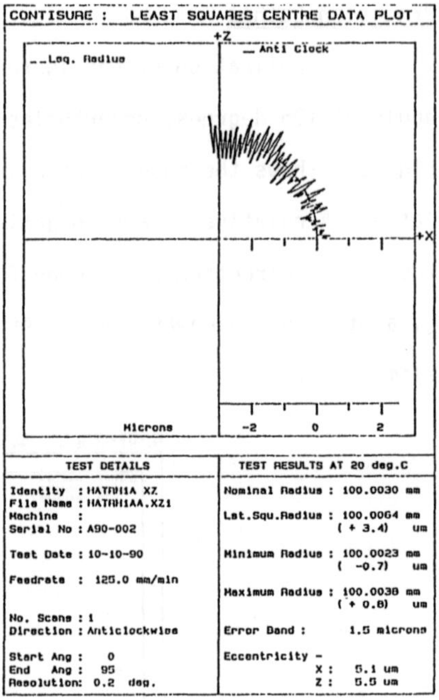

Fig.11/12 Influence of feed rate on high precision lathe

Conclusion

The Contisure system has been shown to be a valuable tool for assessing the accuracy of precision machine tools under contouring conditions.

The hardware is relatively simple in construction and produces instant results which are fully traceable to national standards.

The computer software facilities are extensive and provide the user with flexibility in the analysis and presentation of his results as well as the option to recall and compare with previous calibrations.

The results have great diagnostic significance and eliminate the need to carry out extensive and expensive test piece machining.

Acknowledgements

The Authors would like to acknowledge the helpful suggestions made by users of the Contisure system. These have contributed much to the software development. In particular Philips (Eindhoven), AWE (Aldermaston) and GLF (Arbroath).

References

1. Contisure System - User Manual, C.D. Measurements Ltd, 'Chomlea House', Hadfield, Hyde, Cheshire, SK14 8ER, England.

2. W. Knapp. BIDIM IC Circular Test., P.O. Box 162, CH.2400 Le Locle, Switzerland.

3. J.B. Bryan. A simple method for testing measuring machines and machine tools. Precision Engineering, Vol. 4, No.2, 1982.

An Evaluation of Ultra-precise Machine Tool Contouring Performance: The Low Amplitude Sine Tracking (LAST) Test

Mark Craig Gerchman and David H. Youden

Rank Taylor Hobson Inc.
Keene, New Hampshire 03431 USA

Introduction

Micro-stepping position tests are often used to demonstrate the performance of ultra-precise machine tools. These tests record the ability of a machine tool to precisely move slides over very short distances. To optimize machine performance for these tests it is necessary to detune slide servo system characteristics from typical operating values. Therefore discreet micro-stepping tests are not representative of the dynamic contouring performance of ultra-precise machine tools.

The low amplitude sine tracking test measures the contouring performance of ultra-precise machine tools under typical slide servo conditions. The metrology instrumentation necessary to perform the test is available in many laboratories and the testing configuration is easy to align. Interpretations of the test results provide not only a measure of machine tool contouring performance but also information useful for the proper tuning of slide servo system characteristics. The test as described is not biased to a particular machine tool geometry and parameters that define the test are suitable to machine with various resolutions. Although this is a contouring performance test in two dimensions, generalizations of the test to three dimensional machines are straightforward.

Test Description

The low amplitude sine tracking test, or LAST test, measures the ability of a machine tool to execute a coordinated sequence of motions. The sine wave motions that make up the tracking sequence require low amplitude variations in slide speed that are typical of ultra-precise machine tools. In a thorough LAST test, several tracking sequences are performed in different orientations. Usually, for a two-axis machine tool with an orthogonal slide configuration, three tracking sequences are performed. These include orientations that are parallel, perpendicular, and at 45 degrees to the principal spindle axis. The use of these different orientations provides an evaluation of the motion of each slide including reversals and speed variations.

The tracking sequence of the LAST test, shown in figure 1, is a five part motion. Two axes can be defined in the plane of these motions. They are designated the fast and slow axes; because of the relative speed of coordinate slide motions in these directions. The five contouring motions consist of: (1) a straight move along the fast axis of one-quarter wavelength; (2) a half-wavelength sine wave motion; (3) a reversal in the fast direction and a one and one-half sine wave motion; (4) another reversal in the fast direction and a one-half sine wave motion; and (5) a straight move along the fast direction of one-quarter wavelength that returns to the starting position.

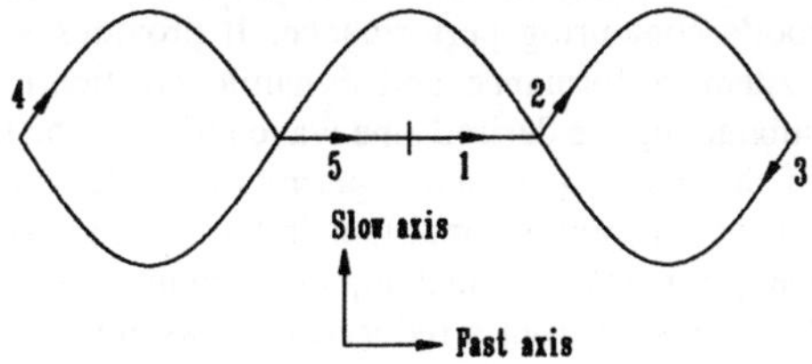

Figure 1. LAST Tracking Sequence

The choice for the sine wave's amplitude and wavelength are related to the machine tool's minimum programming increment and typical operating velocity. For the ultra-precise machine tool measured to date, the amplitude of the sine wave has been 25 minimum programming increments: yielding a total motion in the slow axis of 50 increments. The time to perform the LAST test tracking sequence has been standardized to three minutes. This has helped to reduce temporal drift in the metrology instrumentation and eliminate long term environmental considerations. Given the duration of the tracking sequence the wavelength is determined from a typical machining speed. As an example, a diamond turning lathe with 0.000,01 mm programmability operating at typically 2.5 mm/min has a LAST sine wave with a 250 nm amplitude and 2.5 mm wavelength.

The programming of the tracking sequence is defined using 100 constant speed linear moves per wavelength. These moves are equally spaced along the fast axis. Since the beginning and ending straight line motions are also divided into moves of the same size, this results in a program sequence 300 blocks long. The use of linear interpolation and this spacing provides a bench mark for the comparisons of different machine tools under identical conditions. The adequacy of these parameters for the description of the tracking sequence will be established later. A LAST evaluation using other interpolation schemes and curve fitting can be performed to evaluate and compare machine controller operations.

A preferred technique for measuring and recording the tracking sequence is capacitance gaging. Figure 2 show a typical configuration of the LAST test. By mounting a capacitance probe on one axis and a conductive straightness standard on the other, a measure of the slow axis position in time can be obtained. The measurement should be performed at cutting height in the vicinity of the typical tool-work interaction. Typically the frequency response of the recording system should be adequate to sample at 0.001 of the tracking sequence wavelength. For the example given, this corresponds to a 16.7 Hz filter.

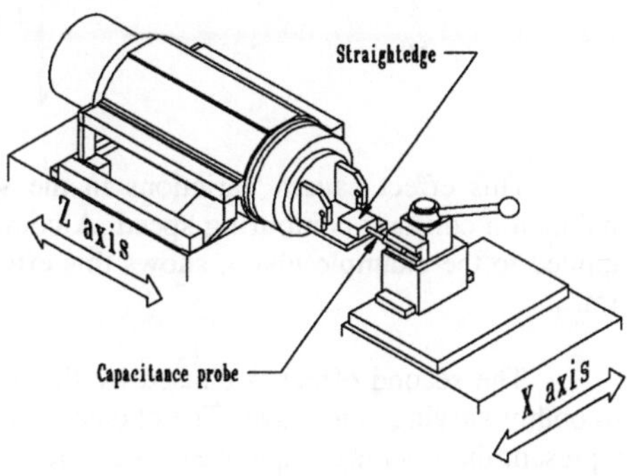

Figure 2. LAST Test Configuration

Plotting the slow axis position in time provides an easy way to evaluate a machine tool's contouring performance. It provides a valuable diagnostic tool for evaluating servo system performance and dynamic positioning accuracy. By digitizing the information an subtracting the desired sine wave motions, the residual errors can be quantified. The symmetry of the tracking sequence permits the effects of gaging electronic drift to be separated from out-of-squareness conditions in the test configuration. The resultant error plot can be analyzed using a number of techniques including harmonic analysis. A contouring position error can be expressed from a reduced trace as either a peak-to-valley or *rms* error.

Tracking Sequence Analysis

Although the representation of the tracking sequence is straightforward, it's description masks some inherent inaccuracies. These inaccuracies are caused by describing a tracking sequence that does not yield the expected sinusoidal output. The errors created are caused by three different factors. Fortunately, because of the large wavelength to amplitude ratio (10,000 to 1 in the example given), these errors can be shown to be below influencing levels.

The first effect is caused by the use of constant speed in the description of the side motions. To describe this effect consider the position along the fast axis as 'x' and the position along the slow axis as 'z'. The description of a portion of the tracking sequences sine wave is shown in equation 1. Where z is a function of x, the amplitude a, and wavelength lambda.

$$z = z(x) = A \cdot \sin\left(\frac{2\pi x}{\lambda}\right) \tag{1}$$

Because of the effects of maintaining constant speed in the contouring mode the output of z as a function of time is not a sine wave. The exact relationship between z and time is given by the differential equation shown in equation 2.

$$\frac{\partial z}{\partial t} = \frac{v}{\sqrt{1 + \frac{\left(\frac{\lambda}{2\pi A}\right)^2}{\left(1 - \left[\frac{z}{A}\right]^2\right)}}} \tag{2}$$

This effect causes variations in the speeds along both axes by the requirement to maintain a constant contouring speed. A computer simulation using Runge-Kutta techniques applied to the example given, shows this effect to produce maximum errors of less than 0.1 nm.

The second effect is caused by the use of straight line segments to approximate the smoothly varying sine wave. The choice of an appropriate number of segments to accurately represent the tracking sequence is also influenced by a third effect which is the requirement that the segment endpoints be integral multiples of the minimum programmable increment. Although a large choice of straight line segments may appear beneficial to better fitting the curve, once the curve fit is less than (plus or minus) one half the minimum programmable increment the total error in representing the curve increases with additional segments. An infinite number of straight line segments whose ends are forced to the nearest programmable increment results in a total error fit of one programmable increment. The selection of 100 line segments per wavelength was selected as a reasonable trade off between these two effects regardless of the tracking orientation used.

Example

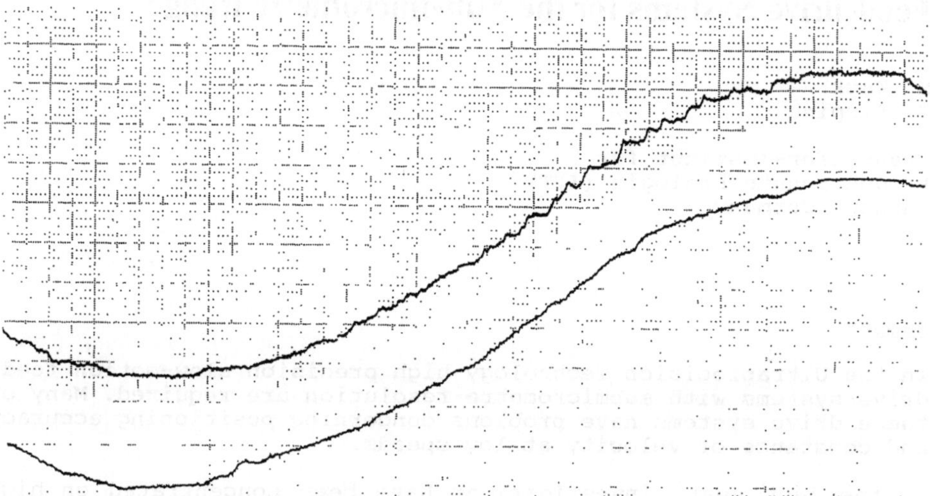

Figure 3. Illustrated here are segments of two traces from LAST tracking sequences. The upper trace shows the performance of a diamond turning lathe with a minimum programmable increment of 10 nm and a feedback resolution of 10 nm. The lower trace shows an identical machine, with the same programmable increment, operating at a feedback resolution of 2.5 nm. In this instance, the LAST test illustrates how the contouring performance of these machine tools has improved with increased feedback resolutions.

Examination of High Precision Slow-motion Feed-drive Systems for the Sub-micrometre Range

M. Weck
T. Bispink

Fraunhofer-Institut für
Produktionstechnologie (IPT),
Aachen, Germany

Summary

In the Ultraprecision Technology high precision slow-motion feed-drive-systems with submicrometre resolution are required. Many of these drive systems have problems concerning positioning accuracy and constance of velocity at low speeds.

In the last years investigations have been concentrated on high dynamic responses and high load capacities with accuracies in the micrometre-range. Properties of feed drive systems, e.g. accuracy or constance of velocity, which are very important in ultraprecision technology, have not been analysed systematically yet. A comparison between drive systems of different machines is very difficult, because of the various influences of slide-, feedback- and control-systems.

Therefore, a test-bench has been built up to make a systematic comparison between different types of feed drive systems under comparable conditions. Special test procedures have been used to analyse the accuracy, repeatability and slow motion behaviour.

This report will give an overview about the results of the examinations of ball- and roller-screws as of friction drive systems.

1. Introduction

The performances of many ultraprecison machines have been tested at the Fraunhofer-Institut für Produktionstechnologie for some years. Fig. 1 shows the example of a typical diamond turnig machine, which was designed and built at the IPT. It is used to manufacture laser mirrors and precision, nonferrous metal parts /1, 2/. The machine is laserinterferometrically controlled with a resolution of 10 nanometres. Highest machine accuracies are required for the production of reflective parts. Aluminium, copper and brass workpieces of flat, conic, spheric or aspheric shape with roughnesses down to the nanometre-range and accuracies in the submicrometre-range can be turned on this machine.

fig. 1: numerically controlled diamond turning machine UPM 1

The accuracy and constance of velocity of the feed drive systems have big influence on the rouhgness and shape accuracy of the produced parts. Special performance tests have been developed to analyse the behaviour of different types of feed-drive-systems. [1]

2. Test-Assembly and Test-Methods

A test bench has been built up to compare different types of high precision feed drive systems under similar conditions, fig. 2.

---

[1]The research project "Examination of High Precision Slow-Motion Feed-Drive-Systems for the Sub-Micrometre Range" is financed by the "Bundesminister für Wirtschaft".

fig. 2: test-bench for the examination of high precision feed-drive-systems

Because of its neglible friction, the aerostatic slide system of the carriage does not have any influence on the measured feed drive behaviour. Pneumatic vibration isolation pads keep ground vibrations out of the test bench. The test- and control system is based on a HP 5527A laser transducer system with a 10936A servo-axis board /3/. The resolution of the system is 10 nm. The position transducer in this configuration is not only been used for the feedback of the motion control, but also for the analysis of the feed-drive behaviour. The advantage of this setup is that no measurement errors caused by the laser position transducer will affect the test results /4/. The control scheme of the test-set-up allows a very flexible testing, fig. 3. A PID controller can be used for the analysis of the different drive-systems. For the tests, nevertheless a proportional control algorithm with a fixed loop gain was used, so that the different drive systems can be compared under similar conditions. The feed forward compensation was used, so that the following error does not overlay the path errors.

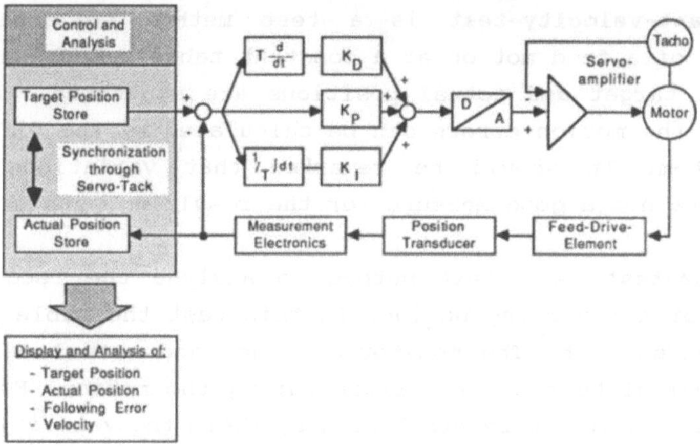

fig. 3: control scheme of the test- and control system

Different test methods have been applied to examine the characteristic behaviour of the feed-drive systems, fig. 4.

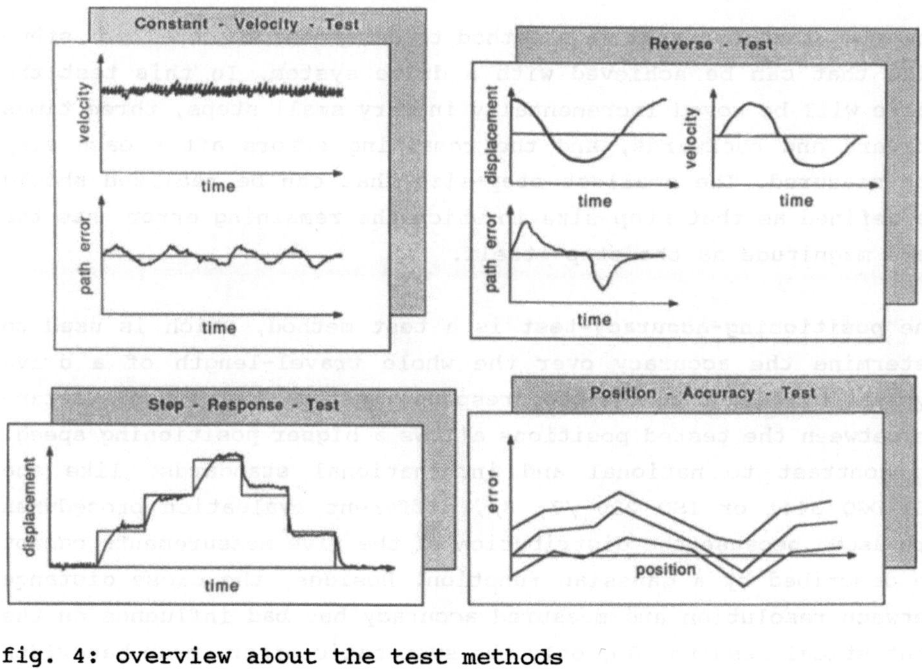

fig. 4: overview about the test methods

The **constant-velocity-test** is a test method to analyse the uniformity of a feed motion at a constant table speed. During the motion the target and actual positions are monitored. Basing on these data the motion errors can be calculated by the differences between them. It should be remarked that variations of the velocity are not a good measure for the resulting path errors.

The **reverse-test** is a test method to analyse the feed characteristics of a reversing motion. In this test the table performs a cosinusoidal path. The monitored target and actual values are taken to calculate the path errors during the motion. Especially the path errors at the reversal points, where the velocity changes it's sign, are of very big interest. This test method is very similar to the circular or ball bar test /5, 6/, but gives only the information of one axis. This makes the analysis of the feed-drive systems much easier. All tests have been performed with amplitudes of 0.1 mm and motion time of 2 seconds, so that the maximal velocity is 18.85 mm/min.

The **step-response-test** is a method to determine the smallest step-size that can be achieved with a drive system. In this test the table will be moved incrementally in very small steps, three times forward and backwards, and the remaining errors after each step are measured. The smallest step-size that can be realized should be defined as that step-size in which the remaining error has the same magnitude as the step itself.

The **positioning-accuracy-test** is a test method, which is used to determine the accuracy over the whole travel-length of a drive system. Different to the step-response-test, a much longer distance between the tested positions allows a higher positioning speed. In contrast to national and international standards, like the VDI/DGQ 3441 or ISO 230 /7, 8/, different evaluation procedures are used, because the distribution of the five measurements cannot be described by a gaussian function. Besides, the close distance between resolution and measured accuracy has bad influence on the statistical results. So, only the systematic error, mean bandwidth and hysteresis will be compared.

3. Analysed Drive Systems

The results of three different drive-systems will be presented in this report. The first system is a ball-screw with a diameter of 20 mm and a pitch of 5 mm. It is driven by an synchronous motor, which has a resolver for the velocity feed-back.
The second system is a planetary roller-screw with a diameter of 15 mm and a pitch of 2 mm. It is driven by a DC-disk-motor, which has an analog tachometer. The third system is a friction drive system /9/ with a high performance direct-drive DC-torque-motor and a high performance analog tachometer, fig. 5.

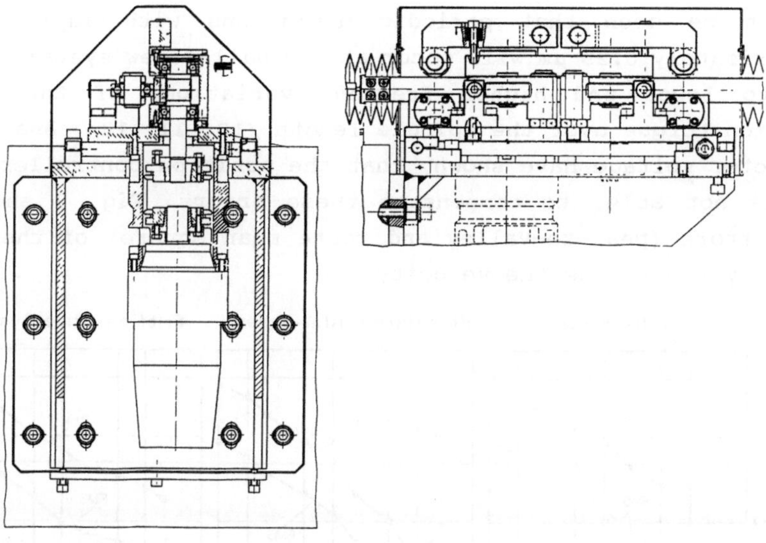

fig. 5: analysed friction-drive-system

4. Measurement Results

At the **constant-velocity-test** the analysed drive elements have been moved with contstant speeds between 0.25 and 100 mm/min. The path errors were plotted versus the measurement time, fig.6.

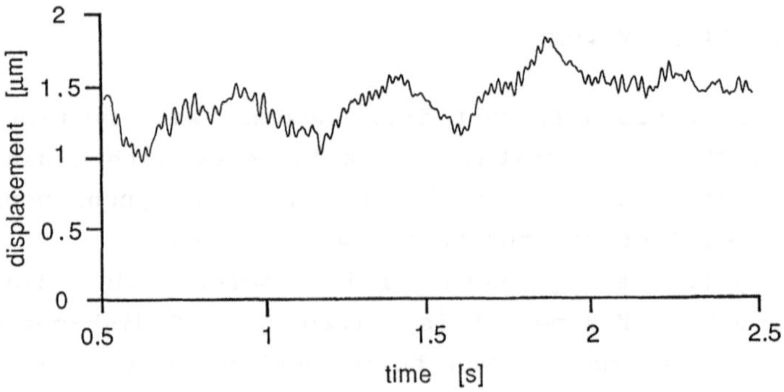

fig. 6: constant-velocity-test of the ball-screw-system

It can be seen that periodic deviations with amplitudes of approximately 0.25 μm will occur at the ball-screw system. This is due to lower and higher frequent variations of the spindle friction torque over the spindle length /10, 11, 12/. Analysis of the motor voltage have shown, that the position controller tries, but is not able, to compensate these errors. Fig. 7 shows the path-errors (peak-to-valley and route mean square) of the tested drive-system versus the velocity.

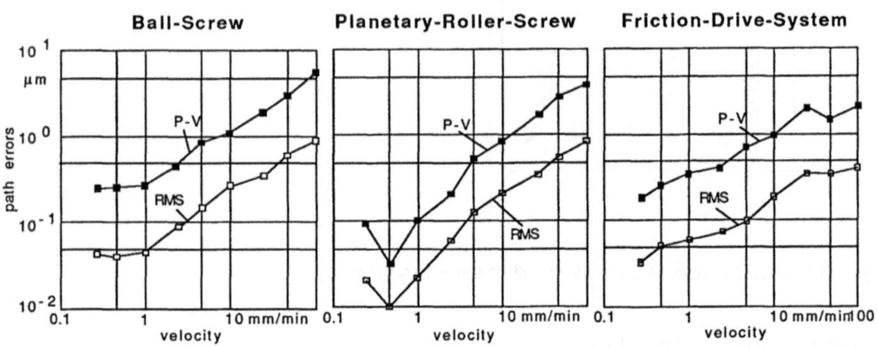

fig. 7: results of the velocity-tests

It can be seen, that at low speeds (smaller than 10 mm/min) the planetary roller-screw system is more accurate than the other systems. At speeds higher than 10 mm/min the friction drive system is more precise. One reason may be the high reduction ratio (2 mm/rot.) of the roller-screw in comparison to 5 mm/rot. of the ball-screw and approximately 100 mm/rot. of the friction drive system.

Fig. 8 shows the results of a **reverse-test** of the ball screw.

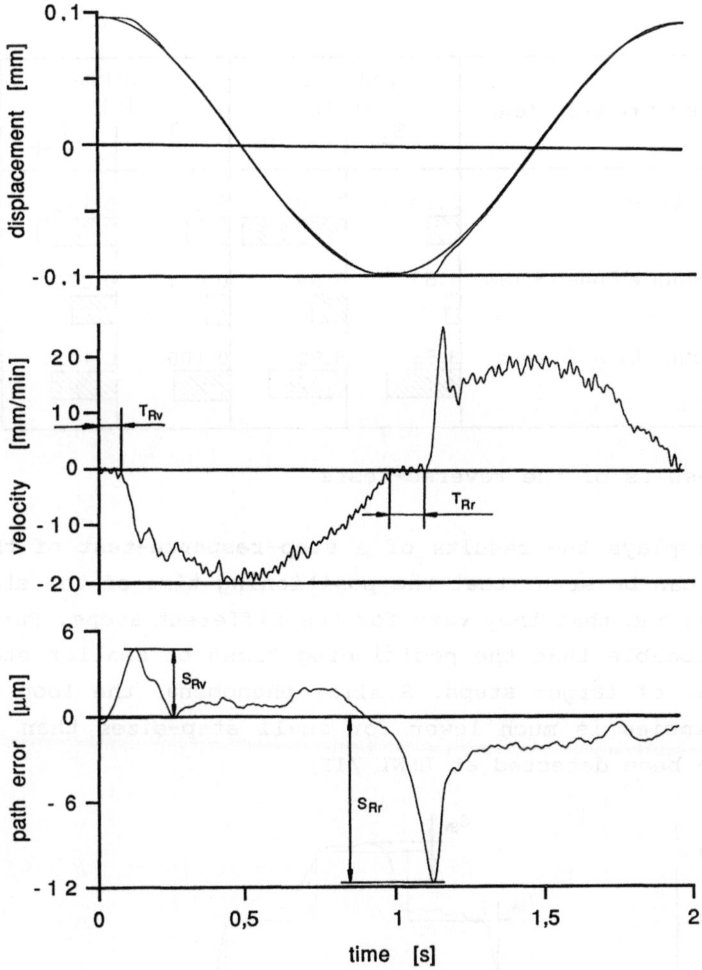

fig. 8: reverse-test of the ball-screw-system

The plot shows that the largest path errors S_R occurs at the reverse-positions. These "starting problems" can be explained by the transition from stick- to glide-friction and result in a dead time T_R. It can also be seen that the system has a non-linear behaviour.

The errors S_{Rv} in the forward direction are smaller than those (S_{Rr}) in the backward direction. The shown behaviour is typical for all the analysed drive systems, differences can only be seen in their magnitudes. In fig. 9 the characteristic errors of the different analysed drive systems are compared.

Feed-Drive-System	path error [μm]		dead time [s]	
	S_{Rv}	S_{Rr}	T_{Rv}	T_{Rr}
Ball-Screw	3.73	12.01	0.074	0.139
Planetary-Roller-Screw	1.53	3.89	0.041	0.079
Friction-Drive-System	8.55	8.93	0.100	0.113

fig. 9: results of the reverse-tests

Fig. 10 displays the results of a **step-response-test** of the ball screw. It can be seen, that the positioning time of the steps are rather long and that they vary for the different steps. Further on it is noticeable that the positioning times of smaller steps are longer than of larger steps. Similar phenomena, the loop gain at low frequencies is much lower for small step-sizes than for big ones, have been detected at LLNL /13/.

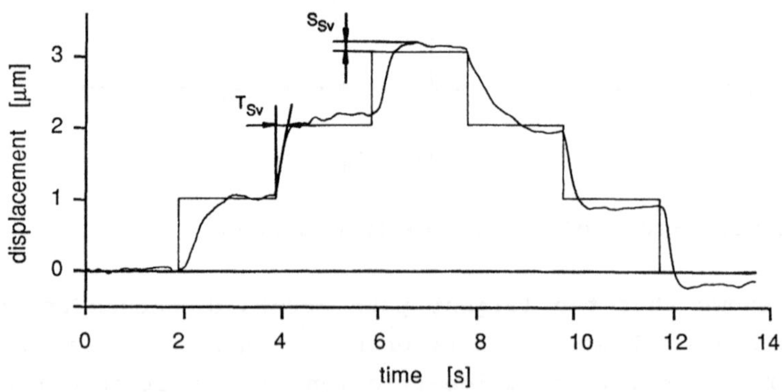

fig. 10: step-test of the ball-screw-system

The best way to reduce the remaining positioning errors is to use an PI-controller. Fig. 11 shows the test results of several feed drive systems.

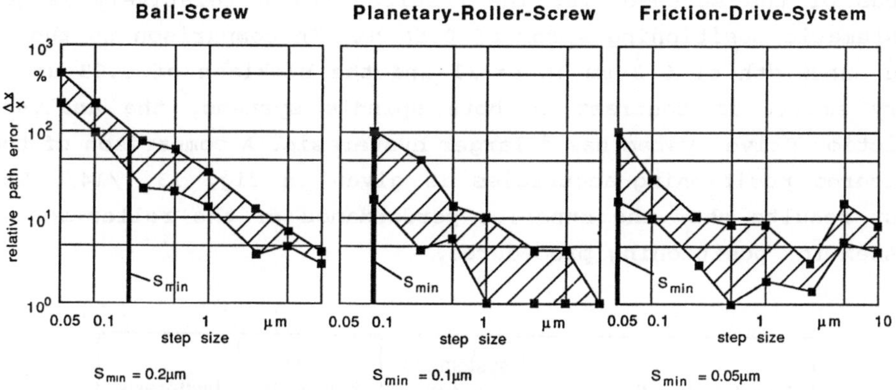

fig. 11: results of the step-tests

The smallest step-sizes S_{min}, picked up by 3 forward and 3 backward measurements (worst case) are approximately 0.2 μm of the ball-screw, 0.1 μm of the planetary roller-screw and 0.05 μm of the friction drive system. This shows the superior performance of this special drive system to conventional systems.

Fig. 12 shows a typical result of a **positioning-accuracy-test** of the ball screw.

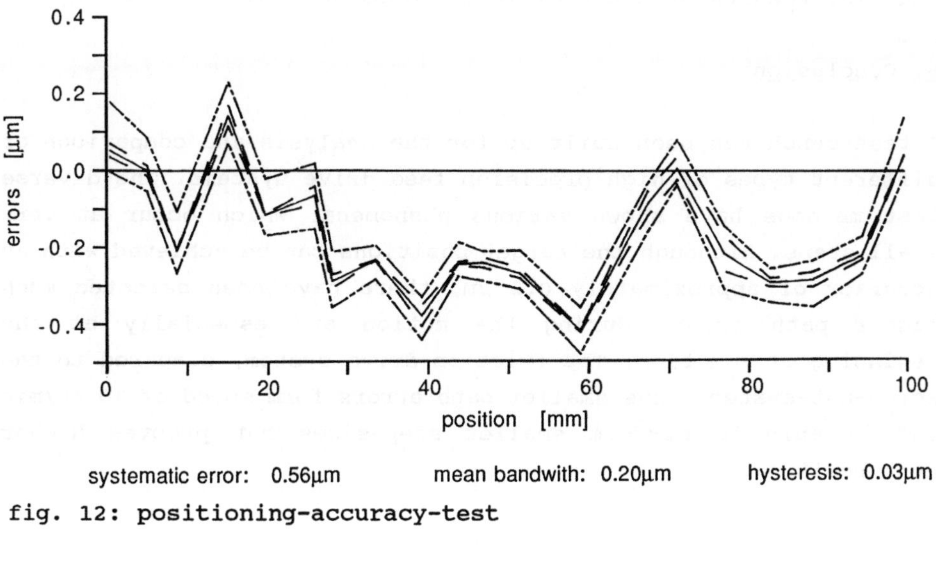

fig. 12: positioning-accuracy-test

The path of the systematic errors are very typical for the ball or roller screws, and cannot be detected in the friction drive system. A reason could be difficulties in the inlet and outlet areas of the screw-balls. The results show a relatively large, systematic positioning error of 0.56 μm. In comparison to those, the bandwidth of 0.2 μm is small and the backlash of 0.03 μm is very small. In contrast to both spindle systems, the analysed friction-drive-system has a larger hysteresis. A comparison of the measured positioning-accuracies is given in fig. 13, /14/. The test results show the superior performance of the roller-screw-system for positioning procedures.

Feed-Drive-System	systematic error [μm]	mean bandwidth [μm]	hysteresis [μm]
Ball-Screw	0.56	0.20	0.03
Planetary-Roller-Screw	0.07	0.02	0.01
Friction-Drive-System	0.04	0.06	0.10

fig. 13: results of the positioning-accuracy-test

6. Conclusion

A test-bench has been built up for the analysis and comparison of different types of high precision feed drive systems. The diverse test methods have shown various phenomena, which occur at very small steps. Although the target positions can be achieved with an accuracy of approximately 0.2 μm, there have been detected much higher path errors during the motion and especially at the beginning of a motion. The friction drive system, compared to the screw-nut-systems, has smaller path errors form speed of 10 mm/min and is able to perform smaller step-sizes but prooves higher hysteresis.

7. References

1. Weck, M.; Hartel, R.; Bispink, T.: Limits of Workpiece Accuracy Caused by the Geometrical and Dynamical Behaviour of Ultraprecision Diamond Turning Machines. Conference "Ultraprecision in Manufacturing Engineering", Aachen, May 1988, Proceedings "Ultraprecision in Manufacturing Engineering", Springer-Verlag, Berlin, Heidelberg, New York, 1988, ISBN 3-540-19241-7
2. Weck, M.; Hartel, R.: Bispink, T.: The Geometrical and Dynamical Behaviour of Ultraprecision Diamond Turning Machines set Limits to the Workpiece-Accuracy. SPIE-Conference, Session "Micromachining Optical Components and Precision Engineering", Hamburg, 1988, SPIE Volume 1015
3. HP 5527 A Laser Position Transducer System - Complete, High Performance Motion Control, Now With Fast, Programmable Servo Tuning. Company-brochure of Hewlett Packard, Santa Clara/USA, March 1988
4. Weck, M.; Hartel, R.; Bispink, T.:Improving the Workpiece Quality in Precision Engineering - a Challenge for the Development of New Contol Systems. International Conference on Advanced Mechatronics, The Japan Society of Mechanical Engineers, Tokyo, May 1989
5. Bryan, J.B. et al.: A Simple Method for Testing Measuring Machines and Machine Tools, Part 1: Principle and Applications. Precision Engineering, 4, 2, Butterworth, London, April 1982
6. Knapp, W.: Ergebnisse bei der dynamischen Prüfung von Werkzeugmaschinen und Koordinaten-Meßgeräten. VDI-Bericht 711, Fertigungs-Meßtechnik und Qualitätssicherung, 1988.
7. Statistische Prüfung der Arbeits- und Positionsgenauigkeit von Werkzeugmaschinen - Grundlagen. Richtlinie VDI/DGQ 3441 VDI-Verlag, Düsseldorf 1977.
8. ISO 230-2 International Standard - Acceptance Code for Machine Tools, Part 2. International Organisation for Standardization Switzerland 1988
9. Weck, M.; Hartel, R.; Bispink, T.: Feinstbearbeitung mit Ultrapräzisions-Drehmaschinen. Technika 14 (Jg. 39), Industrie Verlag AG, Zürich/CH 1990

10. Fuchsluger, G.: Planetenrollen-Gewindetriebe. Antriebstechnik 27 (1988) Nr. 12.
11. Weule, H.; Golz, H.U.: Dynamisches Verhalten von Kugelgewindetrieben. wt Werkstattstechnik 78 (1988)
12. Otsuka, J.; Fukada, S.; Osawa, T.: Fundamental Study of Planetary Screw. Bull. Japan. Soc. of Prec. Engg. Vol 21, No. 1 (Mar. 1987)
13. Hansen, H.J.; Maddux, A.S.: Comparison of Three High Precision Linear Axis Positioning Systems. Report at the 4th International Precision Engineering Seminar. Cranfield, U.K. May 11 - 14, 1987
14. Golz, H.U.: Analyse, Modellbildung und Optimierung des Betriebsverhaltens von Kugelgewindetieben. Dissertation TH Karlsruhe, 1990

The Influence of the Crystallographic Structure of the Substrate Material on Surface Quality and Cutting Forces in Micromachining

W. König
N. Spenrath

Fraunhofer-Institut für Produktionstechnologie
(IPT), Aachen, Germany

Ultra precision cutting is a machining method to manufacture mainly optical components with a surface roughness of a few nanometers and form tolerances which are in the submicron range. Due to continuous developments in the field of machine tools, it has become possible to machine increasingly complex surface contours. Thus, for example, recent efforts lead to the possibility to generate defined modulations on workpiece surfaces within the micron range by means of a piezoelectcrical fast-tool-servo (*figure 1*) [1]. As machining operations are becoming increasingly complicated, it is necessary to investigate the technological processes on the cutting edge more in detail.

Particularly in micromachining mirror-like surfaces, difficulties have been experienced due to the necessity of working with a minimum chip cross section ($a_p x f < 20 \mu m^2$) [2, 3]. As a result of these small chip cross sections, the chip formation takes place overwhelmingly inside individual material crystals ($d_K\{Cu\} \sim 250 \mu m$), characterized by different orientations. As a result, the cutting process can be regarded as a subsequent cutting of monocrystals. It is obviously necessary to investigate the cutting behaviour of monocrystals more closely, in order to understand how the chip formation is affected by material crystals of different orientation.

With regard to the quality of the machining process, the elastic and plastic behaviour of the Cu-monocrystal has to be taken into account. The

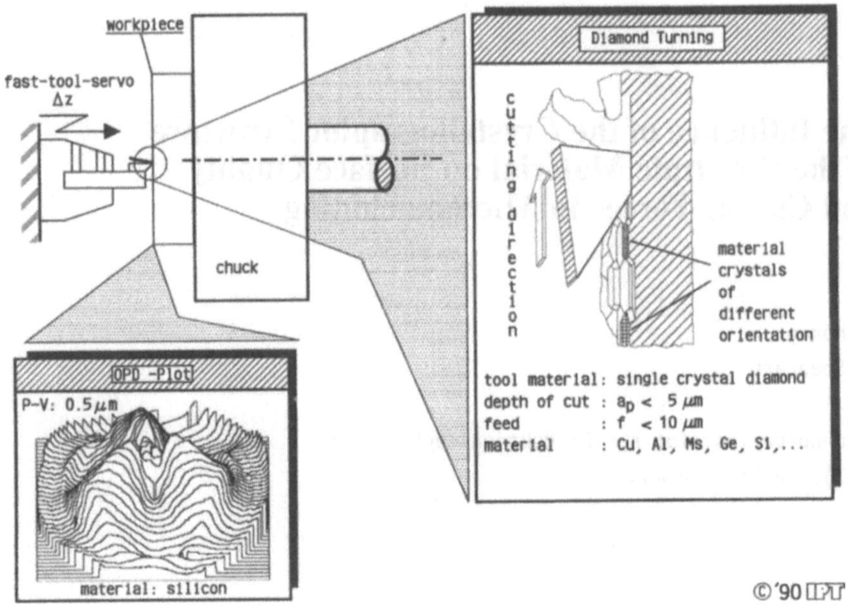

Fig. 1 Non-symmetric laser optics

E-module, which is the major parameter for the elastic behaviour, varies for different crystallographic orientations by the factor 2.9 (*figure 2*).

$$\frac{1}{E}=s_{11}-2(s_{11}-s_{12}-\frac{1}{2}s_{44})*(\alpha^2\beta^2+\beta^2\gamma^2+\gamma^2\alpha^2) \tag{1-1}$$

s_{ik} : constant of elasticity

α,β,γ : direction cosine of the crystrallographic direction

These differences in the E-module are considered to be responsible for the differences in the surface level, which can be observed at the polycrystalline copper substrate after machining the grain edges [8].

However, this is not the only cause for the different cutting behaviour of the individual copper crystals. The plastic material behaviour has an even stronger influence on the surface quality. During plastic deformation, the coefficient of consolidation ϑ, for example, can vary by the factor of 11

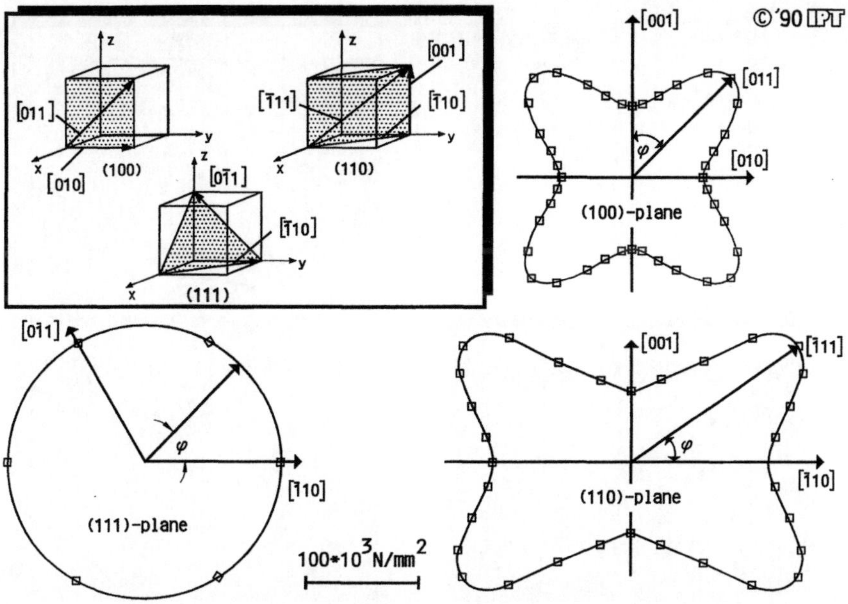

Fig. 2. Direction-specific E-modules in the Cu-monocrystal

for different crystal orientations [4]. In the cutting process, the slipping takes place on defined gliding systems, which are specific to the crystal system. The thickness of these material layers, in which the motion of dislocations occur, amount to be about 30 Å thick [4]. This anisotropy of the material can be determined from the chips of the polycrystalline OF-Cu substrate (*figure 3*). As illustrated on the left hand side of the SEM-figure, the chip lamella show different structures and directions from grain to grain. Such a behaviour can be explained by the different orientations of the individually chipped material crystals. The result is an activation of different gliding systems within the material crystals due to the cutting forces, induced on the material by the cutting edge. A change of the gliding system also leads to a variation of the shear angle [6, 7], and thus, to a variation of the direction of material flow (direction of slip). Following, this leads as well to a variation of the

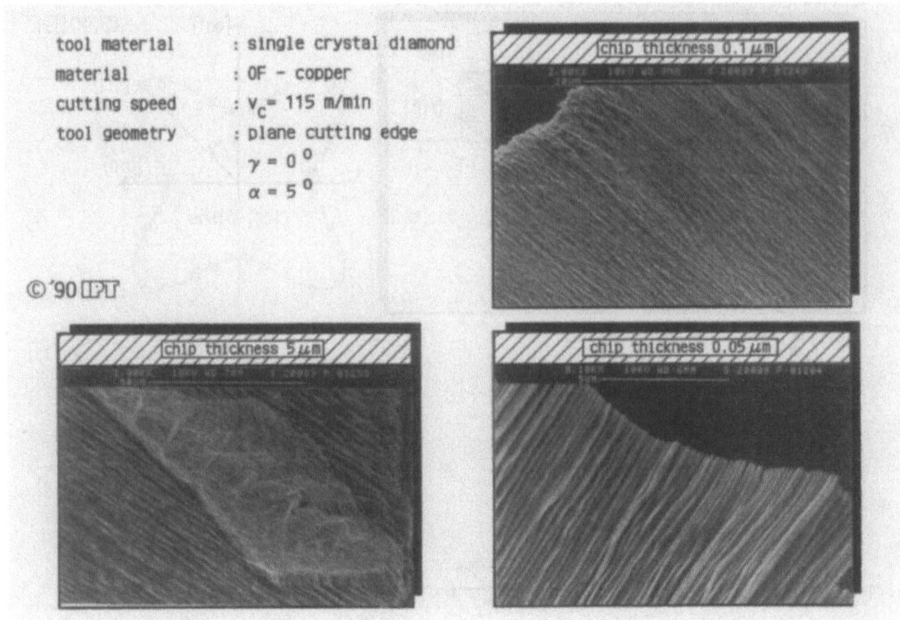

Fig. 3. Chip structure on polycrystalline OF-Cu chips

dynamics of the cutting forces [5]. If the cutting thickness is reduced by means of different cutting parameters, a more homogeneous lamella structure can be seen, as is illustrated in the SEM-pictures on the right hand side. This indicates that with increasing thickness of cut the gliding of the individual chip lamella is more impeded due to higher chip compression.

A closer analysis of the material-specific influential factors on the machining process was based on cutting tests with monocrystalline OF-Cu-substrate materials with {100}-, {110}- and {111}-oriented crystals featuring a diameter of 40 mm. Samples were prepared according to the Bridgman-method, and the crystallographic orientations were verified by means of X-ray structure analysis. The original material for the monocrystals had a purity of more than 99,99%.

At first, an off-axis machining of the material samples was carried out on an ultra precision lathe in order to achieve an almost straight cutting path, so that cutting could be done in defined crystal directions. In further tests, the crystals were also centrically clamped in order to cut cyclically in the different directions of the crystals, and thus, to find out possible process excitations.

The results of the cut in <110>-direction show a continuous, homogeneous course of the cutting force at a low static level (*figure 4*). This corresponds to the good surface quality, which can be achieved for cutting in this direction. However, if the cutting direction is varied 45 degrees towards the <100>-direction, there is a strong dynamic proportion superimposed. The dynamic component of the cutting force is considered to be the cause for the poorer surface quality, since it activates the

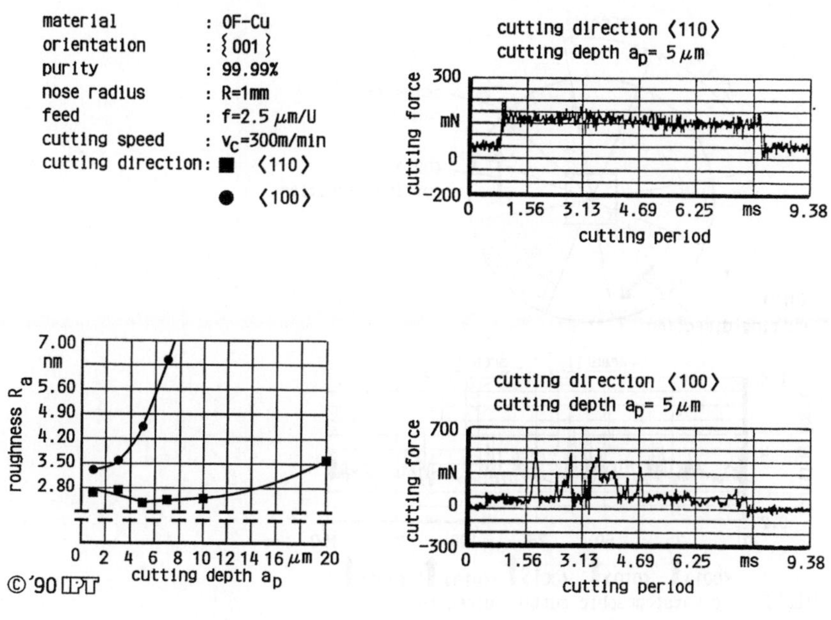

Fig. 4. Effect on the crystallographic cutting direction on cutting force and surface quality at the {100}-Cu-crystal

tool-workpiece-machine system to vibrate at high frequences. With decreasing chip cross sections, it can be observed that test results are increasingly approaching.

Translating these results from off-axis machining to turning with centric workpiece clamping, the cutting process is carried out in alternating sequences of 45 degrees towards the crystallographic directions mentioned above. This leads to a cyclical activation with a quadruple rotational frequency (*figure 5*). The outputs of the dynamometer clearly show two ranges with large (<100>-cutting direction) and small dynamic force components (<110>-cutting direction), which could also be measured in off-axis machining if carried out in this direction.

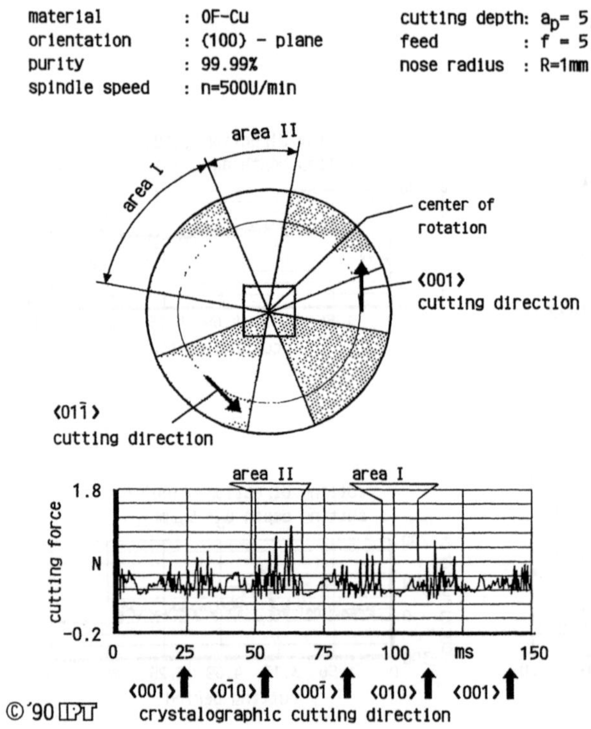

Fig. 5. Cutting force modulation during turning a {100}-oriented OF-Cu monocrystal

As a result, the {100}-oriented material crystal cannot be recommended for precision machining with centric workpiece clamping. Similar excitations can be observed with the {111}-orientation, but, due to crystallographic reasons, with a triple symmetry.

However, the smallest dynamic activations can be observed at the {110}-orientation, although here, too, quite obvious differences in the cutting forces occur. Apart from the small dynamic force components, the {110}-orientation is also characterized by the lowest static force components. Thus, this kind of orientation has to be preferred among the monocrystals. As owing to the low dynamic components of the cutting force, the activation of the machining process is reduced, leading to less technological problems. Additionally, the reduced static force components lead to minimum deformations in the tool-workpiece-machine system.

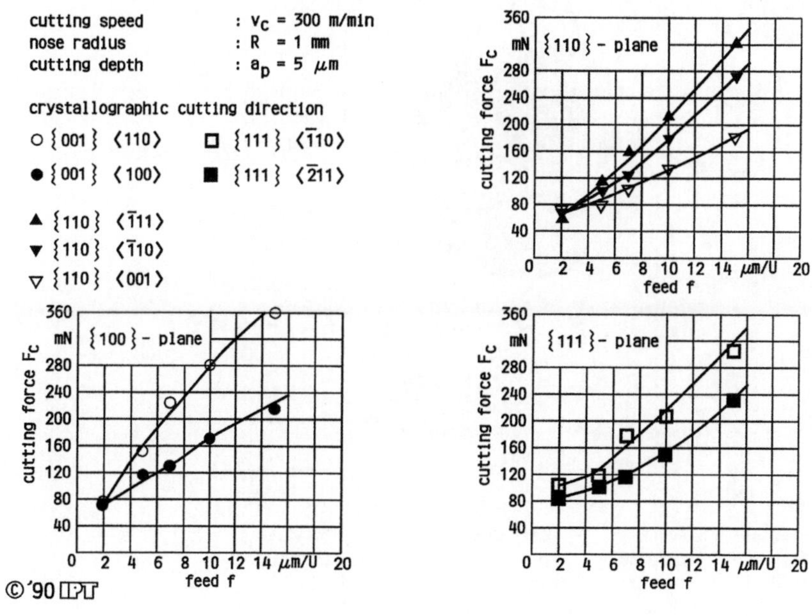

Fig. 6. Static cutting force at different crystallographic cutting directions

In the sequence of the static cutting force levels as function of the crystallographic cutting directions, a strong analogy to yield stress values from one-axis tensile stress tests can be observed. These yield stress values continuously decrease from the <111>-tensile direction via the <110>- to the <100>-tensile direction /4/. Furthermore, the yield stress for the <111>-tensile direction is slightly above that of the <211>-tensile direction. However, this is exactly the sequence in amount of the static cutting forces, as measured on the different crystal surfaces (*figure 6*).

In order to translate the results into the application of polycrystal substrate materials, special ultra-pure copper was employed, whose crystallites had grown vertically through the sample. The medium grain

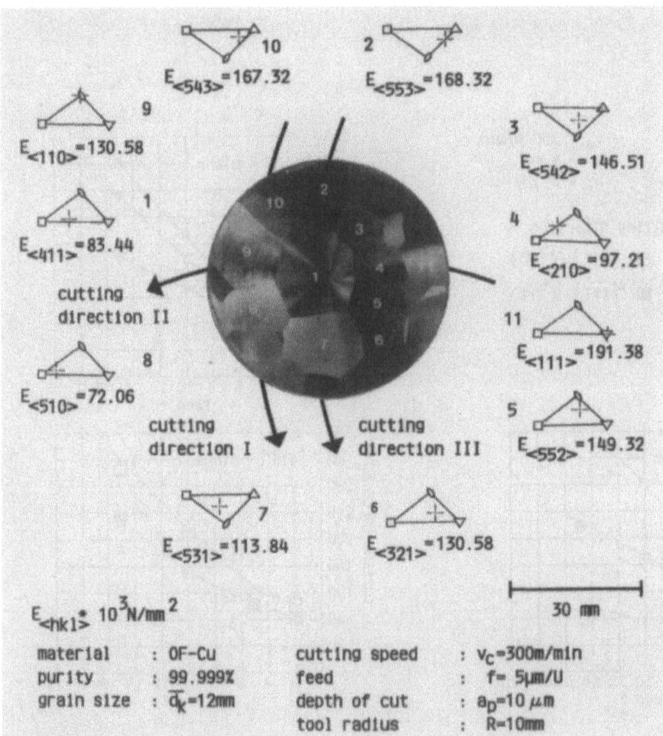

Fig. 7. Orientations and Young's Modulus of a macrocrystalline OF-Cu specimen

diameter of the monocrystals amounts to be about 12 mm (*figure 7*). Due to the vertical growth of the crystals through the sample, it is guaranteed that the process is not affected by other material crystals, which might be close under the surface. Furthermore, the large crystallites render both, the orientation test of the individual crystals and the cutting force measurement inside the individual crystals.

Owing to the big differences in orientation of the individual crystals, these cutting force measurements show clear force alterations. However, it is not possible to show correlations between differences of E-modula and surface steps at the grain boundaries. For instance, there are adjacent grains whose E-modula are much different, but which do not show a different topography, nor a contrary behaviour. Moreover, differences in the cutting

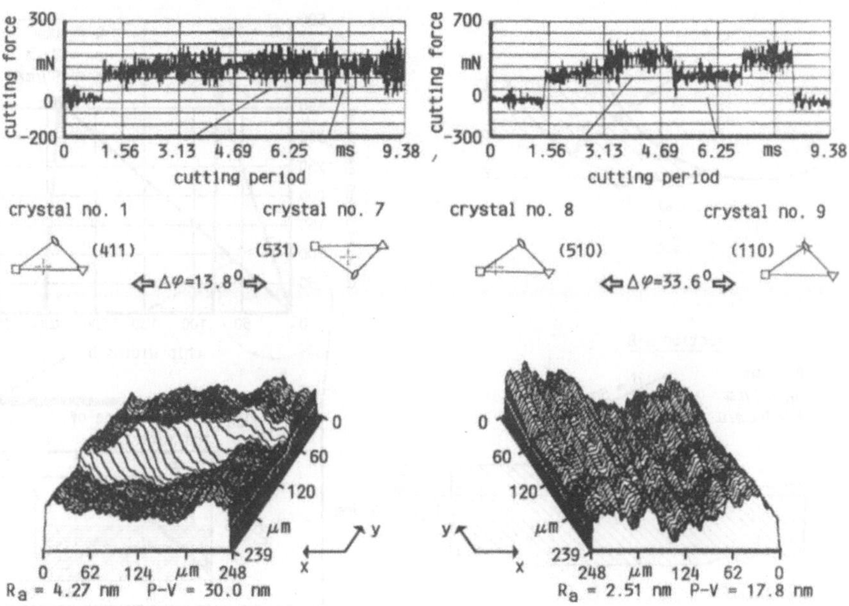

Fig. 8. Differences in the cutting force values do not lead to differences in the surface level at the grain edges

force level between two different grain orientations do not necessarily lead to differences in height at the grain edges (*figure 8*). Neither do grain edge alterations lead to differences in cutting forces.

Thus, the direction-oriented E-module of the individual material crystals cannot be the cause for the measurable surface steps at the grain boundaries. As already described above, the cause is rather to be found in the plastic material properties.

Here, the size of the non-deformed chip thickness h_{cu} has to be seen in comparison to the cutting edge radius of the tool. Recent investigations have shown that the edge radius of mono-crystalline diamond cutting edges amounts to about 40-50nm [9] and, thus, is larger than the chip thickness of that part of the cutting edge, where the surface to be generated will be formed (*figure 9*). That means that the chip removal in the area of the surface to be generated is carried out with a strongly negative rake angle.

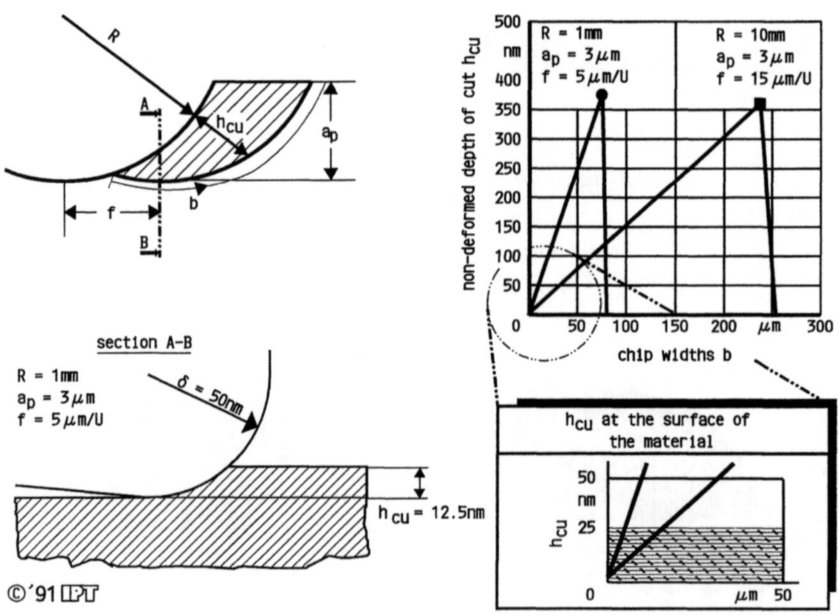

Fig. 9. Comparison of the kinematically calculated chip thicknesses with the cutting edge radius

In particular in the lobe, the area of the chip where the chip thickness approaches zero, the material is rather dislocated plastically than cut off as a chip. However, this process of plastic deformation is, as mentioned above, dependent on the crystal orientation and the direction of cut.

These problems can be possibly avoided when substrata materials with uniform plastic properties are chosen for the cutting process. As far as milling or off-axis-machining of small areas is concerned, a monocrystal with a selected crystal orientation can be employed. In the case of the turning operation, however, this will become problematic. Since the cut has to be carried out through different crystallographic directions, the cutting process will be characterized by strong dynamic activations. Least technological problems will be experienced with the {110}-oriented material crystal, which is therefore most useful for plane surfaces.

From the technological point of view the most ideal substrate material should have an amorphous structure, characterized by identical elastic and plastic properties in all directions (like chemically precipitated nickel).

References

1. Weck, M.; Pyra, M. *Diamond turning of non-symmetric laser optics* ASPE Annual Meeting; Sept. 23-28, 1990 Rorchester, New York, USA
2. Martin, K. *Untersuchung der Zusammenhänge von technologischen Parametern und der erzeugten Oberflächenqualität bei der Mikrozerspanung im Plandrehprozeß* Internal Report, 1989, IPT-Aachen, Germany
3. Mot, E. *Surface roughness and diamond turning tools* Production Engineering Laboratory; Technical University Eindhoven, Netherlands
4. Mitchel, T.E. *Dislocations and plasticity in single crystals of face-centred cubic metals and alloys* Departement of Metallurgy, Case Institute, Cleveland, Ohio, U.S.A. 1964
5. Ueda, K.; Iwata,K.; Nakayama,K. *Chip formation mechanism in single crystal cutting of b-brass* Annals of the CIRP 29(1980)1, pp. 41-46

6. Sato, M.; Kato, Y.; Tuchiya, K. — *Effects of material anisotropy upon the cutting mechanism* Trans. JIM 1978, Vol.19, pp. 530-536
7. Sato, M.; Kato, Y.; Aoki, S. — *Effects of crystal orientation on the cutting mechanism of the aluminium single crystal* Bulletin of the JSME 26(1983)215, pp. 890- 896
8. Stadler, H. Freisleben, B. Heubeck, C. — *Response of metallic Material to micromachining* 4th International Precision Enginieering Seminar; 11.-14. May 1987, Cranfield, UK
9. Asai, S. Taguchi, Y. Horio, K. Kasai, T. Kobayashi, A. — *Measuring the very small cutting-edge radius of a diamond tool using a new kind of SEM having two detectors* Anals of the CIRP 39(1990)1, pp. 85-87

Mirror Surface Grinding by Metal Bonded Super-abrasive Wheel with Electrolytic In-process Dressing

Hitoshi OHMORI*, **Ichiro TAKAHASHI****, and **Takeo NAKAGAWA****
*Institute of Industrial Science, University of Tokyo
**Institute of Physical and Chemical Research (RIKEN)

1. Introduction

The authors have been involved for several years in the development of grinding and processing techniques mainly targeting finish grinding of hard and brittle materials, in particular, in order to effectively utilize cast-iron fiber bonded wheels that are considered to be the toughest among all metal bonded wheels. Those results have realized a reliable dressing method that was absent in previous grinding techniques by employing a controlled electrolytic phenomenon. The application of this dressing method during the course of grinding operation led to the development of **"Grinding Method with Electrolytic In-process Dressing"("ELID-grinding"**, in short) which allows chips or debris to be removed easily, and allows grindability to be maintained over an extremely long period of time[1-11].

Although this dressing mechanism is a phenomenon that one of the authors accidentally discovered during a grinding test of silicon[1], it has grown up to the method which can fully utilize a metal bonded super-abrasive grinding wheel (especially, a micro-grain sized grinding wheel). It is not only effective in the grinding of materials that present difficulties for their efficient machining, but also fundamentally solves such problems as the occurrences of wearing and loading during a grinding operation[1-11]. In particular, there is considerable significance in the successful realization of a mirror surface grinding method (**Table 1**) having superior general-purposed adaptability (to any machine tool and material) using a metal bonded micro-grain super-abrasive wheel with this dressing method. In this report, the authors would like to provide a general outline of the mirror surface grinding technique which has been developed up to the present, and also would like to introduce its properties by showing some typical examples.

Table 1 Features of the Elid-mirror surface grinding method

No	Selection Factor	Factor Item	The Elid-mirror surface grinding	Conventional finish grindings
1	Machine	Feeding accuracy	Insensitive (No special machine is required)	Sensitive (Special machine is required)
		Cutting accuracy	Insensitive	Sensitive
		Machine vibration	Insensitive	Sensitive
2	Wheel	Abrasive grain	Super-abrasives: diamond or cBN	Depends on work materials
		Grain size used	No restriction	Strongly restricted
		Bond material	Cast iron fiber bonded wheel, etc.	Depends on demands
3	Condition	Wheel velocity	Insensitive (No special condition is required)	Sensitive (Machine tools and conditions influence)
		Feeding speed	Insensitive	Sensitive
		Depth of cut	Insensitive	Sensitive
4	Characteristics	Process stability	Maintained semi-permanently	Requires re-dressings
		Produced accuracy	Extremely high	Depends on machine tools
		Produced quality	Mirror surface only by grinding	Mirror surface is difficult
		Wheel wear rate	Extremely low	Depends on conditions
		Work material	No restriction	Depends on wheels, etc.

2. Mirror Surface Grinding Method

The mirror surface grinding method that the authors have developed is **"Grinding with Electrolytic In-process Dressing"** in which processing is performed continuous dressing through the electrolytic effects to a metal bonded super-abrasive wheel (with diamond or cBN abrasive grain). The use of #4000-#8000 abrasive wheel with an average abrasive grain size of 2-4 micron is standard. Differing from electro-chemical dressing methods of the past, this method is not of the type in which electrolysis excessively progresses. In this method, a formation of a isolating coating substance such as iron oxide occurs during electrolysis,and it is indeed a mechanism which controls the protrusion of abrasive grains on the wheel surface;after removing the initial coating substance on the wheel surface by friction between the oxide and the work surface to be ground ,thickness of the oxide layer decreases. Then,electrolysis of the bond material of the wheel restarts because the isolation capacity of the oxide layer reduces,as seen is **Fig.1** . This mechanism is balanced by wear of grains and electrolyzing strength. In the cource of grinding operation,the in-process dressing mechanism proceeds almost constantly. Therefore,mirror surface grinding can be realized with good efficiency and an extremely high grinding ratio,while at the same time having a high degree of stability. Among the conditions for its establishment listed in **Table 2**,a major point that should be noticed is the use of conventional grinding coolant and cast-iron fiber bonded wheel(**CIFB** wheel,in short)(as well as other metal bonded wheels). One of the authors has named this grinding technique the **"ELID-mirror surface grinding method"**. The "ELID" was made from underlined characters of "Electrolytic In-process Dressing".

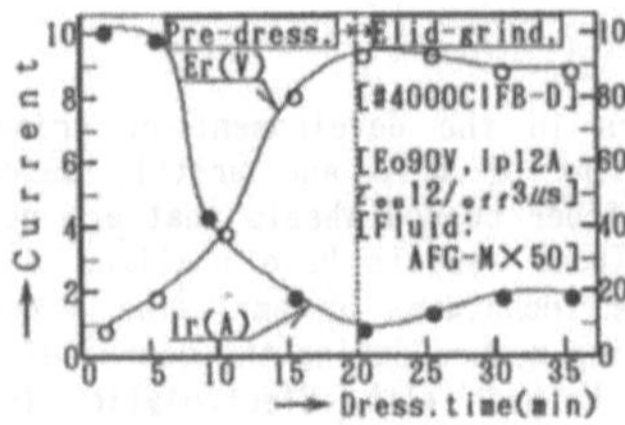

A)Electrical behavior of the Elid

Fig.1 Schematic illustration of the Elid-grinding mechanism

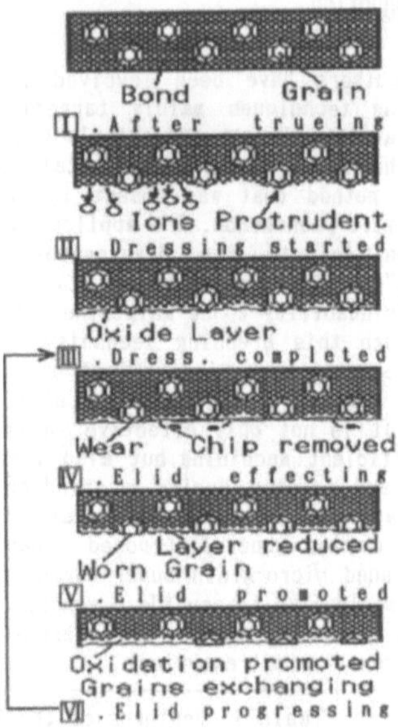

B)The Elid-mechanism illustration

Fig.1 Schematic illustration of the Elid-grinding mechanism

Table 2 Major factors for the establishment of the Elid-mirror surface grinding

No	Factor type	Main factor to be established	Example of standard conditions for Elid-grindings
1	Physical condition	[-]electrode position	Gap of the poles:around 0.1mm
		[-]electrode area	Around 1/6-1/4 of wheel area
2	Electrical condition	Electrolyzing conditions	$E_0$60-120V,I_P10-40A,$\tau_{on/off}$2-5μs
		Coolant conductivity	Around 2200μ S(after mixed)
3	Chemical condition	Coolant components	Requires rust-inhibitors
		Wheel's bond material	Susceptible oxidation:Fe,etc.
4	Mechanical condition	Coolant supplying	Uniform supply into the gap
		Grinding conditions	Wheel wear must match Elid

3. Types of Application for ELID-Mirror Surface Grinding

Various types of application of the Elid mirror surface grinding have been developed based on the authors' research. As is shown in **Table 3**,it is possible to realize mirror surface grinding with various types of system using existing machine tools.

Table 3 Processing machine types available for the Elid-mirror surface grindings

No.	[Processing machine] Machine type (Abbreviated)	Spindle type	[Processing type] Wheel type	Processable surface	Feeding way	Cutting way	Type No.
1	Rotary surface grinder (RG)	Vertical	Cup	Flat	Creep	Constant depth	①
2	Machining center (MC)	Vertical	Cup	Flat	Creep	Condtant depth	②
		Horizontal	Straight (Blade)	Flat / Non-flat	Rapid	Traverse	
				Cylindrical	Creep	Plunge	
			Mounted	Flat / Non-flat	Creep	Traverse	
				Cylindrical outer/Inner		Plunge	
3	Surface grinder (SG)	Horizontal	Straight (Blade)	Flat / Non-flat	Rapid	Traverse	③
				Grooved	Creep	Plunge	
4	In-feed grinder (IG)	Horizontal	Cup	Flat	In-feed	Constant speed	④
		Vertical				Constant pressure	
5	Lap-grinding machine (LG)	Vertical	Lap	Flat	In-feed	Constant pressure	⑤
					Creep	Constant speed	
6	Turning center (TC)	Horizontal	Cup	Cylindrical outer surf.	Creep	Traverse	⑥
			Straight (Blade)	non-cylindrical surf.	Rapid		
				Cylindrical end surface	—	Plunge	
			Mounted	Cylindrical inner surf.	Creep	Traverse	⑦
7	Cylindrical grinder (CG)	Horizontal	Straight (Blade)	Cylindrical outer surf.	Creep	Traverse	⑧
				non-cylindrical surf.	Rapid		
					—	Plunge	

*The mesh portions are typical application types.

(1)Rotary Surface Grinder

The Elid mirror surface grinding method that uses a rotary surface grinder has the background of being used in various grinding tests since the early stages of the authors' research[1-3]. As shown in Fig.2A,together with the mounting of a dressing attachment ([-] electrode) to any area that is not used in processing with the grinding wheel for electrolytic in-process dressing,a brush or similar device is brought in contact with the metal portion of the wheel to supply the electrolyzing power. An illustration of the experimental device is shown in Fig.2B.

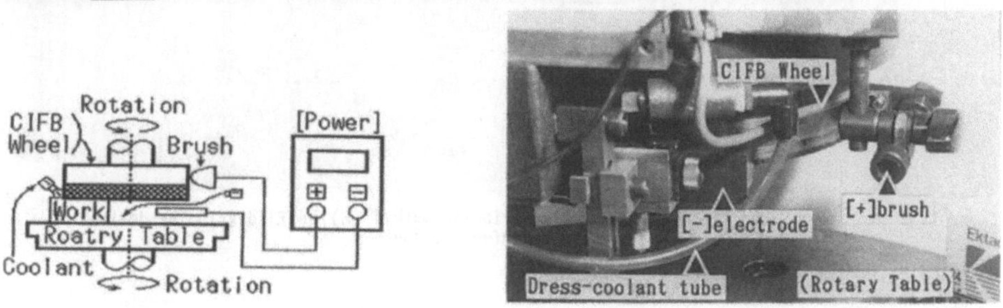

A)The Elid-grinding principle

B)The Elid-grinding device

Fig.2 The Elid-grinding method using a rotary surface grinder

(2)Machining Center

The Elid-mirror surface grinding which uses a machining center is different from that which uses the above-mentioned rotary surface grinder,in its taking some variations of grinding type. Basically,there are two types consisting of that which uses a cup type grinding wheel and that which uses a straight type grinding wheel. The processing principle is shown in

Fig.3A. By utilizing automatic tool changing (ATC) of a cup wheel,a straight wheel,a mounted wheel or a cutting wheel(blade),it becomes possible to realize efficient shaping and mirror surface machining with shucking a workpiece only once. The Elid-grinding is then possible by providing electrolytic dressing electrodes that correspond to each of the grinding wheels' shapes[4]. Fig.3B shows an example of a Elid-device equipped with a straight wheel.

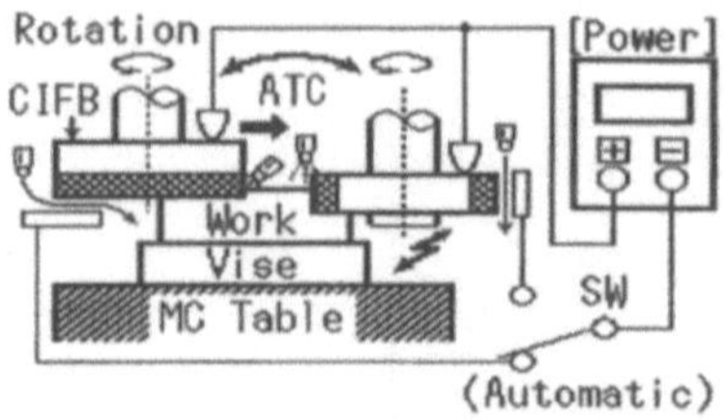

A)The Elid-grinding principle

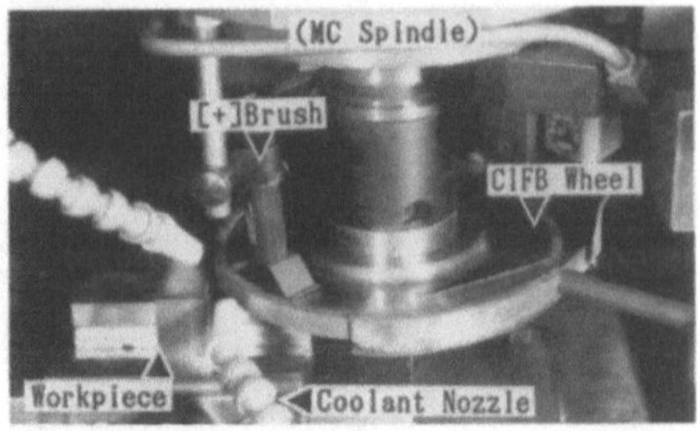

B)The Elid-grinding device

Fig.3 The Elid-grinding method using a machining center

(3)Surface Grinder

The Elid-mirror surface grinding method which uses a surface grinder can be said to be equivalent to the system which used a straight wheel in MC(as above-mentioned). However,if the Elid-mirror surface grinding can be uses in a conventional surface grinder,it could be said that its application type has more practical significance than in MC,because it can be applied more frequently to actual production processes. Based on the principle shown in Fig.4A,the authors have realized mirror surface grinding of hard and brittle materials[8] and many steel materials by applying the Elid-grinding method on a surface grinder. As shown in Fig.4B,a mirror surface grinding machine can be created by simply mounting a [-] attachment or so. In particular,a major advantage of this machine is that it allows easy mirror surface grinding operations of large workpieces (of 100mm square,for example,or larger) with a constant diameter wheel. In the Elid-grinding with MC, a cup wheel needs to be changed according to width of a workpiece.

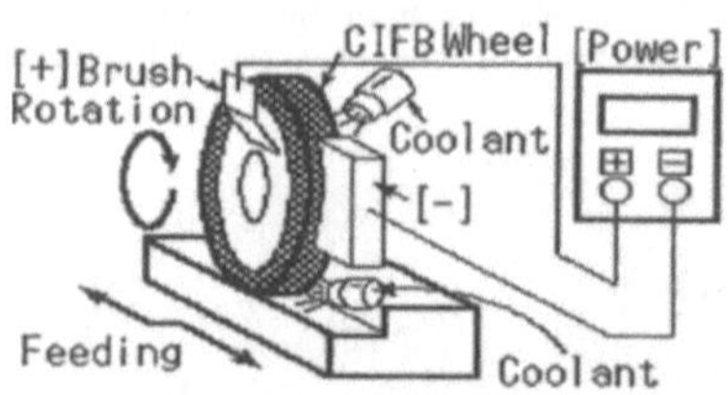

A)The Elid-grinding principle

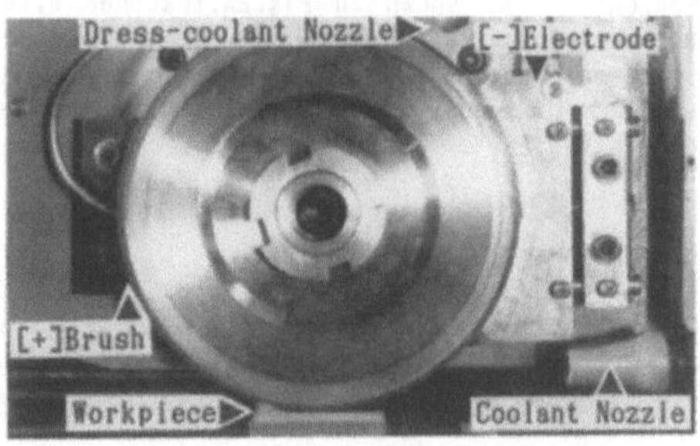

Fig.4 The Elid-grinding method using a surface grinder B)The Elid-grinding device

(4)Turning Center

Turning center(TC) refers to CNC lathe equipped with a mill spindle in its construction,and it has begun to be used in machining of ceramics in recent years in the same manner as MC. Beginning with the realization of outer mirror surface grinding about cylindrically shaped ceramics[5],recently,the authors have reached the development of mirror surface grinding of inner surface of cylindrically shaped workpieces[9]. Fig.5A indicates each of the processing principles,and Fig.5B indicates an outer mirror surface grinding machine. In the case of the inner,although the restriction about interference of a [-] Elid attachment with an inner surface of workpiece can exist,stable mirror surface grinding properties have been obtained by its applying to the cases of larger inside diameter of workpieces than 45mm.

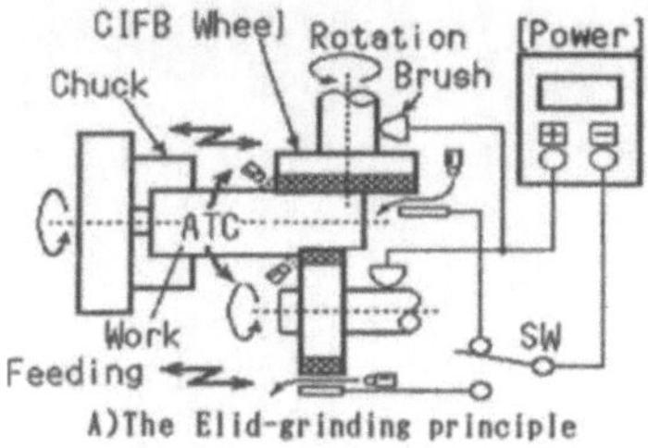

A)The Elid-grinding principle

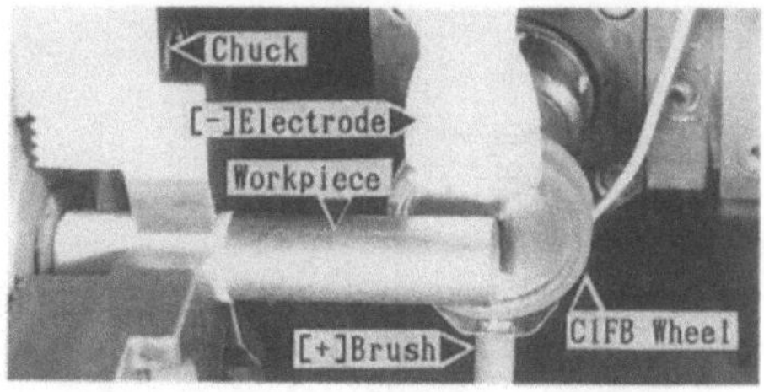

B)The Elid-grinding device

Fig.5 The Elid-grinding method using a turning center

(5)In-feed Grinder

In-feed grinder adopts the processing system of feeding the grinding wheel into the surface of the workpiece while rotating of both the wheel and the workpiece. It is most suited to the processing of circular workpieces and can be easily applied over a range from coarse grinding to precise grinding with a same grain size of abrasive wheel by simply controlling the feeding speed,thereby resulting in a high level of processing efficiency. In this system,since the grinding surface area remains constant,a satisfactory degree of processing flatness can be obtained. It also has the characteristics of allowing adjustment of the shape of the produced surface(either concave or convex) to be made easily according to the setting of the relative angle between the two rotating spindles (of wheel and workpiece). The use of the processing principle indicated in **Fig.6A** allows successful surface grinding of a wide variety of materials including silicon wafers[6]. **Fig.6B** shows an example of an experimental Elid-mirror surface grinding machine that uses this principle.

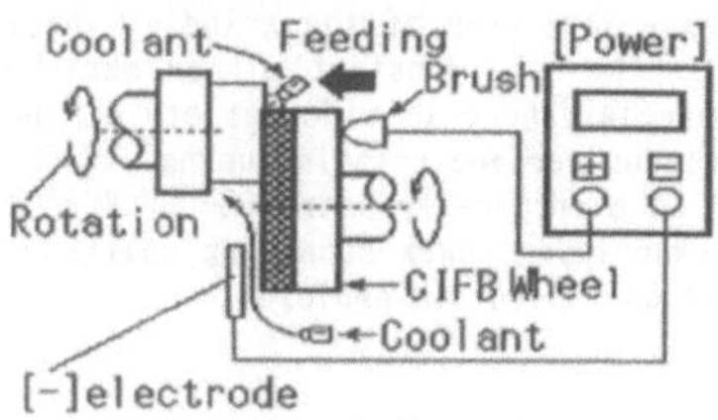

A)The Elid-grinding principle

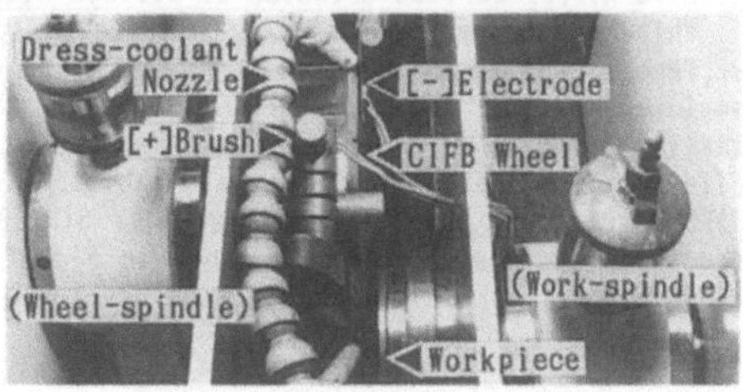

B)The Elid-grinding device

Fig.6 The Elid-grinding method using a in-feed grinder

(6)Lap-Grinding Machine

Lap grinding machines refer to processing machines that perform constant pressure grinding using bonded abrasive lapping plates such as cast iron fiber bonded super-abrasive plates. As shown in the diagram of **Fig.7A** indicating the principle of the Elid-grinding in **Fig.7A**,if an area for electrolytic in-process dressing is provided in a portion of the lapping plate,it is possible to prevent disturbance of the properties about the produced surface quality,which has been due to loading by debris which had inhibited that practical application of bonded abrasive lapping in the past. In addition,since flatness of produced surface by this grinding type naturally tends to be satisfactory,the Elid-mirror surface grinding of this machine type is expected to realize more easily both surface accuracy and roughness than other Elid-grinding types. One of the authors named this Elid-grinding type "Elid lap-grinding method" from two reasons;The first is that the machining way is like that of conventional lapping. And the second is that the machining rate is comparable to that of grinding. The Elid lap-grinding method is especially suitable to some materials that are easily susceptible to brittleness and crushing in the case of grinding with constant depth of cut such as alumina[7]. As is shown in **Fig.7B**,a larger electrode and a higher capacity power supply are required for this Elid system for maintaing electrolysis over the entire surface of the lapping plate utilized . As a practical usage of the Elid lap-grinding,there are some methods proposed employing combination of electrolytic and mechanical dressing.

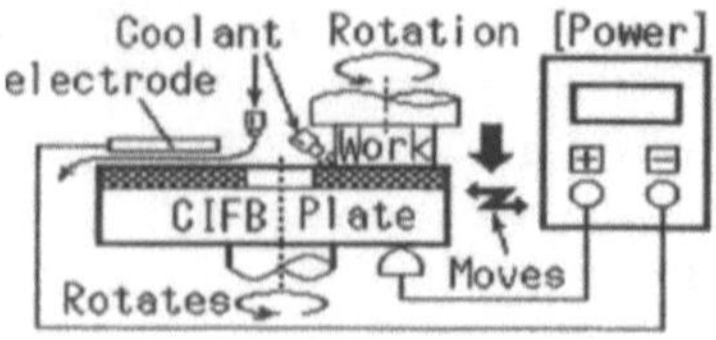

A)The Elid-grinding principle

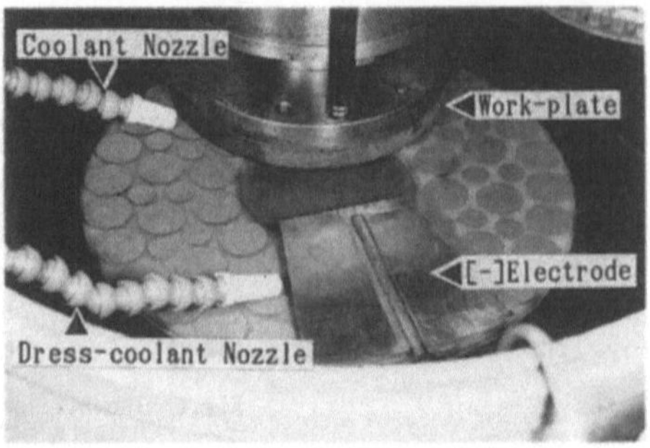

B)The Elid-grinding device

Fig.7 The Elid-grinding method using a lap-grinding machine

4. Examples of Elid-Mirror Surface Grinding for Ceramics

In this section, we will give some examples of mirror surface grinding for some typical fine ceramics using the various machine types described in the previous section.

(1)Elid Mirror Surface Grinding of Silicon Nitride(Si_3N_4)

While on the one hand silicon nitride has poor machinability, it is easy to obtain a glossy surface even with conventional grinding. However, due to excessive wear of the grinding wheel employed, one has almost no choice but use the Elid-method for the constant mirror surface quality grinding of this material. For processing this material, there is wide variety of the Elid-grinding application types available. An example of mirror surface grinding using a #4000 CIFB-D grinding wheel (manufactured by SINTOBRATOR Ltd.) on a surface grinder (GS-CHF:KURODA Precision Industries Ltd.) in shown in Fig.8[6]. In this example, an power supplying unit(EDD-04S:SINTOBRATOR Ltd.) which has been developed for the Elid-operation was employed.

CIFB-D=Cast Iron Fiber Bonded Diamond Wheel

[Grinding conditions]		[Elid conditions]	
Wheel used	#4000CIFB-D	Voltage	Eo120V
Velocity	v1200m/min	Current	Ip5A
Feed-rate	f20m/min	Pulse-width	τon 5μs
Depth of cut	d1μm P0.4mm		τoff1.7μs

A)Mirror surface grinding conditions

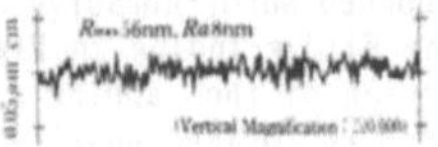

B)Surface roughness pattern

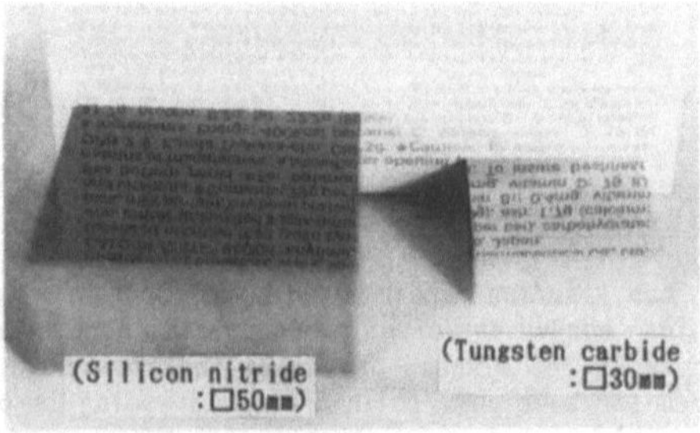

C)View of produced mirror surface samples

Fig.8 Example of silicon nitride produced by Elid-grinding (application type : ③)

Next, examples of mirror surface grinding of outer cylindrical surface is shown in Fig.9, which examples contain a silicon nitride ceramics'. That is an example of traverse grinding using a cup wheel(#4000 CIFB-D Wheel:SINTOBRATOR Ltd.), which is mounted on the wheel spindle in the direction perpendicular to the work spindle, on a turning center(QT-10N:YAMAZAKI MAZAK Corp.)[5]. A conventional power supply for an electro-discharge machine(MAKINO Milling Machine Co. Ltd.) was utilized as a dressing power supply.

[Grinding conditions]		[Elid conditions]	
Wheel used	#4000CIFB-D	Voltage	E o60V
Peripheral speed(m/min)		Current	I p10A
Wheel	Work	Pulse-width	τ on1. 5μs
vt471	vw30		τ off1.5μs
Feed-rate	fz60mm/min	Depth of cut	dz2μm

A)Mirror surface grinding conditions

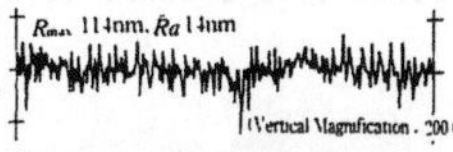

B)Surface roughness pattern

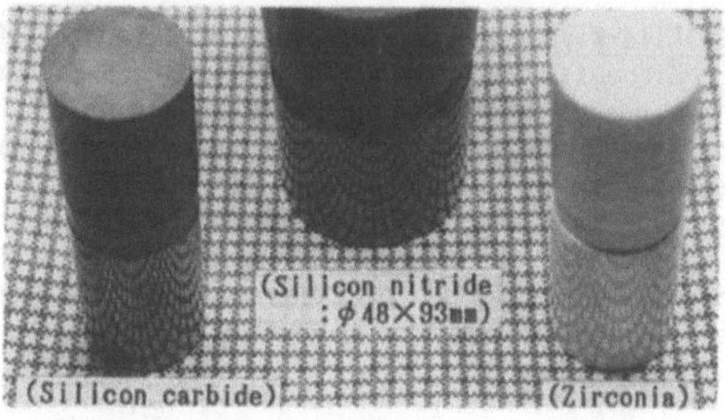

C)Veiw of produced mirror surface samples

Fig.9 Example of cylindrical ceramics produced by Elid-grinding (application type : ⑥)

(2)Elid-Mirror Surface Grinding of Silicon Carbide(SiC)

Although the machinability of silicon carbide is not as poor as that of silicon nitride, it is susceptible to the occurrence of rapid wheel wearing. This material's processing can be also selected from the wide range of the Elid-grinding types. An example of using a submicro-grain abrasive wheel on an in-feed mirror surface grinding system is shown in Fig.10[10]. A horizontal in-feed surface grinder(SHONAN Eng. Co. Ltd.) was used along with a cast iron fiber bonded submicro-grain diamond abrasive wheel(#120000:SINTOBRATOR Ltd.). In addition, an electrolytic dressing power supplying unit(Stanley Electric Co. Ltd.) was employed for the Elid-grinding.

[Grinding conditions]		[Elid conditions]	
Wheel used	#120000CIFB-D	Voltage	E o60V
Velocity/Rotation		Current	I p30A
Wheel	Work	Pulse-width	τ on 5μs
vt337m/min	Sw600rpm		
Feed-rate	fd10μm/min		τ off5μs

A)Mirror surface grinding conditions

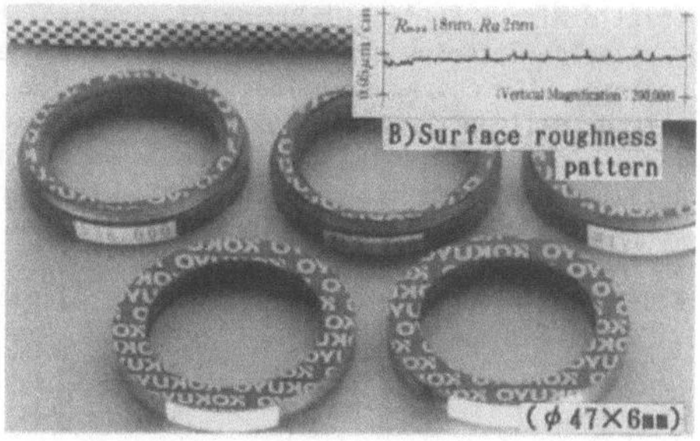

Fig.10 Example of silicon carbide produced by Elid-grinding (application type : ④)

C)Veiw of produced mirror surface samples

(3)Elid-Mirror Surface Grinding of Tungsten Carbide(WC,WC-Co)

Although tungsten carbides(WC,WC-Co) are metal materials, they are also ceramics as far as they are sintered bodies.

Rather than wheel wearing, this material causes susceptibility to loading due to debris, and usually requires frequent dressings in conventional grindings. In the sense of confirming the effectiveness of Elid-grinding, in-feed grinding(using the same machine as the above-mentioned) was attempted with a #120000 CIFB-D wheel, a submicro-grain abrasive wheel. The result yielded stable mirror surface grinding without any loading[10]. The produced surface roughness was 14nm in Rmax, 2nm in Ra, indicating that this method is able to realize the level of **"grinding process"**, which could be said to be the highest in the world (Fig.11).

[Grinding conditions]		[Elid conditions]	
Wheel used	#120000CIFB-D	Voltage	E o60V
Velocity/Rotation		Current	I p30A
Wheel	Work	Pulse-width	τ on 5μs
vt337m/min	Sw600rpm		
Feed-rate	fd10μm/min		τ off5μs

A)Mirror surface grinding conditions

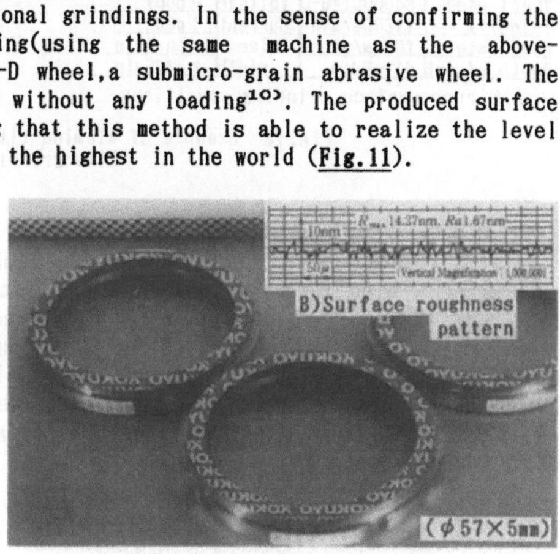

Fig.11 Example of tungsten carbide produced by Elid-grinding (application type : ④)

C)View of produced mirror surface samples

(4)Elid-Mirror Surface Grinding of Zirconia(ZrO_2)

Zirconia has comparatively favorable machinability allowing mirror surface grinding with a high degree of efficiency. Together with silicon nitride and silicon carbide,a wide variety of the Elid-mirror surface grinding types can be selected. In Fig.12,an example produced by creep feeding mirror surface grinding using a rotary surface grinder is shown[3]. A rotary surface grinder having hydrostatic pressure spindles(RGS-60:NACHI-FUJIKOSHI Corp.) was employed for the processing,and an electrolytic dressing power supplying unit manufactured by Stanley Electric Co. Ltd. was employed. Of course, a cup type #4000 CIFB-D wheel (SINTOBRATOR Ltd.) was used for the processing.

[Grinding conditions]		[Elid conditions]	
Wheel used	#4000CIFB-D	Voltage	Eo60V
Velocity	v1000m/min	Current	Ip30A
Feed-rate	f60mm/min	Pulse-	τon 5μs
Depth of cut	d1~2μm	width	τoff5μs

A)Mirror surface grinding conditions

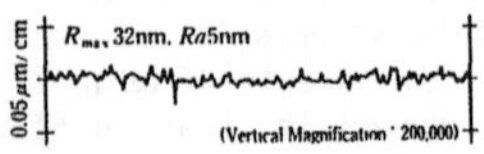

B)Surface roughness pattern

C)View of produced mirror surface samples

Fig.12 Example of zirconia produced by Elid-grinding (application type : ①)

(5)Elid-Mirror Surface Grinding of Alumina-Titanium Carbide(Al_2O_3-TiC)

Alumina-Titanium carbide is one of fine ceramics that presents favorable machinability. In contrast to characteristics that in the fine grinding of such materials as silicon nitride,silicon carbide and zirconia,the produced surface properties are plastic fashioned,this material can be observed to be finished by extremely fine brittle fracture. Notwithstanding,since the scale of this brittle fracture is extremely fine,the produced surface easily shows a mirror quality. An example of creep feeding grinding on arotary surface grinder(RGS-60:NACHI-FUJIKOSHI Corp.) using a #4000 CIFB-D wheel is shown in Fig.13[3]. A dressing power supply manufactured by Stanley Electric Co. Ltd. was employed.

[Grinding conditions]		[Elid conditions]	
Wheel used	#4000CIFB-D	Voltage	Eo60V
Velocity	v1200m/min	Current	Ip30A
Feed-rate	f60mm/min	Pulse-	τon 5μs
Depth of cut	d1~2μm	width	τoff5μs

A)Mirror surface grinding conditions

B)Surface roughness pattern

C)View of produced mirror surface samples

Fig.13 Example of alumina-titanium carbide produced by Elid-grinding (application type : ①)

(6)Elid-Mirror Surface Grinding of Sialon

Sialon typically demonstrates its poorer machinability than silicon nitride's. As such,since a high tendency towards wheel wearing is unavoidable during the grinding,it is necessary to stabilize the Elid-mirror surface grinding by increasing the dressing current. The example shown in Fig.14 illustrates the creep-feed ground one by a machining center(VQC-15/40:YAMAZAKI MAZAK Corp.) using a #4000 CIFB-D cup wheel. In the Elid-grinding,an power supply(SUE-87) for an electro-discharge machine manufactured by Sodick Co. Ltd. was utilized. In the case of this material,mirror surface grinding could be performed easily even with a #2000 CIFB-D wheel.

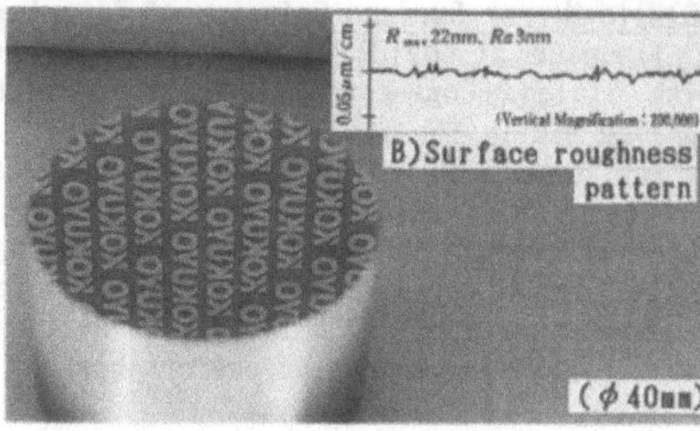

[Grinding conditions]		[Elid conditions]	
Wheel used	#4000CIFB-D	Voltage	Eo120V
Velocity	v1200m/min	Current	Ip24A
Feed-rate	f60mm/min	Pulse-	τon12μs
Depth of cut	d1~2μm	width	τoff3μs

A)Mirror surface grinding conditions

C)View of produced mirror surface sample

Fig.14 Example of sialon produced by Elid-grinding (application type : ②)

(7)Elid-Mirror Surface Grinding of Aluminium Nitride(AlN)

Aluminium nitride,together with alumina,is well known to have poor mirror surface finishability. Although its machinability is comparatively favorable,voids tend to remain on the ground surface due to its porous nature. Therefore,the above-mentioned lap-grinding or in-feed grinding at a low feeding speed is considered to be effective as an available mirror grinding method. In the example indicated in Fig.15,a submicro-grain,#30000 CIFB-D wheel was used with the latter system. An in-feed surface grinder(SHONAN Eng. Co. Ltd.) was employed for the processing and a electrolytic dressing power supply(by Stanley Electric Co. Ltd.) was employed.

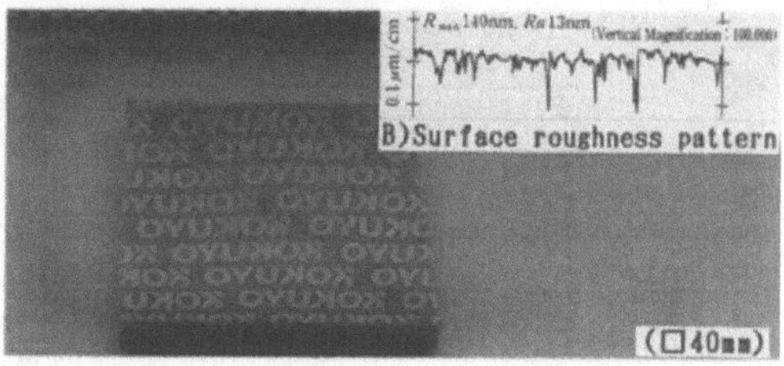

[Grinding conditions]		[Elid conditions]	
Wheel used	#30000CIFB-D	Voltage	Eo60V
Velocity/Rotation		Current	Ip30A
Wheel	Work	Pulse-	τon 2μs
vt337m/min	Sw600rpm		
Feed-rate	fd20μm/min	width	τoff2μs

A)Mirror surface grinding conditions

C)View of produced mirror surface sample

Fig.15 Example of aluminium nitride produced by Elid-grinding (application type : ④)

(8)Elid-Mirror Surface Grinding of Alumina(Al_2O_3)

Alumina is extremely diverse as its nature,and mostly tends to be subject to brittle fracture removal quite easily. Therefore,its mirror surface finishing is not easy. In constant feeding grinding methods employing constant depth of cut,obtaining of the mirror quality surface is quite difficult to be achieved,because the relation between the cutting depth and the abrasive grain size utilized tends to cause its brittle fracture destroying. However,a favorable finishing was realized by applying the above-mentioned lap-grinding method which uses constant pressure feeding (Fig.16). This feature has been confirmed by a use of lap-grinder(revised BAL-18S:SINTOBRATOR Ltd.) and a #4000 CIFB-D lapping plate[7]. The applied pressure for its processing was 1-3 kgf/cm². In this example,the suitable shape and the surface area of the electrode were decided from experiences in order that the electrolytic dressing could be effected uniformly on the entire surface of the lapping plate utilized.

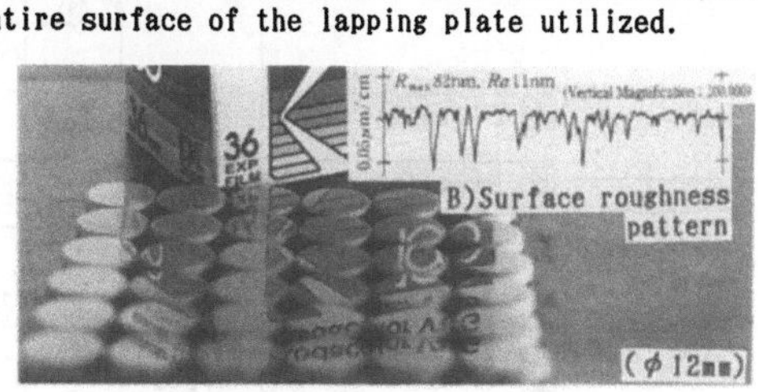

[Grinding conditions]		[Elid conditions]	
Wheel used	#4000CIFB-D	Voltage	Eo60V
Velocity/Rotation		Current	Ip60A
Wheel	Work	Pulse-	τon 2μs
vt42m/min	Sw30rpm		
Pressure	Fz1~3kgf/cm²	width	τoff2μs

A)Mirror surface grinding conditions

C)View of produced mirror surface samples

Fig.16 Example of alumina produced by Elid-grinding (application type : ⑤)

(9)Elid-Mirror Surface Grinding of Titanium Di-boride(TiB_2)

Titanium di-boride is considered to be one of "Difficult-To-Be-Ground-Materials" along with Sialon. Conversely,since it is electro-conductive,electro-discharge machining can be applied easily for its efficient shaping.

The example shown in Fig.17,was produced by applying inner mirror surface grinding which had been difficult in conventional way,mainly because of the poor machinability of the material. The Elid-mirror surface grinding was performed on a turning center(QT-10N:YAMAZAKI MAZAK Corp.) using a #4000 CIFB-D mounted wheel(30mm in diameter)[9]. An electro-discharge machine power supply(MGN-15W:MAKINO Milling Machine Co. Ltd.) was utilized for the Elid-mirror surface grinding operation,comfortably.

[Grinding conditions]		[Elid conditions]	
Wheel used	#4000CIFB-D	Voltage	Eo60V
Peripheral speed(m/min)		Current	Ip30A
Wheel	Work	Pulse-width	τ on1.5μs
vt283	vw20		τ off1.5μs
Feed-rate	fz3mm/rev	Depth of cut	dx4μm

A)Mirror surface grinding conditions

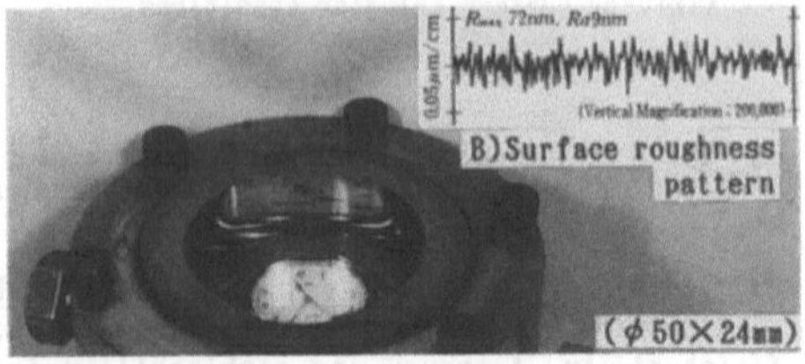

B)Surface roughness pattern

Fig.17 Example of titanium di-boride produced by Elid-grinding (application type : ⑦)

Table 4 lists a summary of the mirror surface ground examples' roughness data,along with the employed devices.

Table 4 Summary of ceramics samples' surface roughness produced by using Elid-grindings

Processed fine ceramics Name (abbrev.)	formula	type No.	Machine power supply	Wheel used	Surface roughness
1.Silicon nitride	Si_3N_4 (Hardness :Hv1700)	①	RGS-60 Stanley	#4000 CIFB-D	Rmax 32nm Ra 5nm
		②	VQC-15/40 SUE-87	#4000 CIFB-D	Rmax 36nm Ra 5nm
		③	GS-CHF EDD-04S	#4000 CIFB-D	Rmax 56nm Ra 8nm
		④	Shonan Eng. MGN-15W	#6000 CIFB-D	Rmax 58nm Ra 8nm
		⑥	QT-10N MGN-15W	#4000 CIFB-D	Rmax114nm Ra 14nm
2.Silicon carbide	SiC (Hv2700)	①	RGS-60 Stanley	#4000 CIFB-D	Rmax 48nm Ra 5nm
		②	VQC-15/40 MGN-15W	#8000 CIFB-D	Rmax 24nm Ra 4nm
		③	GS-CHF EDD-04S	#8000 CIFB-D	Rmax 36nm Ra 5nm
		④	Shonan Eng. Stanley	#120000 CIFB-D	Rmax 18nm Ra 2nm
3.Tungsten carbide	WC(-Co) (Hv1700)	①	RGS-60 Stanley	#4000 CIFB-D	Rmax 52nm Ra 6nm
		②	VQC-15/40 SUE-87	#4000 CIFB-D	Rmax 54nm Ra 8nm
		③	GS-CHF EDD-04S	#4000 CIFB-D	Rmax 76nm Ra 11nm
		④	Shonan Eng. Stanley	#120000 CIFB-D	Rmax 14nm Ra 2nm
4.Zirconia	ZrO_2 (Hv1350)	①	RGS-60 Stanley	#4000 CIFB-D	Rmax 32nm Ra 5nm
5.Alumina-titanium carbide	Al_2O_3-TiC (H_{RA}94)	①	RGS-60 Stanley	#4000 CIFB-D	Rmax 34nm Ra 5nm
6.Sialon	Sialon (Hv1500)	②	VQC-15/40 SUE-87	#6000 CIFB-D	Rmax 22nm Ra 3nm
7.Aluminium Nitride	AlN (Hv1250)	④	Shonan Eng. Stanley	#30000 CIFB-D	Rmax140nm Ra 13nm
8.Alumina	Al_2O_3 (Hv1550)	⑤	BAL-18S MGN-15W	#4000 CIFB-D	Rmax 82nm Ra 11nm
9.Titanium di-boride	TiB_2 (Hv2700)	⑦	QT-10N MGN-15W	#4000 CIFB-D	Rmax 72nm Ra 9nm

At present, it is becoming possible to use a very wide range of grain sizes(shown in **Table 5**) as a cast-iron fiber bonded (CIFB) micro-grain abrasive wheel. A wide range of abrasive grain sizes can be utilized in order that properties of a material being ground and its processing system match in balance with its produced surface accuracy and efficiency. In addition, examples of combination of different grain sized wheel, through the course of processing leading to mirror surface finishing, are indicated in **Table 6**. A combination is selected mainly from characteristics of a material to be ground and its initial surface accuracy, of course, also from its purposed surface quality and efficiency.

Table 5 Micro-grain and submicro-grain sized CIFB wheels for various Elid-grindings

No.	Mesh number	Average grain	diameter	Elid-grinding application available: ①～⑦ in Table3	Example of condition
1	#1200		11.60 μm	Either from ① to ⑦	d ≦ 10 μm/pass
2	#2000	Micro-grain	6.88 μm	Either from ① to ⑦	d ≦ 6 μm/pass
3	#4000	Micro-grain	4.06 μm	Either from ① to ⑦	d ≦ 4 μm/pass
4	#5000	Micro-grain	3.28 μm	Either from ① to ⑦	d ≦ 4 μm/pass
5	#6000	Micro-grain	3.15 μm	Either from ① to ⑦	d ≦ 3 μm/pass
6	#8000	Micro-grain	1.76 μm	Either from ① to ⑦	d ≦ 2 μm/pass
7	#10000	Micro-grain	1.27 μm	Either from ① to ⑦	d ≦ 2 μm/pass
8	#20000	Sub-micro-grain	0.51 μm	①,②,④ have been tested	fd < 30 μm/min
9	#30000	Sub-micro-grain	0.31 μm	①,②,④ have been tested	fd < 20 μm/min
10	#60000	Sub-micro-grain	0.17 μm	④ has been tested	fd < 10 μm/min
11	#120000	Sub-micro-grain	0.13 μm	④ has been tested	fd < 10 μm/min

*The character "d" means depth of cut,
and the character "fd" means feeding speed for cutting.

Table 6 Grain size combination examples from rougher grinding to mirror surface grinding

No.	Rough grinding →	Intermediate finishing →	Mirror surface finishing
1	#100～#325	#600～#1200	#2000～#8000
2	#325～#600	#1200～#2000	#4000～#10000
3	#600～#1200	#2000～#10000	#20000～#120000

5. Examples of Elid-mirror surface grinding of the others

(1)Elid-mirror surface grinding of silicon

It could be said that the Elid-mirror surface grinding has been established during processing silicon wafers in the authors' early research for developing finish grinding techniques. **Fig.18** shows a finished example of a silicon wafer utilizing the Elid-grinding method. In this case, a #4000 CIFB-D wheel and a turning center as an in-feed grinder were employed. Quite low surface damage exists in the ground surfaces by the method. A specialized for silicon wafer's efficient production using the Elid-grinding has already been developed.

Fig.18 Example of silicon wafer produced by Elid-grinding (application type : ④)

(2)Elid-mirror surface grinding of ferrite

Ferrite is also the material in which processing tests many advantages of the Elid-grinding have been confirmed as well as silicon's. As shown in Fig.19, the surface quality produced by a #4000 CIFB-D wheel is very good, together with its chipping-free edges.

Fig.19 Example of ferrite produced by Elid-grinding (application type : ①)

(3)Elid-mirror surface grinding of glass

What is called "**glasses**" contain many kinds of material. In conventional grindings, therefore, it is difficult to gain mirror surface quality of glasses of many kinds in similar conditions. The Elid-grinding method has experienced many glasses, mostly succeeded in obtaining mirror surfaces, as shown in Fig.20. The ground surface using Elid in the left is transparent, clearly different from the conventionally ground right one (the both are BK7). Through applying the Elid-method, the authors have realized efficient lens-processing(Fig.21).

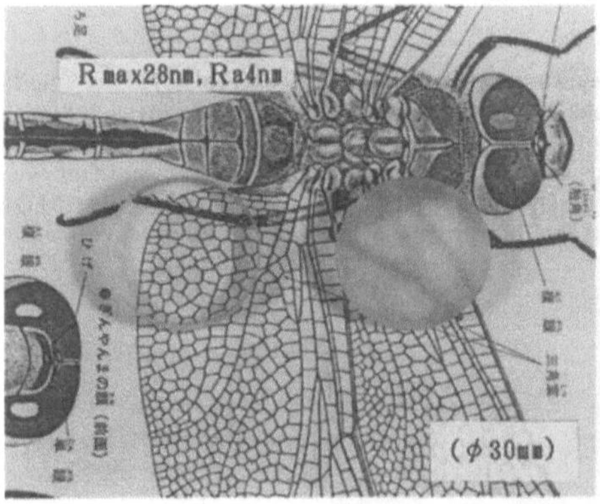

Fig.20 Example of silicon glass produced by Elid-grinding (application type : ①)

Fig.21 Example of optical lens produced by Elid-grinding application

6. Conclusion

The above provided an outline of the Elid-mirror surface grinding techniques and its various types of application, mainly focusing on ceramics finishing. Needless to say, the more other materials including steels which are not mentioned in this paper have been confirmed to be finished by applying the Elid-grinding method mostly using CIFB-cBN wheels in the same manner as the above-mentioned. In addition to metal bonded fine grain wheels and electrolytic dressing power supplies specialized for the Elid-mirror surface grinding method,specialized mirror surface grinding machines(**"Elid-mirror surface grinders"**) have already begun to appear on the market in Japan. These active proliferation about the Elid-devices has even reached the stage where the techniques presented are being put to use at production sites widely and easily,and actually the techniques have started being contributed to manufacturings.

This research should be continued from the viewpoints of practical application resulting in advanced developments,and of the creation about even more superior ductile-mode grinding (**"nano-grinding"**) technology,along with academic researches for analyzing the method to be fully controlled. The authors would like to present many other results of the Elid-grinding techniques with finding opportunities,also efficiently!

Acknowledgements

The authors wish to thank NACHI-FUJIKOSHI Corp.,SINTOBRATOR Ltd.,YAMAZAKI MAZAK Corp.,KURODA Precision Industries Co. Ltd.,Mitutoyo Corp.,NIKON Co. Ltd.,SHIN-ETSU HANDOTAI Co. Ltd.,MAKINO Milling Machine Co. Ltd.,Stanley Electric Co. Ltd.,TANAKA Eng. Ltd.,Sodick Co. Ltd.,NORITAKE Company Ltd.,MARUTO Instruments Ltd.,HITACHI Corp.,Nippon Pillar Packing Co. Ltd.,YKK YOSHIDA KOGYO Co. Ltd.,Olympus Optical Co. Ltd.,MITSUI SEIKI KOGYO Co. Ltd.,and MATSUSHITA Electric Works Co. Ltd. for their supports with respect to experimental equipments,and also Mr. K.Park(Graduate-course,University of Tokyo) for his assistances.

References

[1] H.Ohmori,T.Nakagawa,Grinding Process of Silicon Using Cast-iron Bonded Diamond Wheel(3rd Report),Preprint of Autumn Conf. of JSPE,1987,687-688

[2] H.Ohmori,T.Nakagawa,Finishing Method of Ferrite by Mirror Surface Grinding,Preprint of Spring Conf. of JSPE,1988,519-520

[3] H.Ohmori,T.Nakagawa,Mirror Surface Grinding of Hard and Brittle Materials Using Cast-iron Fiber Bonded Wheel,Preprint of Autumn Conf. of JSPE,1988,355-356

[4] H.Ohmori,E.Yamada,et al.,Grinding Process of Steel Materials with Machining Center(3rd Report),Preprint of Autumn Conf. of JSPE,1988,43-44

[5] H.Ohmori,Y.Takata,et al.,Cylindrical Mirror Surface Grinding with Turning Center,Preprint of Autumn Conf. of JSPE,1988,353-354

[6] H.Ohmori,K.Toyama,et al.,Grinding Process of Silicon Using Cast-iron Bonded Diamond Wheel(5th Report),Preprint of Autumn Conf. of JSPE,1988,715-716

[7] H.Ohmori,T.Narita,et al.,Complexed Finishing Process by Bonded Abrasive Lapping(3rd Report),Preprint of Autumn Conf. of JSPE,1988,203-204

[8] I.Takahashi,H.Ohmori,et al.,Highly Efficient Grinding of Hard and Brittle Materials(2nd Report),Preprint of Spring Conf. of JSPE,1989,355-356

[9] K.Park,H.Ohmori,et al.,Mirror Surface Processing of Cylindrical Inner Surfaces by Grinding Method Using Electrolytic Dressing,Preprint of Autumn Conf. of JSPE,1989,899-900

[10] H.Ohmori,T.Nakagawa,Grinding Method Using Submicro-grain Wheel,1989,903-904

[11] H.Ohmori,T.Nakagawa,Mirror Surface Grinding of Silicon Wafers with Electrolytic In-process Dressing,Annals of the CIRP Vol.39/1/1990,329-332

Results of Non-linear Mixing in Precision Machining

R.Haberland, G.Pfeifer, Feinwerktechnik - Universität Kaiserslautern

In every kind of machining, the machine virbation is a limiting factor of the generated surface quality. The nonlinear mixing of frequency components during cutting are creating different surface shapes. Using a computer-simulation the relationship between the machine vibration and the generated surface can be described. The theoretical calculation of surface shape is shown in an example for fly cutting. Surface amplitudes and wavelengths are calculated and compared with real machined surfaces shapes.

1 INTRODUCTION

Ultraprecision machining as a well established technology is almost limited by the disturbance of machine vibration. This vibrations are generating a surface texture which is typical for the vibration mode of the machine.
The vibration amplitude can be decreased by increasing the machine stiffness. This often involves considerable costs.
Always an easier way of optimization is found in optimized speed and feed velocities. The knowledge of the relationship between surface shape and vibration conditions enables the user to optimize the machine.

2 GENERATION OF SURFACE SHAPE

The generation of surface profile is depending on the relation between the relative movement of the tool and its rotating frequency. The number of complete cycles of the tool-oscillation during one rotation of the tool is specified by

$$\frac{\omega_a}{\omega_s} = a + b$$

ω_a : relative movement frequency
ω_s : spindle rotation frequency
a : integer part of frequency ratio
b : fractional part of frequency ratio

By this means the parameter **a** represents the integer part, and **b** the fractional part of the frequency relation. If b = 0, the location of the maximum an minimum heights on the surface are not changing from one to another rotation.
Normally the fractional part is different from zero. This causes a constant phase shift per revolution.

$$\delta = \omega_s \cdot b \cdot T_a$$

b: fractional part of frequency relation
ω_s : spindle rotation frequency
T_a : period of vibration frequency
δ : phase shift per revolution

Caused by this phase shift, the position of each point located at a defined high level is moving on an cycloid. The coordinates of this position can be calculated

$$x(n) = x \cdot n + r \cdot \cos(-b \cdot \omega_s \cdot T_a + \omega_s \cdot T_a)$$

$$y(n) = r \cdot (1 + \sin(-b \cdot \omega_s \cdot T_a + \omega_s \cdot T_a))$$

s : feed rate
n : number of spindle revolutions
r : distance tool - rotation axis
ω_s : spindle rotation frequency
T_a : period of vibration frequency

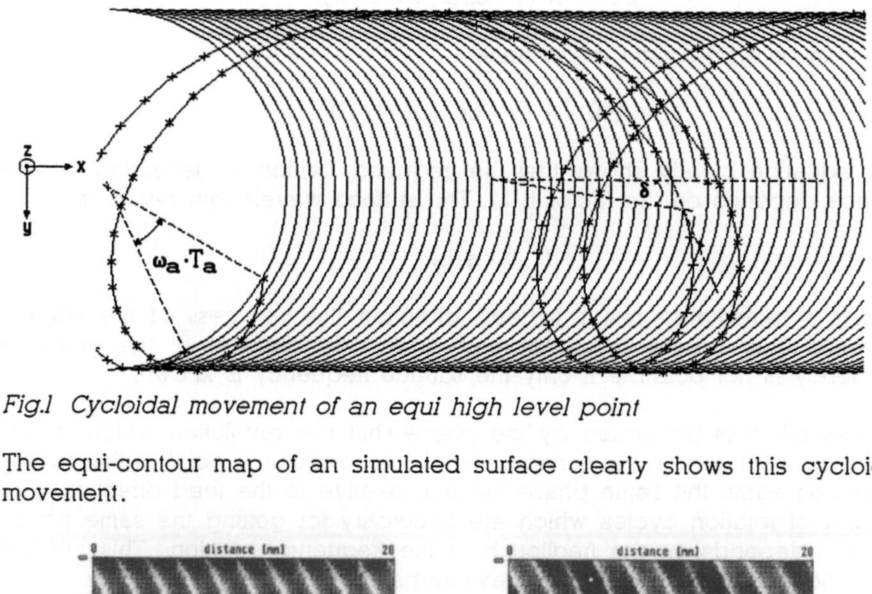

Fig.1 Cycloidal movement of an equi high level point

The equi-contour map of an simulated surface clearly shows this cycloidal movement.

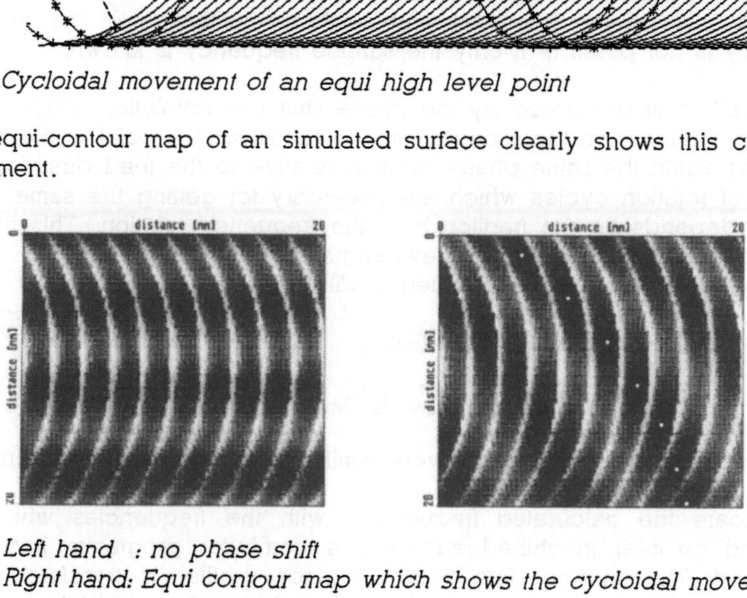

Fig.2 Left hand : no phase shift
Right hand: Equi contour map which shows the cycloidal movement of a point of constant height

3 GENERATION OF SURFACE FREQUENCIES

In feed direction, the profile of the surface can be calculated.
The parametric equation of the movement of one point of the cutting edge

$$k(t) = (s \cdot t/T_a + r \cdot \cos(\omega_s \cdot t) \ , \ r \cdot \sin(\omega_s \cdot t) \ , \ A \cdot (1-\cos(\omega_a \cdot t))$$

$$= (x),(y),(z)$$

with

$$\omega_s \cdot t = \varphi_o + 2 \cdot n \cdot \pi$$

allows to calculate the surface profile in feed direction.

$$z(x) = A \cdot (1-\cos(2 \cdot \pi \cdot b \cdot x/s))$$

This can be written time dependent:

$$z(t) = A \cdot (1 - \cos(\omega_s \cdot b \cdot t))$$

This equation clearly shows that the frequency which is generated on the surface depends on the factor $b \cdot f_s$. The surface wavelength results in

$$\lambda = s/b.$$

This is a remarkable result. In feed direction the waviness of the surface depends only on the fractional part b. An identification of the vibration frequency is not possible if only the surface frequency is known.

Another effect is generated by the phase shift per revolution which results in a long waviness on the surface. After $2 \cdot \pi/\delta$ rotations, a point of the cutting edge has again the same phase location relative to the feed direction. The number of rotation cycles which are necessary for getting the same phase location depends on the fraction **b** of the frequency relation. This number of rotations specifies the lowest wavelength on the surface.
So normally existing surface frequency will be

$$f = b \cdot f_s$$

another one will be

$$f_{low} = f_s/n_{min}$$

As this can be very different or very similar depending on b and n, very different surfaces can exist.
To compare the calculated frequencies with the frequencies which are contained on real machined surfaces, a computer programm has been developed. The surface data from a surface profiler is transfered to a computer. A Fourier analysis of the profile shows the containing surface frequencies.

4 GENERATION OF SURFACE AMPLITUDE

If no disturbing vibration is present during machining, the tool radius is generating a regular periodic surface. The approximation of the tool radius by a parabol results in

$$p(x) = x^2/(2\cdot\sqrt{R^2-s^2/4})$$

for the surface profile in feed direction.
The Fourier-Transformation of this parabolic tool shape has the coefficients

$$z_n(x) = \frac{s^2}{24\cdot\sqrt{R^2-s^2/4}} + \sum_{n=1} \frac{s^2}{2\cdot n^2\cdot \pi^2\cdot\sqrt{R^2-s^2/4}}\cdot \cos(n\cdot \frac{2\cdot\pi}{s}\cdot x)$$

The coefficients show clearly, that the amplitudes are decreasing as the square of the frequency.
The fourier spectrum of a real machined copper sample, with no disturbing vibration during cutting, is represented in the next figure. The amplitude of the second order surface frequency should be only about 1/4 of the first order frequency. This is in a good agreement with the value from the real surface.

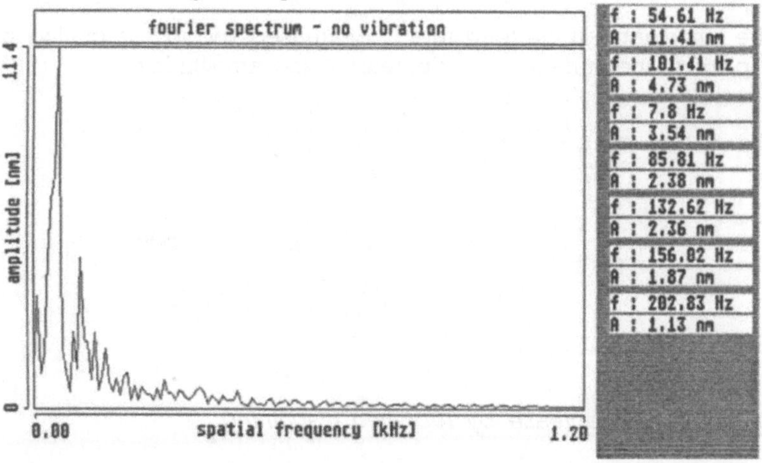

Fig. 3 Fourier spectrum of a surface profile, machined without vibrations

Machining under real conditions is not free of vibration. If the feed per revolution is high enough, the amplitude of this vibration is found unchanged on the surface (but the frequency shifted as explained before).

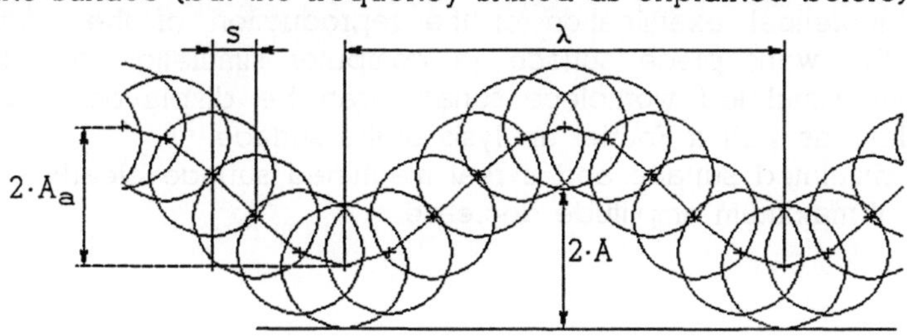

Fig.4 Vibration amplitude (A_a), tool radius and generated surface (λ,A)

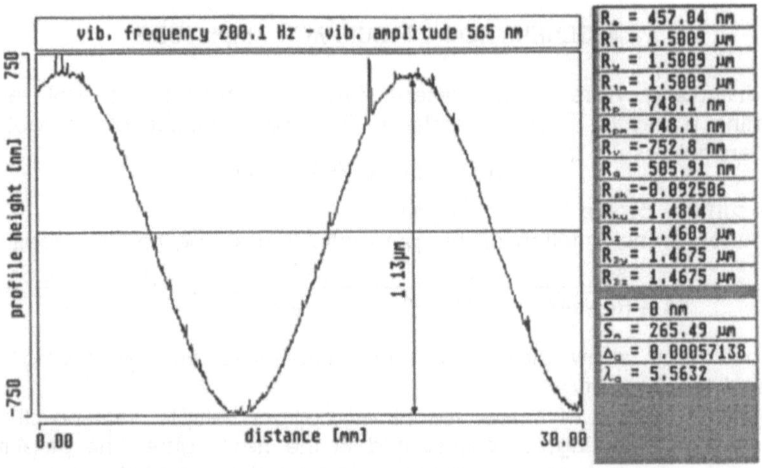

Fig.5 Unchanged vibration amplitude (real machined sample)

On the other hand, if the feed rate is decreased, the tool is cutting away the high parts of the surface, thus decreasing the amplitudes.

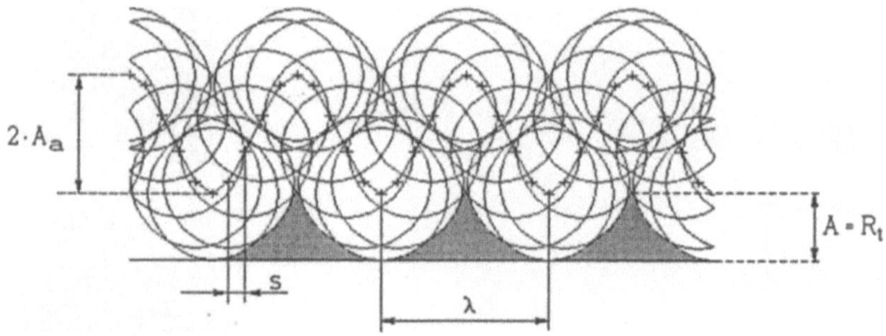

Fig.6 Amplitude decrease by repeated machining

The maximum of amplitude decrease can be calculated. As a result, the maximum Peak to Valley (R_t) value of the surface frequency $f = b \cdot f_s$ is

$$R_t = \frac{s^2}{8 \cdot R} \cdot \left(\frac{f_s}{f}\right)^2$$

To test the theoretical examination of the reproduction of the machine vibration on the work piece surface, a computer simulation has been developed. The simulated workpiece surface can be displayed and the program displays as well a Fourier analysis of the surface.
Comparing a simulated surface and a real machined surface clearly shows the influence of maximum amplitude decrease.

For example:

vibration frequency	: 512.5 Hz	surface amplitude	: 40 nm
spindle frequency	: 50 Hz	surface frequency	: 12.5 Hz
tool radius	: 20 mm	surface wavelength	: 80 μm
feed rate	: 1 mm/s		

(Right-hand columns: calculated parameters:)

Table 1 parameters for simulation and real machining

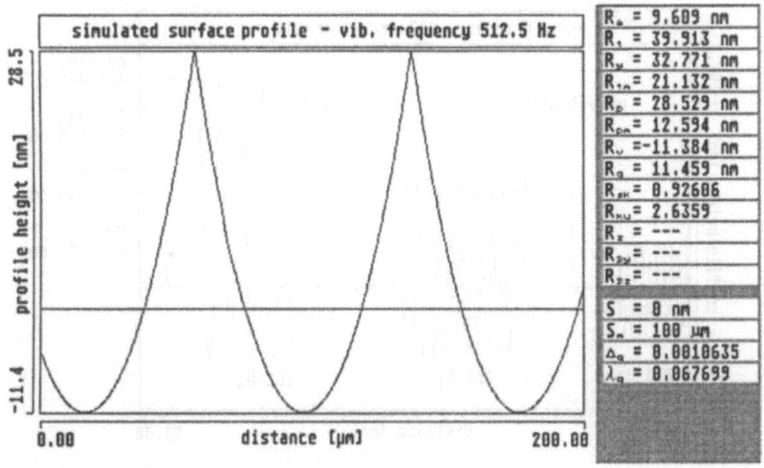

Fig.7 Simulated surface, maximum amplitude decrease , calculated according to previous equation (parameters see Tab. 1)

The calculated Peak to Valley value for the parameters from Fig.7 is 40nm.

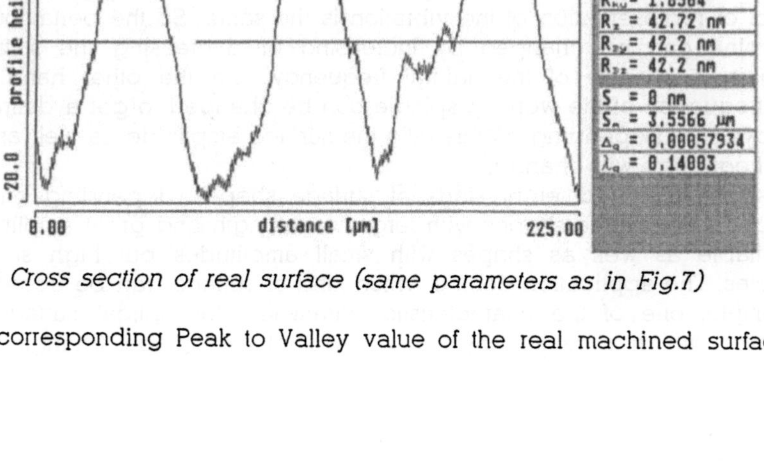

Fig.8 Cross section of real surface (same parameters as in Fig.7)

The corresponding Peak to Valley value of the real machined surface has

been 45nm. The difference of 5nm is caused by the non ideal sample material (grain boundaries), not ideal sharpness of the tool and possibly some small additional vibration amplitudes.
The decrease of the surface amplitude depends on the vibration frequency (on condition that the rotation frequency is constant). This effect can be demonstrated by machining a sample with constant vibration amplitude and constant feed velocity, only changing the vibration frequency.

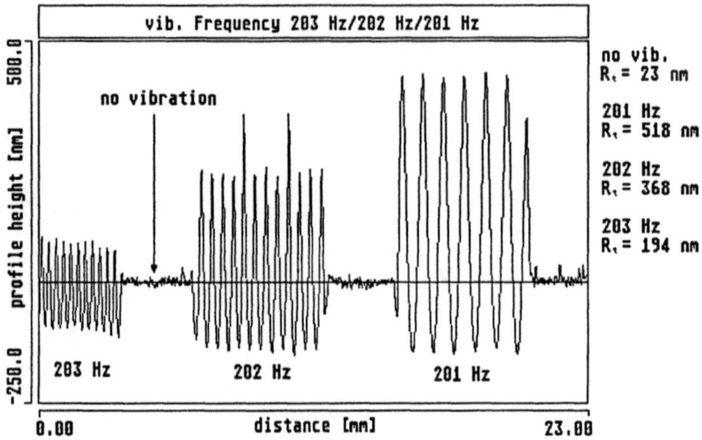

Fig.9 Machined sample. Constant feed velocity (1mm/s), constant vibration amplitude (0.6 μm), constant tool revolution frequency (50 Hz), changing vibration frequency

5 OPTIMIZATION OF FLY-CUTTING PROCESS

Choosing the ratio of rotation frequency to vibration frequency, it's possible to change the resulting surface amplitude. In real production machines there is no forced vibration amplitude as in the described experiments. Instead vibrations of the machine structure itself will disturb the working process, but the effect of representation of the vibration is the same. So the behaviour of the machine can be changed by increasing or decreasing the stiffness, resulting in a change of the natural frequency. On the other hand, the rotation frequency of the working spindle can be changed to get a definitive frequency ratio. Depending on this ratio the surface amplitude as well as the surface frequency will change.
It is possible to get different kinds of surface shapes, depending on the frequency ratio. Surface shapes with large wavelength and great amplitudes are available as well as shapes with small amplitudes but high surface frequencies. The application decide which kind of surface can be accepted. For example, one of the characteristic parameters for optical surfaces is light scatter, which is caused by the waviness of the surface. Changing the

surface characteristics by variation of the frequency ratio, is resulting in different scatter characteristics. A large diffraction angle and a small scatter intensity or a small diffraction angle and a large scatter intensity can be realised.

6 CONCLUSIONS

Knowing the effects of disturbing frequency, rotation frequency of the tool-spindle, feed velocity and tool radius, the characteristics of the resulting surface can be affected. By small changes in disturbing frequency or spindle rotation frequency the resulting surface shows quite different properties. The surface frequency with the most dominant amplitude depends on the fractional part of the ratio f_a/f_s. The clear identification of the vibration frequency is not possible if only the surface frequency is known.

6 REFERENCES

/1/ Shuhei Takasu; Influence of Study Vibration with small Amplitude upon Surface-Roughness in Diamond Turning; Annals of the CIRP Vol.34/1/1985

/2/ A.V. Oppenheim,A.S. Willsky; Signal and Systems; Prentice Hall Inc. Englewood Cliffs, New Jersey

/3/ R.Haberland, G.Pfeifer; Generation of Surface Topology by Interactions of Vibration, Spindle Speed and Feed Velocity; SPIE Proceedings 1990

/4/ S.M. Pandit; Characteristic Shapes and Wavelength Decomposition of Surfaces in Machining; Annals of the CIRP Vol. 30/1/1981

Micropositioning for Submicron Electron Beam Lithography

Harald Bohlen and Werner Kulcke

IBM German Manufacturing Technology Center, Max-Eyth-Str.6 7032 Sindelfingen, Germany

Abstract

Positional accuracy is the most important parameter in future lithography for integrated circuits. With shrinking dimensions it is a great challenge to meet the associated overlay requirements. Electron Beam Proximity printing is described as a lithography for future integrated circuit fabrication. A unique correction feature was developed for this exposure method which is applicable for many tool and mask related pattern distortions. Digital signal processor cards have been developed which can handle the large amount of data for corrections in real time during the exposure. Results of this exposure method on test devices are presented.

Introduction

Lithography is the most essential manufacturing process of integrated circuits and especially of dynamic random access memories (DRAM). Since the first circuits were developed in the sixties a remarkable trend still continues that a new generation of chips appears on the market about every three years which then contains four times more bits than the previous one. It is difficult to predict when this increase in complexity will come to a stop.

DRAMs with a memory capacity of 4 million bits are presently fabricated in mass volume. Experts believe that first 1 Gigabit chips will be available at the end of this century. Mainly two factors have supported this enormous progress in technology over more than 20 years:

- Reduction of the smallest geometries (or lithography resolution) on the chips
- Increase of the chip area

Some technical data about present and future DRAM chip generations are given in table 1. As one can see in table 1 the minimum linewidth shrinks by a factor $\sqrt{2}$ per generation. At the same time the chip area is increased by about the same factor and last but not least the overlay in lithography has to be improved at the same pace. The overlay error is defined as deviations between two exposures. Furthermore, each individual exposure produces positional errors which under optimum conditions are caused by the mask and by the tool to equal amounts. Therefore, tool errors must be kept negligible.

Optical lithography tools are the workhorse of today's IC manufacturing. It is the subject of a variety of publications how far the optical technology can be extended. It is obvious that smaller wavelengths will be needed in order not to offend basic physical rules of imaging. Experts agree that 0.35 μm linewidth will still be governed by optical tools.

DRAM	Minimum Linewidth	Chip Area	Overlay (3σ)
1 Mbit	1.0 μm	$\approx$60 mm^2	350 nm
4 Mbit	0.7 μm	$\approx$90 mm^2	250 nm
16 Mbit	0.5 μm	$\approx$130 mm^2	175 nm
64 Mbit	0.35 μm	$\approx$180 mm^2	125 nm
256 Mbit	0.25 μm	$\approx$250 mm^2	90 nm
1 Gbit	0.17 μm	$\approx$360 mm^2	60 nm

Table 1: Chip generations and the lithographic requirements

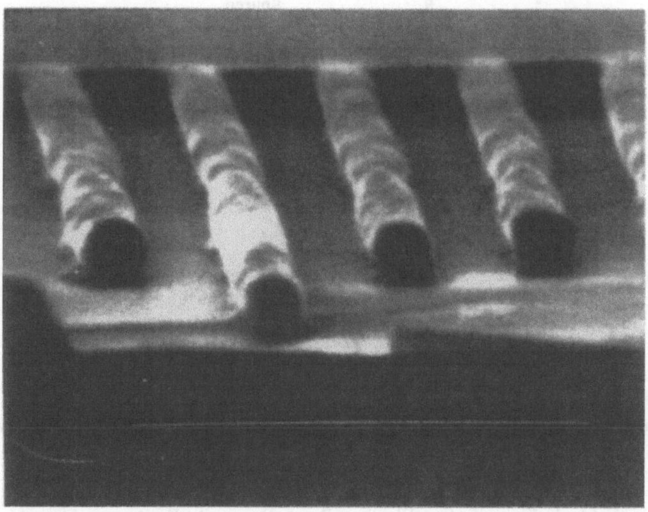

Figure 1: Lines and spaces of 0.15 μm printed with EBP into PMMA resist

X-ray lithography is being developed for applications after the optics era. Many groups around the world are engaged in this field and made progress in many areas. The main problem of X-ray lithography is to achieve the required positional accuracy.

This paper deals with a new electron beam lithography technique which has been demonstrated to offer new solutions to the challenge in positional accuracy, the most problematic parameter for future progress in this area.

Electron beam proximity printing (EBP) has been developed by our group as an alternative with unique features to meet the tight requirements in positional accuracy of the future IC lithography. Its high resolution capability is demonstrated in fig. 1. The SEM shows a single mask slit several times exposed and accurately stepped to produce the structure of 0.15 μm lines and spaces in PMMA resist. The stepping accuracy is, apparently, in the range of 13 nm which is the resolution of the laser interferometer (see below).

Basic principle of EBP

EBP employs shadow projection of a chip sized mask by means of 10 keV electrons. Shadow projection uses a mask in 1:1 scale which is not limited in field size. However, a mask in the scale 1:1 is required. Electrons at 10 keV have a de Broglie wavelength of 0.15 Å so that diffraction effects at submicron linewidths can be neglected. Thus a relatively large distance (e.g. 500 μm) between mask and wafer can be used. A schematic block diagram of an EBP tool is shown in figure 2.

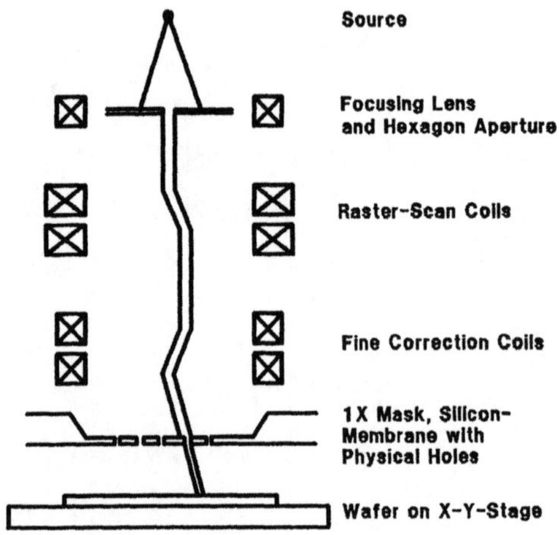

Figure 2: Schematic block diagram of an Electron Beam Proximity Printer

The main part of the tool is the electron-optical column. A gun emits the electrons which are accelerated to 10 kV. A magnetic lens produces a nearly parallel beam which is shaped to a hexagonal cross-section of about 1 mm diameter. The beam then passes some deflection yokes whose functions will be explained later. The beam illuminates a transmission mask on top of the wafer. The mask is supported by a rigid frame and can be rotated by means of a stepping motor at very small rotational increments ($7x10^{-6}$ rad.). The wafer is carried by a x-y stage which is driven by linear motors. The stage position is monitored by a laser interferometer at three axes to track the x and y positions as well as the table yaw error. The resolution of the laser interferometer is $\lambda/48 = 13$ nm. As an electron beam can only be produced in vacuum the system is installed into a vacuum chamber.

The correction feature of EBP

Since resolution is not a problem for electron beam exposure systems the main challenge is to meet the extremely tight positional requirements for sub-half- micron lithography applications. Some unique correction methods have been developed for this kind of exposure tool which are important for the goal of high positional accuracy. Important for that is the choice of mask illumination. The beam with its diameter of 1 mm is scanned across the pattern area in a raster scan mode. Therefore, only small parts of the mask are illuminated at a time and shadow projected to the wafer. A so-called fine correction unit enables to tilt the beam slightly with a pivot point in the mask plane. The result is that the illuminated mask area is shifted on the wafer surface relative to its position at vertical beam incidence. (Fig.3)

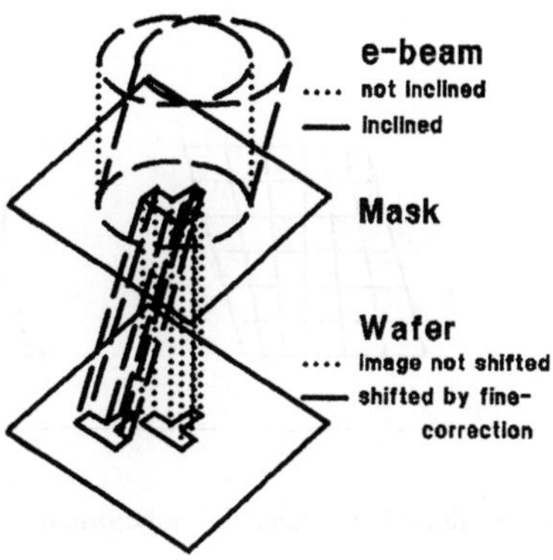

Figure 3: Principle of fine correction in EBP

The inclination angle of the beam can be varied during the raster scan motion across the mask. Therefore, different mask areas can be shifted differently. It should be noted that the variation in beam inclination during the exposure sequence produces some kind of pattern blur due to the finite size of the beam. The maximum tolerable blur has been discussed in /1/. For practical use this limitation is larger than needed.

Although each one of the following discussed error sources may be small the sum of all may exceed the small tolerable error budget of future lithography applications. Table position correction is the simplest use of beam tilt. A constant beam tilt angle during the raster scan illumination corrects for this error. The information about the amount of tilt is gained from the interferometer reading. Also, table yaw errors which rotate the wafer below the mask can be corrected by a change of the beam tilt during the raster illumination. It is obvious that even table oscillations are tolerable within certain limitations because corrections can be applied in real time.

In the same manner as table mispositioning also mask pattern distortions can be corrected, provided that they are known or can be measured. This is a typical application for non-linear corrections by means of beam tilt. The present EBP tool calculates polynomials up to the 4th order for a best fit correction of known mask distortions. Some polynomial corrections are illustrated in fig. 4. Also, combinations of the various polynomial corrections can be applied.

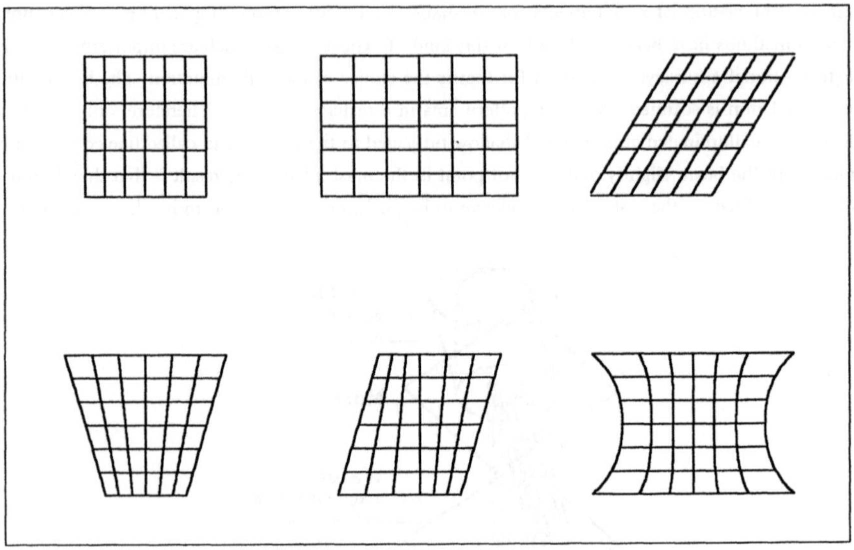

Figure 4: Some simple corrections achievable by means of polynomials.
The original pattern is top left, others are possible modifications.

This feature was successfully applied in a device exercise in order to improve the overlay of the layers. Typically, overlay errors were reduced by a factor of about three.

Detailed investigations of the exposure tool yielded additional causes which affect the overlay accuracy. One is, e.g., that the gap between mask and wafer may vary because the wafer is bent or has the cross-section of a wedge.

Registration mark detection

The manufacturing process of integrated circuits requires the precise superposition of several exposure levels. A prerequisite is that marks on the wafer can be detected with high precision.

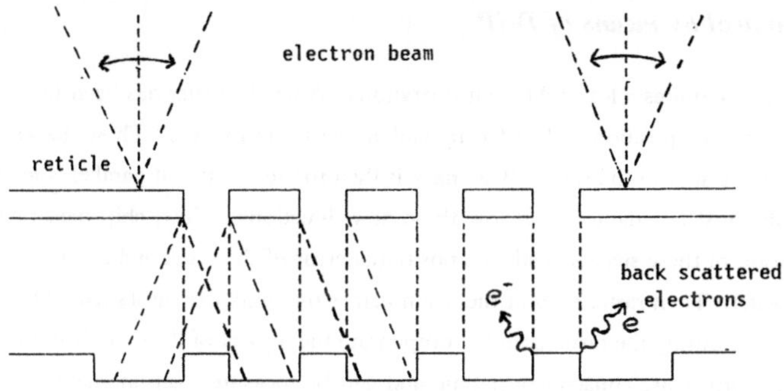

Figure 5: Principle arrangement for mark detection in EBP

For this purpose the electron beam illuminates small holes of e.g. 1 μm x 1 μm size in the mask. (Fig. 5) The same structure is etched into the wafer surface. The small square beams produced by the holes are now scanned across the structure on the wafer by tilting the beam. When the beam hits the walls of the structure the amount of backscattered electrons changes and so the absorbed beam current in the wafer. The change in absorbed current is detected by an amplifier connected to the wafer. In order to get a strong signal several thousands of holes are arranged in a registration mark with a pitch of about 3 μm. The overall size of the mark is typically 200 μm x 600μm.

Such registration marks are placed at each corner of the chip. This permits to derive corrections from the 4 marks so that the printed chip exactly fits into the area defined by the marks. Thus lateral or rotational offsets as well as magnification changes can be compensated which may have been occurred during the print of the previous structures on the wafer. The occurrence of such errors especially happens when a in a mix-and-match exposure strategy previous layers were exposed in a tool of different kind. Also, changes of the overall size of the wafer can be compensated which originate from stresses introduced during the fabrication processes.

The described chip registration procedure allows to determine the distance between mask and wafer at every mark position. The tilt of the beam is increased until it hits the next structures located 3 μm apart. The distance between mask and wafer can be calculated from the required tilt. A precise knowledge about the mask/wafer gap is needed as some of the corrections are gap dependent, others not.

For this chip registration procedure it is a prerequisite that the position of the marks has been prealigned to better than 1 μm. Otherwise an error of multiples of the 3 μm registration pattern pitch could occur. To avoid this, a so-called global align procedure is employed every time when a wafer is introduced into the tool. The principle for detection of the global marks is similar to the above. The global align mark pattern, however, is different and a hill climbing procedure permits to detect two marks at the wafer rim very precisely. (For more detail compare /2/).

Exposure cycle control by means of DSP

The exposure of a wafer requires a lot of different operations. After the wafer has been loaded into the tool its lateral and rotational position is detected by global registration marks. These marks determine a coordinate system associated to the wafer. The mask is then rotated so that it parallels the wafer axes. The table movements must be adapted to follow these new directions. Now, chip registration marks are detected. Evaluation of these signals yields the positional error of the wafer and the width of the gap between mask and wafer. Then raster scan of the beam across the mask can be started. The correct tilt of the beam is calculated during the raster scan to compensate the impact of the described different error sources. Some of the errors are constant over time and can be measured 'off-line' and stored in data tables. However, they may depend on the beam position in the scan field due to e.g. non-linearities of the raster scan or remanent magnetic fields varying with the stage position. The correction of these kind of errors corresponds to a tool calibration. Dynamic errors must be measured and evaluated in 'real time'. A programmed algorithm calculates the data for compensation. Typical examples are the lateral and rotational table errors, the registration mark detection and mask/wafer gap measurement.

In addition to the normal tool operation the tool control system must be capable to handle the information in real time for:

- Correction of tool errors
- Correction of stage errors
- Optimization of pattern placement
- Compensation of mask distortions

In practice a lot of data has to be handled very fast for determining the correct beam tilt during the exposure scan. Digital signal processors (DSP) have been selected as the best choice for this task. Its design allows parallel processing. Especially, the very powerful ability to handle compound instructions at a cycle time of 100 ns offers excellent capabilities and the flexibility to perform the different control tasks needed in EBP.

A tool control system employing special digital signal processor cards has been developed. Five identical DSP-cards are used in the tool controller. They work in parallel and are hosted by a PC-AT. The tool control system, therefore, has a speed of up to 200 million arithmetic instructions per second and a communication rate of 6 MBytes/s between DSP's. The following data operations are needed for example for each beam tilt up-date and each beam axis:

- 34 Multiplications
- 35 Additions
- 7 Logical operations

The DSP's allow tilt up-date every 8 μsec or every 0.25 mm of beam scan.

EBP transmission masks

As electrons are scattered even in very thin layers transmission mask for EBP must contains physical holes at transparent areas The fabrication steps of such masks are similar to processing steps used in the fabrication of integrated circuits. A cross-section of a mask is shown in fig. 6:

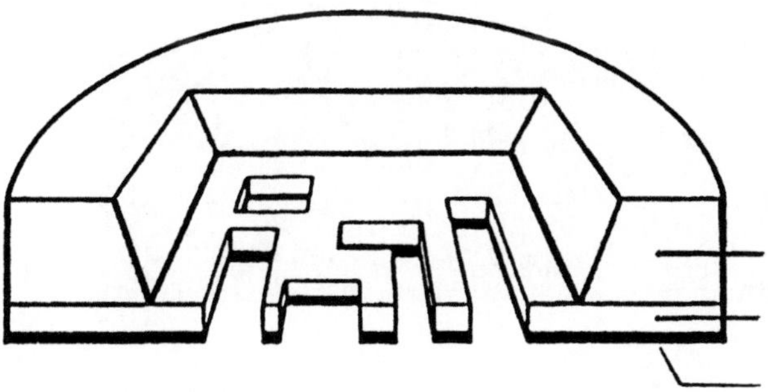

Figure 6: Schematics of a EBP transmission mask

A silicon wafer is heavily boron doped. The wafer is then wet etched in an EDP (ethylen-diamine-pyrocatechol-water mixture) solution through a window on the bottom side which corresponds to the desired mask pattern size. The boron doping acts as an etch stop so that a membrane with a thickness of about 2 to 3 μm is left in the center of the wafer.

In a next step the mask pattern is delineated by an electron beam pattern generator into a resist layer on the membrane. The hole pattern in the resist is transferred into the silicon membrane by reactive ion etching steps. A thin metal layer of gold or tungsten on top of the membrane is used to have a high amount of reflected electrons on opaque areas so that the temperature increase remains moderate. A SEM detail of a transmission mask is shown in fig. 7.

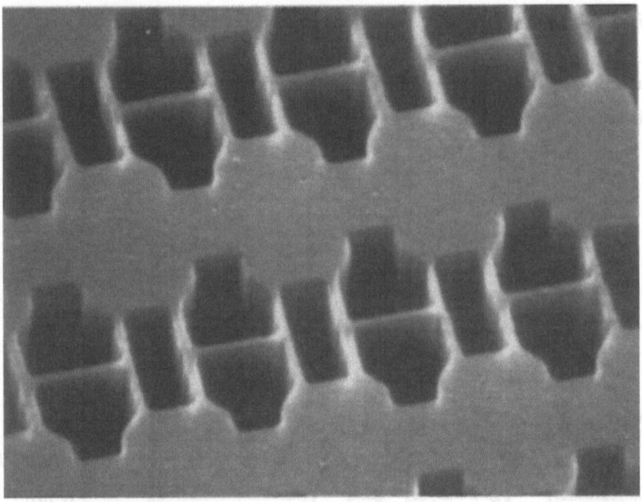

Figure 7: SEM micrograph of a transmission mask containing a typical DRAM pattern. Smallest dimension is 0.35 μm.

The requirement of physical holes in the mask excludes the use of ring shaped structures. To allow for ring shaped structures the pattern is split into two complementary half masks which are placed side by side on one mask substrate. Fig. 8 shows a simple example of a complementary pattern split.

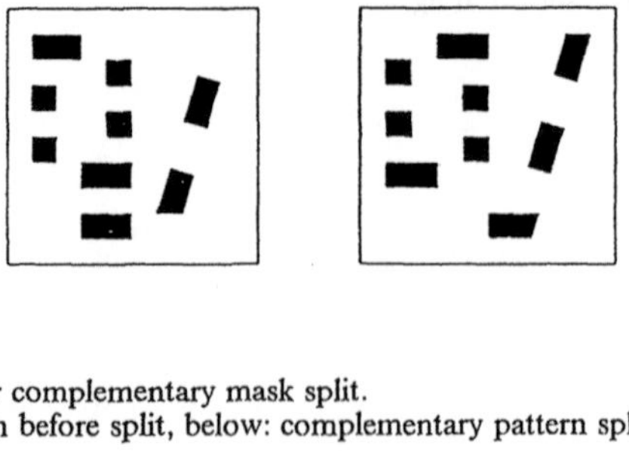

Figure 8: Example for complementary mask split.
Top: Pattern before split, below: complementary pattern split.

The distance between the two complementary patterns in the mask is equal to the distances of chips on the wafer. Thus, the two patterns can be exposed to adjacent chip positions at one table step.

Mask heating

An electron beam current of 10 μA has been chosen for the exposure. The absorbed beam energy is about 40 mW which is dissipated by heat conduction only. The temperature increase during the exposure is about 10 K. Finite element analysis was used to determine the thermally induced distortions across the mask area. Maximum distortions are about 40 nm. Fortunately, it was found that this kind of distortions is (nearly) pattern independent so that a simple correction program can be installed to reduce thermal distortions to negligible amounts.

Throughput estimate

Throughput is a very important aspect as it determines the economy.

For an EBP manufacturing tool the following throughput specifications were derived for 200 mm wafers with chip sizes expected for 256 Mbit chips ($\approx$ 2.5 cm^2). (Table 2) About 90 chips are assumed on a wafer.

Wafer load and unload:	30 sec
Global align:	6 sec
Exposure time (90 x 0.5 sec):	45 sec
Chip registration (90 x 0.05 sec):	4.5 sec
Table stepping time (110 x 0.2 sec):	22 sec
Total time per wafer:	107.5 sec

Table 2: Cycle times for wafer exposure

This estimate results in a throughput of about 33 wafers/h.

Positional accuracy of EBP

Test devices of integrated circuits with 4 exposure levels have been exposed in the EBP lab tool. One device was designed to test the capabilities of the tool at 0.5 μm minimum linewidth. An important criterion of an exposure tool is its capability in achieving level to level overlay.

The overlay error (O/L) can be calculated from its main contributors mask error (M), offset error (O) between the registration mark and the chip, tool errors (T) and alignment errors (A). Each contribution is independent from the other so that the result is given by the ATOM-formula, /3/:

$$O/L = \sqrt{2}\sqrt{A^2 + T^2 + O^2 + M^2} \quad [1]$$

The contributions A, T, O, M were determined separately: (Table 3)

Alignment error	A = 17 nm
Tool errors	T = 10 nm
Offset error	O = 10 nm
Mask error (mean)	M = 22 nm

Table 3: Contributors to overlay. (All numbers are 1 σ values.)

The ATOM-formula (1) with these values results in an overlay error O/L = 44 nm. The actual measured results on the exposed wafers are listed in table 4:

Levels	x-value	y-value
Poly/ROX	34 nm	43 nm
Contact/ROX	30 nm	36 nm
Contact/Poly	27 nm	32 nm
Metal/Contact	47 nm	45 nm

Table 4: Measured overlay results. (1 σ values)

The errors of pattern placement in the physical mask were measured on a NIKON2i measurement system. These measurements yielded a (1 σ) error of 73 nm. A computer program was used to simulate the correction feature of the EBP tool. The simulation indicated that the correction feature would reduce the average mask distortion error by a factor of three down to 22 nm - the value used in table 4. The excellent agreement between the measured values in table 4 and the result of the ATOM-formula indicates that the tool correction must have worked correspondingly. Without the mask correction our computer simulation predicted only a (1 σ) overlay value of 102 nm. When the device exercise was finished we found out that during the Metal level exposure the tool error corrections were erroneously not used. This explains partly the larger values for Metal/Contact overlay in table 5.

Conclusions

A key aspect for future lithography for integrated circuit fabrication is its capability to achieve the required positional accuracy. Electron beam proximity with its unique correction capability has been demonstrated. A variety of tool, mask or process based error sources have been identified which can be compensated by the correction scheme. Therefore, it is been regarded as an promising approach to meet lithography groundrules after the optics era.

References

/1/ J. Keyser, W. Kulcke: Pattern Distortions in EBP Stencil Masks, Microelec. Engin., Vol 11, 1990, pp 363-363.

/2/ H. Bohlen, W.Kulcke: Evaluation of Level-to-Level Overlay in Lithography Systems, Solid State Techn., Feb. 1990, pp. 59

/3/ P. Nehmiz, U. Behringer, H. Bohlen, and M. Kallmeyer: Registration mark detection in electron beam proximity printing J. Vac. Sci. Technol.B 1 (4), Oct.-Dec. 1983, p. 1023

Refinements in Postprocess Gaging with Feedback in the Production of Diamond Turned Optics

J. W. Roblee, Y. L. Chen, K. B. Becker, and K. H. Fiedler

Carl Zeiss
D - 7082 Oberkochen

Over the past ten years Carl Zeiss has developed and operated two similar diamond turning machines[1-2]. These three-axis machines are designated "A400" because they were designed specifically for manufacturing aspheric optics, and their x and z slides each have **400** mm of travel. The unique configuration of the machine, shown in Fig. 1, has several advantages. The two slides are stacked one upon the other, with the upper slide carrying the air-bearing workpiece spindle. This slide configuration does not sacrifice stiffness, however, because the upper slide is supported by hydrostatic bearings riding directly on the large granite base of the machine. In addition, the guideways on the lower slide are made up of very stiff, opposed-pad, hydrostatic bearings. Finally, the two slides are driven by a stiff hydrostatic leadscrew with a pitch diameter of 103 mm. Furthermore, the hydrostatic nut is made self-aligning by means of a spherical hydrostatic bearing. Both lead screws are supported by hydrostatic bearings and driven by high torque, DC servo motors.

This stacked slide arrangement permits the third, rotary axis to be mounted directly to the granite base. It is also guided by hydrostatic bearings, and it carries the tool post. With the rotary table, the tip of the tool can be maintained normal to the machined surface. This minimizes the machine's sensitivity to imperfections of the cutting edge of the diamond tool. It also facilitates contour machining with large, negative rake angles. The open configuration of the machine allows easy access to the workpiece and enhances the versatility of the machine by providing space for custom tooling and fixturing. One A400 can swing workpieces 700 mm in diameter while the other can swing 1200 mm diameter workpieces, and both can accommodate 400 mm lengths. Perhaps the greatest advantage of this machine configuration is that it permits the use of a simple and very accurate metrology system. By employing two Zerodur straightedges, kinematically mounted to the upper slide, the x and z straightness and position errors can be compensated for, without any Abbe offset, by only two laser interferometers directly in line with the tool. Both interferometers operate with a 10 nm resolution.

Other important features of the machines include its support on pneumatic isolators with 2 Hz natural frequencies vertically and horizontally. Also, the bearings are pressurized with a low pulsation oil supply that is temperature controlled to 0,01 °C. The oil escaping the bearings is then allowed to flood over the granite base. Furthermore, each machine is located in individual rooms with a vertical air shower having 0,01 °C temperature control. Both machines are then operated from outside their rooms, with monitoring via a television camera, so the heat from the control electronics and the operator do not influence the machine's stability.

Both machines are located side by side in a closely integrated facility dedicated to the production of optics. Next door is located a flycutting-type, diamond-turning machine and an optical projection device for setting the diamond tools in their mounts before going on the machine. With this system, a tool can be quickly located on the machine's precalibrated tool post with a tolerance of 2 μm. Completing the facility are two more adjacent rooms that are also temperature controlled for the metrology equipment they contain. The range of available equipment plays a crucial role in assuring the quality of the manufactured optics.

In our facility the metrology equipment has also become an integral part of the machining process. Using a procedure known as Postprocess Gaging with Feedback[3] (PPGWF), the accuracy of the manufactured optic can be improved while simplifying the diamond turning operation. The PPGWF process entails the measurement of a workpiece after an initial cut and the utilization of the measured surface errors to modify the control data for the turning machine. The modified control data is then used to recut the initial workpiece, as well as subsequent workpieces in a series production. All of the repeatable errors of the turning process can then be compensated for. Not only are the machine's error motions compensated, but any workpiece specific errors are compensated for as well. Examples include those repeatable errors resulting from fixture induced distortions of the workpiece, from deformations due to the weight of the fixture and workpiece and due to cutting forces, and errors from mislocation of the diamond tool. As an integral part of the production process, PPGWF permits the use of faster and simpler techniques for workpiece fixturing and tool mounting. If a specified procedure is rigourously followed, even errors due to thermal growth of the workpiece spindle can be partially compensated with PPGWF. This can shorten production time by decreasing the necessary warm up time for the spindle.

At Carl Zeiss PPGWF is an everyday process that is often performed by the machinists themselves. With a closely integrated metrology capability the process is rapid (usually less than two hours) and reliable. By regularly involving the machinists in PPGWF, they become more aware of the health of their machine and the success of different machining techniques. Therefore, PPGWF improves their skills and provides a monitoring function on the state of the machine. The alternative to PPGWF is the precalibration of the turning machine. This requires that all the machine's errors be carefully measured before machining, which is not a simple task. Furthermore, precalibration cannot compensate for workpiece specific errors, and it must be repeated periodically as the machine ages.

In our optics production facility we have three different metrology instruments that are dedicated to supporting PPGWF on our A400s. These instruments were carefully selected for this task, and they have complementary capabilities for different types of workpieces. This flexibility allows us to select the most cost effective approach to a particular machining problem. An important consideration for PPGWF is the spatial frequency of the errors that need to be corrected. Fortunately, with the utilization of straightness references on the A400s, the largest error sources have a large spatial wavelength, such as the rotary table motion, squareness, parallelism, fixturing distortions and tool location. Oftentimes, the error correction can be adequately represented by a fourth order polynomial. As few as 40 measurement points on a radial profile of the workpiece are all that is needed to be able to average noisy data and to identify outlying measurement values (i.e. from dirt).

The easiest to use, most general purpose and the fastest metrology instrument in our facility is the UMM 500, 3D coordinate measuring machine which is sold by Carl Zeiss. With a 3D touch probe and 0,2 μm resolution, this instrument can measure absolute size to an accuracy of 0,5 μm P-V within a working volume of 200 x 300 x 500 mm^3. This level of accuracy is adequate for the majority of infrared optics produced in our facility. Programming, setting up the machine running, a measurement and evaluating the data is fast and easy with the software packages available on its integrated computer system.

To evaluate more accurate workpieces, a high resolution (10 nm) coordinate measuring machine was specially developed at Carl Zeiss. This M400 machine[4], with a 1D contact probe, uses five laser interferometers to measure workpieces in a volume of 400 x 600 x 200 mm^3. It can operate in a scanning mode, without stop-and-starts, and it has a repeatability of 30 nm P-V. Because this is a special purpose machine, it is more time consuming to program, set up and operate. It is also used as a cross-checking device when other measuring means give anomalous results.

When making spherical optics, or if the number or accuracy requirements of an aspheric optic warrant the expense of a custom-made null lens, an interferometer is the metrology instrument of choice. Once set up, it is fast and easy to use. The interferometer we currently use was custom made by Carl Zeiss. With a CCD camera interfaced to a personal computer, the interferometer's software can fit Zernike polynomials to the measured fringes. The accuracy of the instrument is better than 30 nm P-V. After 5 minutes of processing time, the interferometer can generate correction data for the A400s from the low-order Zernike polynomials. In the near future, our production facility will take delivery of a new product from Carl Zeiss, the Direct 100 interferometer system. With this more versatile and easier to use system, we expect to obtain measurement accuracies better than 5 nm P-V.

The range of measuring possibilities that we have readily available in our optics production facility has profoundly affected the operation of our diamond turning machines. They promote a culture dedicated to precision that would not be possible without the frequent feedback of information from these devices. With the close coordination that is possible in an integrated facility and with the cross training of machinists and technicians, we have made PPGWF an efficient, accurate and reliable process. Consequently, we believe that we have overcome many of the labor-intensive and time-consuming aspects that have plagued PPGWF in the past[3].

References

1. Becker, K.: Ultrapräzisionsmaschinen in der Fertigung der Optischen Industrie. 6. Internationales Braunschweiger Feinbearbeitungskolloquium, Technische Universität Braunschweig, FRG, Sept. 1990.

2. Becker, K.; Heynacher, E.: Bearbeitung optischer Flächen mit Einzeldiamantwerkzeugen. Feinwerktechnik & Meßtechnik 95 (1987) 4.

3. Thompson, D. C.: Postprocess Gaging with Feedback. Technology of Machine Tools, Machine Tool Task Force, Lawrence Livermore National Laboratory, Livermore, CA, USA, UCRL-52960-5, Oct. 1980.

4. Becker, K.; Heynacher, E.: M 400 - A Coordinate Measuring Machine with 10 nm Resolution. SPIE, Vol. 802, Den Haag, NL, 1987.

Fig. 1 Zeiss A 400 diamond turning machine. 1 - spindle housing, 2 - tool post, 3 - hydrostatic leadscrew.

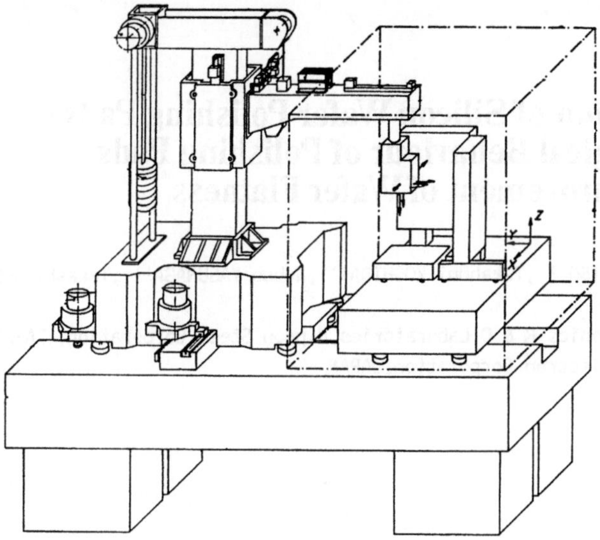

Fig. 2 Zeiss M 400 high resolution coordinate measuring machine.

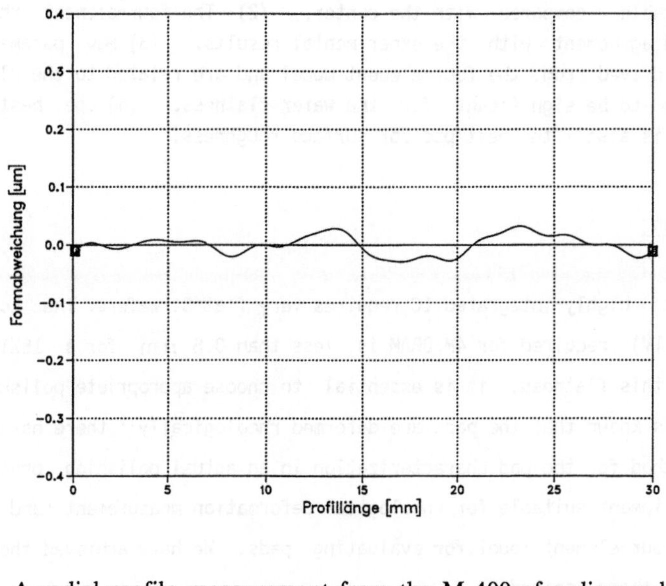

Fig. 3 A radial profile measurement from the M 400 of a diamond-turned optic made on an A400 after a PPGWF correction.

Evaluation of Silicon Wafer Polishing Pads – Rheological Behaviour of Polishing Pads and Improvement of Wafer Flatness

Yamato SAMITSU[1)] ,Takahumi YOSHIDA[2)] ,Nobuo YASUNAGA[1)] ,Takashi OHMOTO[2)] ,Sadamu HORIE[2)]

1)Electronics R & D Laboratories,Nippon Steel Corporation, JAPAN
2)NSCE Electron Corporation,JAPAN

Summary

We have developed some new easy-to-use equipment for measuring quantitatively the rheological deformation characteristics of the polishing pads.The main results, using the measuring equipment, are as follows : (1)With the processing time, the value of th elastic deformation of the pads becomes higher at the peripheral region of the polishing table compared with the center, (2) The four-element rheology model showed good agreement with the experimental results. (3) New parameter α and β, which are drived from the four-element model and are related to the elastic modulus, were found to be significant for the wafer flatness. (4) the best pad for the flatness, is also the best pad for surface roughness.

Introduction

The recent highly-integrated IC requires very flat Si wafers. The Local Thickness Variation(LTV) required for 4M DRAM is less than 0.8 μm for a 15X15mm chip size. To achieve this flatness, it is essential to choose appropriete polishing pads. Although it is known that the pads are deformed rheologically[1)],there has been no reproducible method for the pad characterization in an actual polishing process. We report a new equipment suitable for rheological deformation measurement and new parameters based on four-element model for evaluating pads. We have achieved the flatness for 4M DRAM by these methods.

Experimental Procedures

Rheological characteristics measurement.[2)]

The set-up of the equipment for measuring the rheological characteristics of the

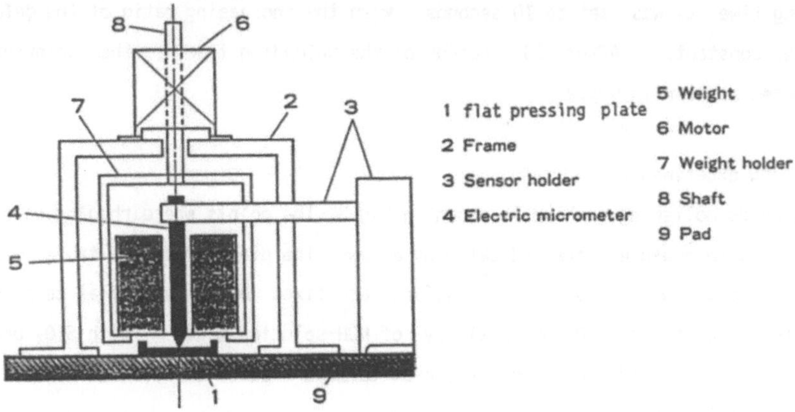

Fig.1 Set up for rheological chracteristics mesasurent

polishing pads is shown in Fig.1. Weight ⑤ is set to move vertically by the weight holder ⑦ which is driven by a motor ⑥. When the weight is set on a flat pressing plate① which is placed on the pad ⑨, the pad deforms. The pad deformation can be detected, measuring the plate's displacement by electric micrometer④.

Pad deformation process.

The typical pad deformation process is shown by a solid line in Fig.2. In this experiment, the step load of 190g/cm² was applied to the pressing plate. The weight

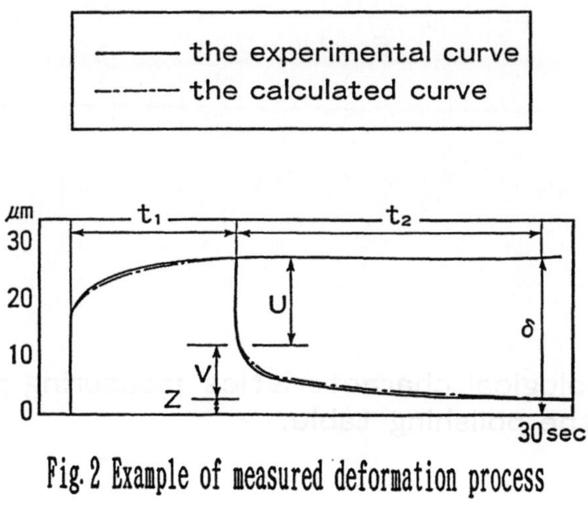

Fig.2 Example of measured deformation process

loading time t_1 was set to 10 seconds, when the increasing ratio of the deformation becomes constant. After 20 seconds of the unloading time t_2, the deformation is saturated as seen in Fig.2.

Polishing experiments

The 4 heads polishing machine is shown in Fig.3. The points where rheological characteristics were measured, are indicated by arrows. The diameter of the table is 1200mm ϕ. At each head, 5 pieces of $5''$ wafers are fixed by wax on a flat ceramic plate. The table rotated at 60 rpm. A slurry of KOH-solution mixed with SiO_2 powder was used. The pH value of the slurry was kept to 10.5 ~11.0. It took 20 minutes for each batch of polishing.

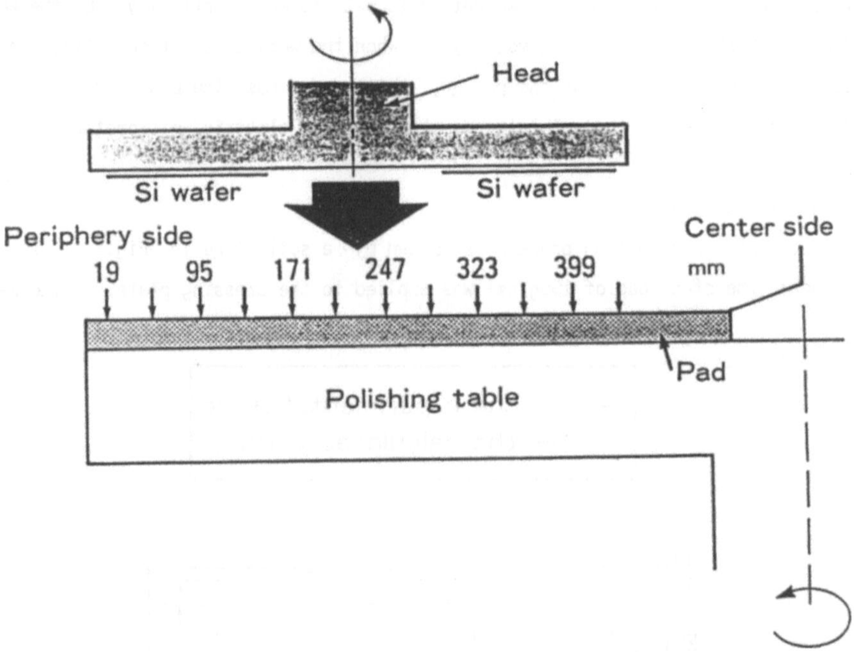

Fig.3 Pheological characteristics measuring position on the polishing table.

Results and Discussion

Rheological characteristics of 4 different kinds of pads after 10 batches of polishing were are shown in Fig.4. U,V,and Z indicate instantanous elastic deformation retarded elastic deformation and permanent deformation,respectively. Total deformation δ (=U+V+Z) is largest in pad B,and becomes smaller in order,pad A, pad D,and pad C. The differences of δ at the peripheral side and at the center are large in pad A, pad B and pad C. These differences are clearly influenced by U .

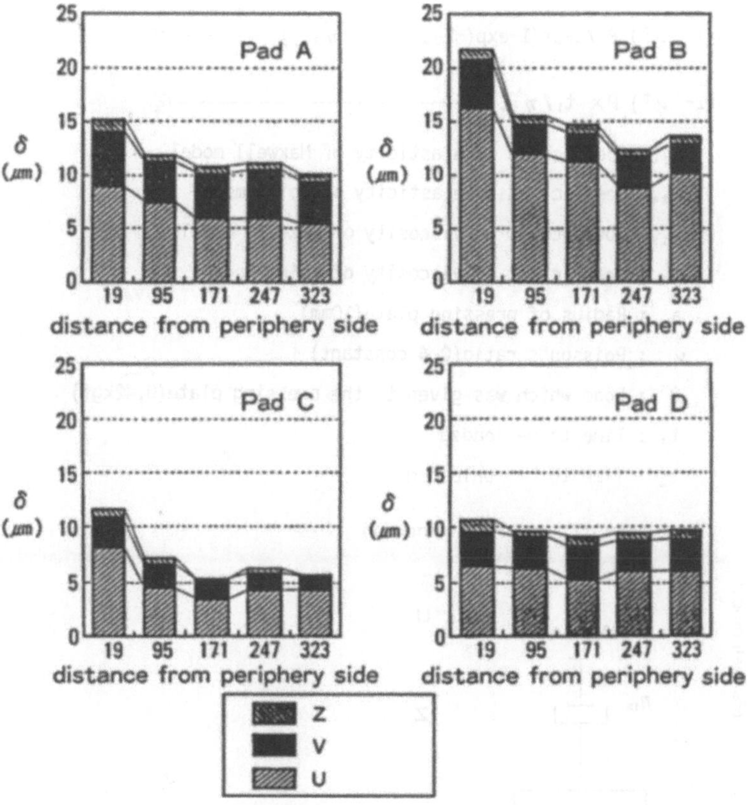

Fig.4 Rheological characteristics of 4 different kinds of pad.
(comparison between periphery side and center side)

Four-element model.

The solid line in Fig.2 shows a typical rheological phenomenon which can be approximately represented by the four-element model, which consists of the combination of the Maxwell and Voigt models as indicated in Fig 5. We can obtain the coefficients of elasticity and viscosity of pads from this model and Timoshenko's theory of elasticity, as follows:

$$\delta = U + V + Z \qquad ①$$

$$2aU = (1-\nu^2)\,P/E_{1*} \qquad ②$$

$$2aV = (1-\nu^2)\,P/E_{2*}\times(1-\exp(-E_{2*}\times t_2/\eta_{2*})) \qquad ③$$

$$2aZ = (1-\nu^2)\,P\times t_1/\eta_{1*} \qquad ④$$

where E_{1*}: Coefficient of elasticity of Maxwell model

E_{2*}: Coefficient of elasticity of Voigt model

η_{1*}: Coefficient of viscosity of Maxwell model

η_{2*}: Coefficient of viscosity of Voigt model

a : Radius of pressing plate(10mm)

ν : Poisson's ratio(0.4 constant)

P : Load which was given to the pressing plate(0.42kgf)

t_1 : Time to be loaded

t_2 : Time to be unloaded

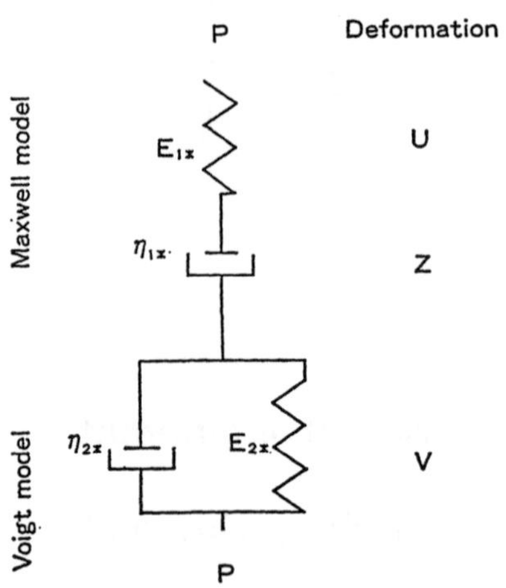

Fig. 5 Four-element model

We can obtain E_{1*} by substituting the value of U measured from Fig.2 into equation ②. When t_2 is infinite, equation③ can be written as follows:

$$2aV = (1-\nu^2)P/E_{2*} \quad ⑤$$

Fig.2 shows that V becomes almost constant, when t_2 is 20 seconds. Therefore E_{2*} can be obtained from equation ⑤ using the value of V at $t_2 = 20$ sec. η_{2*} is obtained by substituting t_2 and V into equation③ , after E_{2*} is calculated from equation ⑤. Z can be read off from Fig.2. Consequently η_{1*} is obtained from equation④. The deformation which was calculated from these processes is shown as a dotted line in Fig.2. It agrees well with the solid line obtained from the experiment.

Discussion

We have calculated the various coefficients for each pad according to the equations ①~⑤. The values of E_{1*} are shown in Fig.6. The E_{1*} of pad D is relatively large and constant, compared with the others. The E_{1*} of pad C is large at the center, but is small on the peripery, that is, E_{1*} varies widely across the table.

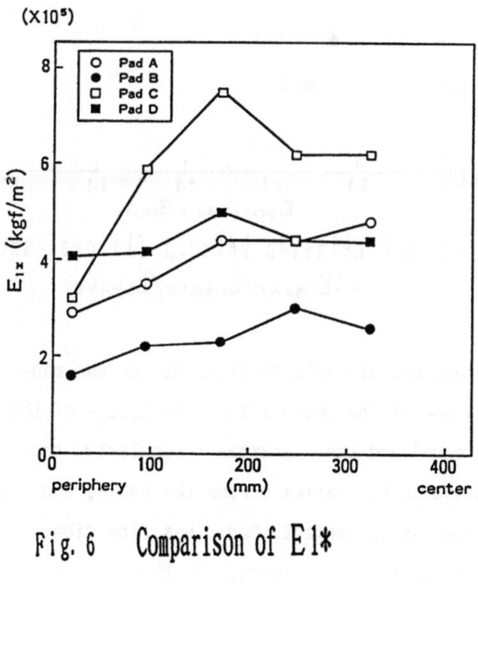

Fig. 6 Comparison of E1*

Here ,we have introduced a new parameter α, which is suitable for evaluating the wafer flatness, shown as follows:

$$\alpha = E_{1*MAX} / E_{1*MIN} / E_{1*AVE}$$

where E_{1*MAX} and E_{1*MIN} are the maximum and minimum value of E_{1*} respectively. E_{1*MAX} / E_{1*MIN} indicates the degree of scattering of E_{1*}. E_{1*AVE} is the average value of E_{1*}.

The relation between α and the flatness is shown in Fig.7. The Non-linear Thickness Variation(NTV) decreases when the value of α decreases.

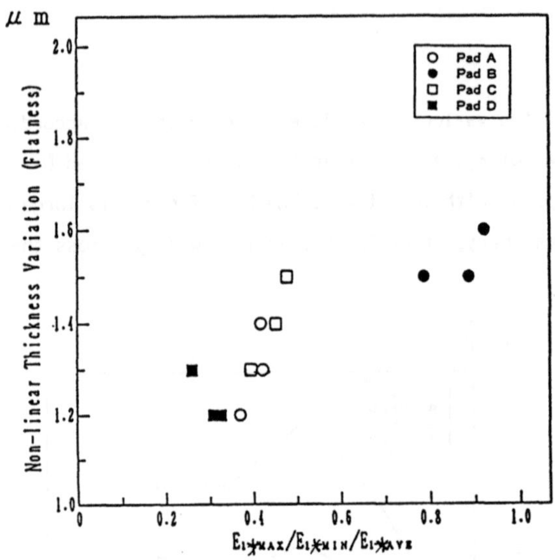

Fig. 7 Relation between flatness and α : $E_{1*MAX}/E_{1*MIN}/E_{1*AVE}$

Photo.1 shows SEM photographs of pad's section. We can see many fibers peeped out of on the pad B. This is one of the reasons that the E_{1*AVE} of the pad B is small, and fibers'length is not equal and they scatter irregularly, therfore, the pad's spring effect, which is relative to E_{1*} varies across the table, and E_{1*MAX} / E_{1*MIN} becomes large, consequently the α is largest. Pad c has fine fibers on the surface. As the same reason with the pad B, it's α is large. On the other hand, almost no fibers are seen on the surfaces of pad A and pad D. Their surface's seem uniform. Therfore, E_{1*MAX} / E_{1*MIN} are small and the values of α are small.

The parameter α is not sufficient to evaluate the defference of the pad A and pad C. Because it is consist of only Maxwell model. So another parameter β has been introduced. This parameter has relation between Voigt model and Maxwell model.
The β is shown as follows:

$$\beta = E_{1*MAX} / E_{2*MAX}$$

$$\beta' = E_{1*ave} / E_{2*ave}$$

When E_{1*MAX} is large,pad deformation is instantaneous and small. When E_{2*MAX} which is the maximum value of E_{2*},is small, pad deformation is retard and large. This means that the high E_{2*}? pad has good cushion. The relation between β and the flatness is shown in Fig.8. The flatness is improved, when β decreases. We have calculated also β'and got almost same result as the β. However the differnce of pad A and pad D is not clear.
As to the other coefficients η_{1*} and η_{2*} ,no clear relation with flatness was found.

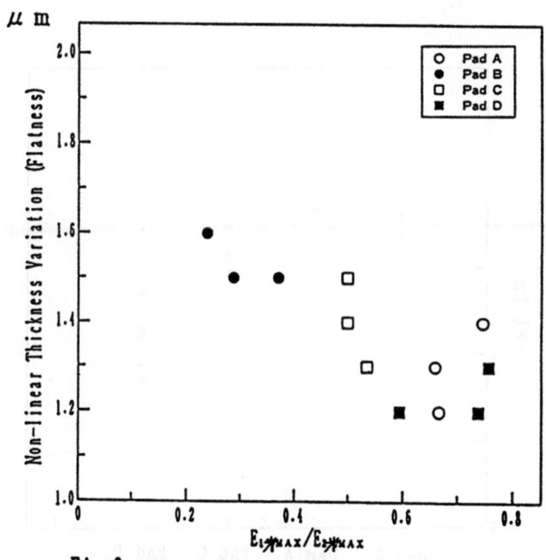

Fig.8 Relation between flatness and β : E_{1*MAX} / E_{2*MAX}

Surface roughneses of Si wafers polished are shown in Fig.9. (Rt is average of ten peak points.) Pad D showed smallest roughness of the four .

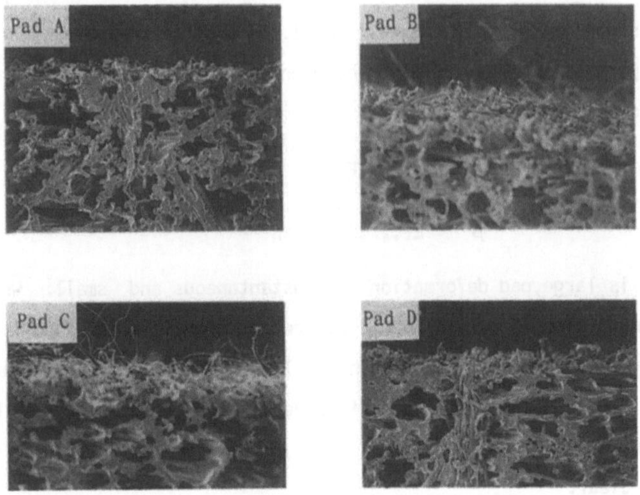

Photo. 1 SEM photographs of pad section (×60)

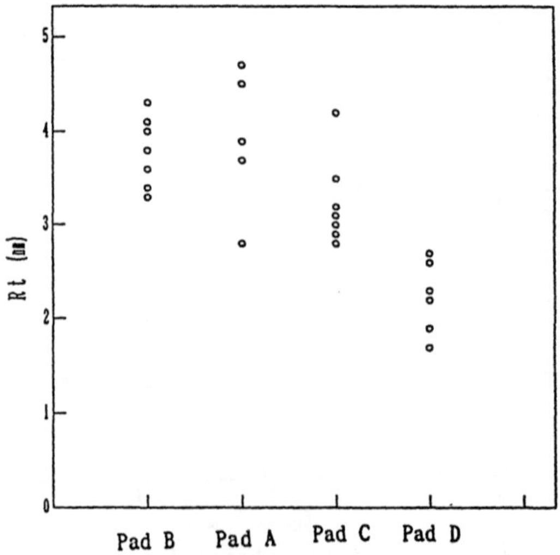

Fig. 9 Comparision of surface Roughness

Conclusions.

We have developed a new rheological characteristics measuring equipment. By analyzing the pad's deformation with a four-element model, suitable parameter α and β which are significant for evaluating Si wafer flatness, were found.
The pad, which gives better flatness, gives better surface roughness.

References

1) T.kurobe et al., Journal of Japan Soc.Prec.Engg.,p55(1983) Sept.
2) T.Yoshida et al.,Proc.Spring Meeting of Japan Soc.Prec.Engg.,p211(1989)

High Aspect Ratio Laser Machining of Composite Ceramics

R. T. Kerth
IBM
Storage Systems Product Division
San Jose, California USA

Summary

High aspect ratio laser machining of composite ceramics has been demonstrated using a pulsed laser wavelength that is largely transparent to one of the composite constituents of the ceramic and highly absorbed in the other. Machined slots 50 microns wide and 400 microns deep are possible as the result of the deep penetration of the laser radiation due to the "semi-transparent" nature of the substrate. This work is presented in two parts. The first is a study of the single pulse interaction of the laser with pure samples of Alumina and separately Titanium Carbide which are the two constituents found in the machined sample. These results are compared to the single pulse interaction with the composite. Finally, results of the high aspect ratio machining on the composite are presented and a probable mechanism is proposed.

Introduction

With the development of reliable high power lasers came their application in various machining activities. In general, the machining application would match the laser wavelength to a strongly absorbing region of the material to be machined (1,2). This matching for highest absorption, results in the energetics of the laser ablation process being concentrated in the top few 1000 Angstroms of the material. In these cases the machining depth does not increase linearly with the laser fluence but saturates at some point with any increase in pulse energy merely increasing the velocity of the escaping ablated material.

If narrow (on the order of 50 microns) but deep machined grooves are desired, the near surface absorption of the material interferes with the machining operation in the following way:

Multiple laser pulses are overlapped with a small displacement of each succeeding pulse to form a groove. As the groove becomes deeper the ablated material does not exit from the top of the groove but deposits on the wall. The next pulse is absorbed by this debris on the side wall and is redeposited elsewhere. Of course, some of the material escapes on the subsequent pulses but the process efficiency is degraded.

In the approach presented here the laser wavelength is chosen such that the energy is absorbed in the Titanium Carbide which represents 30% of the material volume. The remaining 70% of the ceramic volume is Alumina which is transparent to the laser wavelength. The laser pulse is absorbed by the TiC which ablates and in so doing removes the Alumina with it. In the process of making grooves as described above, each pulse penetrates several microns into the surface of this composite for efficient material removal. In addition any wall debris is now largely transparent to the next incoming pulse as it is largely Alumina and devoid of TiC.

Experimental Apparatus and Procedure

The following experiments were completed on an Electro Scientific Industries Model 44 Laser Trimming tool. This tool incorporates X-Y scanning of a pulsed YAG Laser beam . The X-Y stage has positional accuracy of 2.5 microns in both directions and can deliver a laser pulse at integral spacings of the 2.5 microns. The laser is a Q-switched Nd-YAG at 1064 nm. with random polarization delivering up to .5 mJ/pulse with a less than 35 nsec full width, half max duration.

The procedure for the initial materials investigation was to focus a 150 mJ pulse to a 35 micron spot and ablate lapped samples of Alumina, TiC, and 70%-30% "AlTiC" with single and double pulses. The reason for the double pulse was to minimize the possibility of reflection from the lapped surfaces which nominally had a one nano-meter average roughness. These pulses were then profiled, sectioned and photographed.

With the same laser conditions, a sample of the composite material was exposed to single train of pulses spaced 2.5 microns apart. Multiple passes over the same groove were made without changing conditions. The samples were SEM'ed and sectioned.

Experimental Results

The results of the single and double pulse exposures are shown in Figures 1-3. The pulses shown are 75 microns apart and nominally 30 microns in diameter. The TiC ablation in Figure 1, showed non-uniform etching with a large amount of material spatter. This may have been due to the fact that this TiC has 3 to 4% porosity which may have allowed deeper than expected penetration of the laser. Further investigation is needed.

In Figure 2, the Alumina sample tested here was hot isostatically pressed during manufacturing. This sample is very representative of the Alumina in the final composite material. This Alumina was processed under the same conditions as the TiC above. As shown in Figure 2, the Alumina did not ablate on every pulse. Those areas that did show laser ablation were shallow and very smooth compared to previous materials tested. The ablation of some spots but not others is probably due to absorbing impurities near the surface of the Alumina. Generally, the fact that the Alumina did not ablate in some areas is consistent with it being transparent to this wavelength.

Finally, the ablation of the "AlTiC" exhibits uniform etching, as shown in Figure 3, with well controlled deposition of the ablated material around the top of the hole. The result is somewhat a cross between the two other materials. There is not the spattering, characteristic of the TiC but the etch pocket is not as uniform as the Alumina ablation.

From the profiled traces of each etched spot, the volume of material was calculated. The results of these measurements are shown in Figure 4. The final column calculates the volume of TiC removed from the AlTiC by simply multiplying the total volume removed by the ratio of the TiC in the material (30%). The two numbers are within 35% of each other, which might also be accounted for by the high porosity of the TiC leading to a higher etching than lower porosity material.

The final experiment was to show deep narrow trenching of the composite "AlTiC" material and the ability to remove debris from a very deep trench. Figure 5 shows SEM photographs of two trenches and the debris that was removed. The left trench was created using 2.5 micron spaced pulses and 25 passes were made. The right hand trench was the same condition but 5 passes were made. Figure 4 shows the sectioned grooves created in this experiment. A calculation of the area of the grooves, which is proportional to the volume removed per unit length, shows that volume of material removed per pass is reduced by only 12% for the deeper groove. This reduction may be due somewhat to side wall debris but also due to the defocusing of the beam as the ablated surface moves down into the part. In this experiment no attempt is made to re-focus the beam down into the part during the ablation.

Conclusions

Efficient machining of composite ceramics is possible by choosing a laser wavelength that is absorbed strongly by only one component of the composite. The rate of material removal of the composite is close to the rate of the pure absorbing component divided by it's volume percentage in the composite.

Acknowledgements

The author wished to extend his appreciation to M. Knight, B. Auser, Dr. W. Leung, and Dr. S. Lewis for their assistance in the preparation and analysis of the parts.

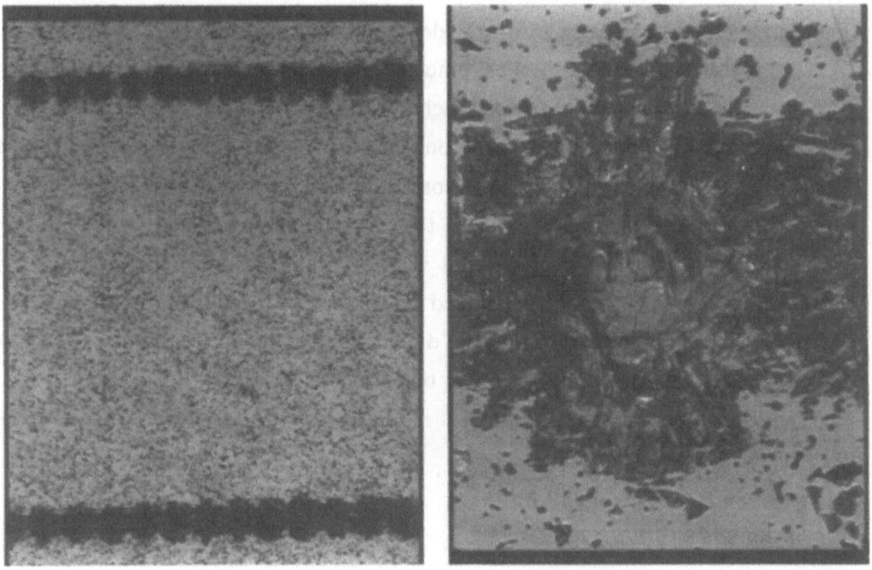

Figure 1. Laser ablated Titanium Carbide showing spattering around a 30 micron spot that is 1.8 microns deep.

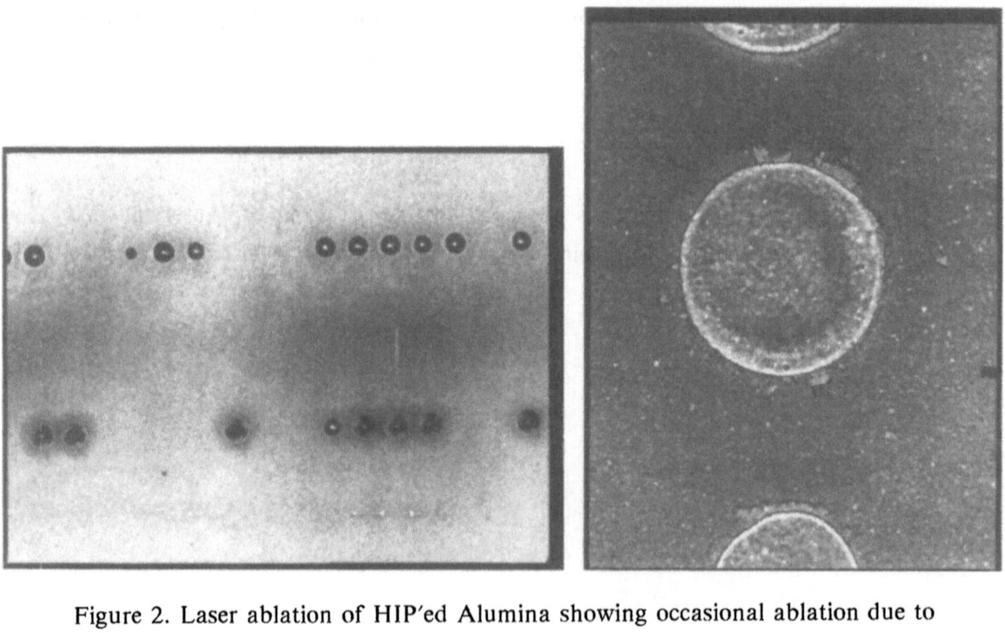

Figure 2. Laser ablation of HIP'ed Alumina showing occasional ablation due to transparent nature of the substrate.

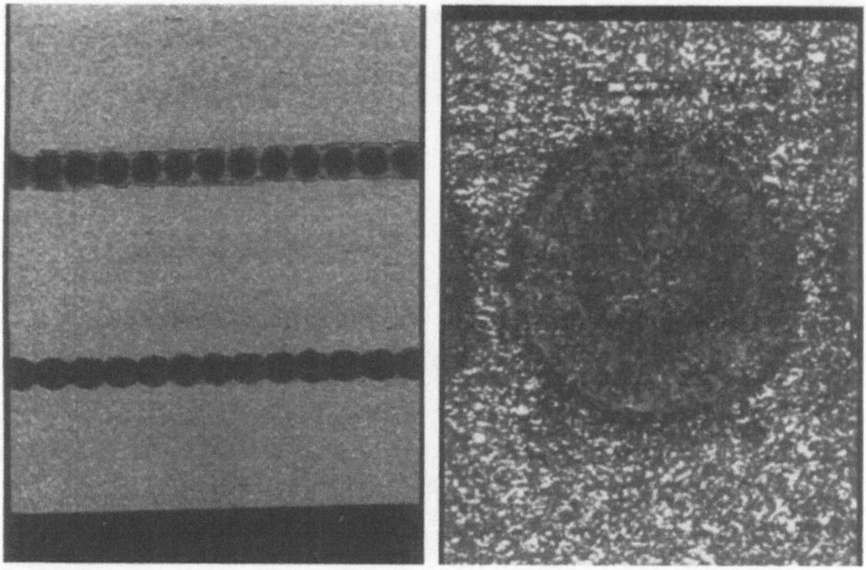

Figure 3. Laser ablated composite ceramic (30% TiC, 70% Alumina) showing 2.5 microns of material removal with little spatter.

Sample	Depth (micron)	Width (micron)	Volume Removed	Volume TiC Removed
TiC	1.8	30	424	424
Alumina	1.0	32	N/A	
"Al-TiC"	2.5	45	1047	314

Figure 4. Comparative ablation of three ceramic samples. Volume of ceramic ablated is calculated.

Figure 5. SEM photograph of high aspect ratio grooves and sectioned parts showing wall angle and depth.

References

1. Chryssolouris, G; Bredt, J.; Kordas, S.; Wilson, E.; Theoretical Aspects of Laser Machine Tool, Jour. of Engineering Industry; Feb. 88, Vol. 110.

2. Ito, S; Nakamura, M; Kanematsu, W.; Machining of High Performance Ceramics, Bull. Japan Soc. of Prec. Eng.; Sept. 87, Vol. 21, No. 3.

Particle Beam Figuring – An Ideal Tool for Precision Figuring of Optics?

M. Weiser, C. Kübler, K.H. Fiedler and K.F. Beckstette

Carl Zeiss, Postfach 1380, D-7080 Oberkochen, FRG

Abstract

Since several years Carl Zeiss manufactures high tech aspheric optical elements quite successfully by means of a CNC polishing mashine. Nevertheless mechanical finishing methods of surfaces find their limits in modelling the edges of the workpieces, in the amount of subsurface damage as well as contaminations of the surface with abrasives and in poor convergence values. One approach to overcome these limits is the use of a particle beam as a tool. Parameters like removal function and removal rate now are tunable almost independently. Convergence values as high as ten or more seem to be within reach. The Carl Zeiss experimental set-up will be presented and first results will be discussed with respect to those of the Computer Controlled Polishing.

Introduction

The aim of the polishing processing is fourfold. Firstly removing the damage-layer due to grinding, secondly reducing the surface micro-roughness in the nm or sub-nm range, thirdly shaping the surface of the workpiece in total in the μm and sub-μm range, and fourthly correcting the workpiece surface locally.

The first three requirements are met by traditional optical techniques. However, for high tech optical elements the fourth gets more and more necesssary due to the demands of now-a-days specifications. One approach, the Computer Controlled Polishing (CCP), is based on experiences of conventional techniques. The obviously new feature of a competing approach, the Particle Beam Figuring (PBF), is the avoidance of any physical contact between tool and workpiece. As CCP allready is a well established methode /1-3/, some properties of PBF - such as IBF (i.e. *Ion* Beam Figurung /4/) - will be outlined. These properties have to be illuminated with respect to the requirements.

Empirical considerations

First of all, for a better understanding of the scenario, the process has to be described on an atomic scale (Fig.1), (further information is available in literature /5/): Sputtering is defined as the removal of atoms from a solid surface due to energetic particle bombardment. The sputtering yield, Y, is defined as the number of ejected atoms per incident ion. This yield is a function of the properties of the system ion and the substrate. The most prominent parameter for the sputter yield function is the energy level per incident ion, which usually gives the baseline for operating a process under consideration:

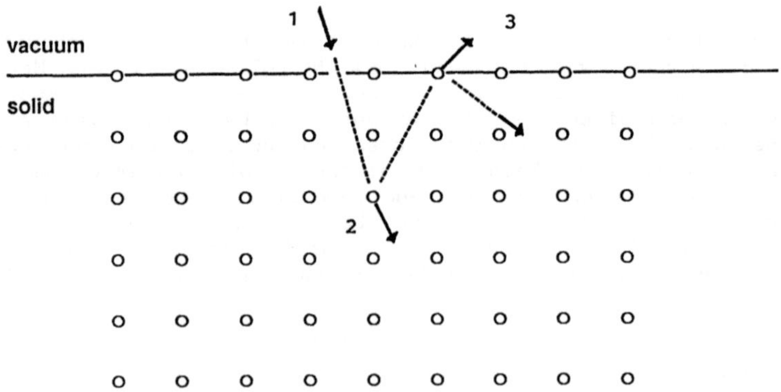

Fig.1 Possible emission mechanism at low particle energy ('few collision approach'). (1) incoming ion, (2) backscattering event: creating of a hole - interstitial (*Frenkel*) - pair, and (3) collision of the primary ion with a surface atom, the latter one will be ejected from the surface, i.e. sputtered.

The energy-yield-relation can be described as follows (we begin at the low energy end): The threshold - energy, E_{thresh}, where sputtering starts, depends on the surface binding energy or sublimation energy of the substrate[1]. The threshold energy typically is about 50 eV. This is one order of magnitude

[1] E_{thresh} depends strongly on the angle of incidence, too. For the sake of clarity we inspect further on the energy dependence of the normal incidence only.

higher than sublimation energy values, because a big portion of the incoming particle's impetus is deflected at the lattice. Within the low-energy interval - above 50 eV and up to some 100 eV - Y is increasing with increasing energy from zero to the order of 0.1 /5,6/. The broad maximum of the yield curve with values of the order of ten is reached at energy levels between 10 and 100 keV depending on the beam-target-combination. At even higher ion energies (MeV) the yield curve is gradually dropping towards zero again. Obviously the dependence of the yield on the ion energy follows approximately the nuclear stopping power, S_n /7,8/[2].

For figuring of high tech optics, the requirements to be met, give kind of an energetic baseline and a frame for other operation conditions. Their utility have to be prooved experimentally.

Because of the demand of production efficiency we look for (1) a predictable, fast, and 'smooth' process, with a high convergence value (to reach the final desired conture with low iteration numbers). The optics' surfaces should stay free of (2) contaminations and (3) subsurface damage. Changes of the working conditions must not affect the removal function (4) (e.g. processing from the middle to the edge of a workpiece or vice versa; see below).

ad (1) The highest possible yield is wanted. This can be reached by chosing a suitable ion energy level, a high current value and a stable ion current density distribution for a defined area size. The resulting tool function convoluted with a relative scanning motion of the workpiece with respect to the ion beam gives a predicted etching of the surface, similar to a CCP process /9/.

ad (2) This condition should be fulfiled already with the measure, well approved at coating facilities, of facing down the workpiece and a coresponding beam direction from bottom to top.

ad (3) In one respect this calls for minimizing the particle beam power /10/: As the depth dependence and total amount of damage production - for example measured as microroughness - by a particle beam impinging on a solid depends on its energy, latter should be as

---

[2] The nuclear stopping power describes the process of slowing down of an ion in matter by elastic impacts with target atoms.

low as possible. This claim opposes to (1). On the other hand, due to the physical sputter-process itself, non-normal incidence leads to higher yield values: There exist preferential directions of the material removal process, at the expense of a worsening of the microroughness /9, 11-13/. In addition, for the stopping process of the particle in the substrate is non-thermal[3], redistribution processes including phase-transitions may occur due to local excess heat.

ad (4) The influence of the electrical properties of the workpiece is growing important for low energy levels, because an increasing amount of time of interaction between charges of the same sign results in increasing deviations of the trajectories due to electrical repulsion forces. This effect is twofold when figuring insulators: the charges of impinging ions stay at the surface and will be accumulated until a discharge to the next defined electrical potential, usually ground potential, occurs. This leads to a reduction of the average potential difference between ion source and the surface to be etched, i.e. ion's energy. Besides bending of the potential lines towards the edges of the insulating surface, in addition the ions are slowed down, leading to extreme defocussing and not predictable removal rates. An established way out is balancing the charges with electrons added to the ion-beam.

State of the art

In the following a practical solution for machining fused silica with a low energy Argon ion beam is proposed: The amount of applied energy for ejecting one atom from the surface of the target by means of a particle beam should be minimum between about 100 and 500 eV. Firstly at very low energies (E_{thresh} up to about 100 eV) the probability of accumulating enough energy for ejecting one stated surface atom is reduced. As a consequence the number of *impinging ions* has to rise strongly, as does the contributed total energy. Secondly, in

---

[3] The stopping of a particle usually lasts about 10^{-13} sec after impact, whereas the energy dissipation through, e.g., phonon-assisted processes is setting in at 10^{-11} to 10^{-9} sec /14/.

the higher energy ranmge (E > 500 eV) the amount of *target atoms* involved rises, because the energy transfer process for stopping the incoming ion is more and more directed into the target. Subsequently only processes of second or even higher order, e.g. the production of cascades of elastic recoils, will lead to ejected atoms. I.e. rising the ion's energy produces a higher yield, but at expense of over-heating and damaging the target.

With the measure of employing an energy band width of 100 - 500 eV the demands #1 and #3 (predictable, fast and damage-reduced) seem to be fulfiled. Here the yield is about 0.3 atom/ion or 20 nm/min at a current density level of 1 mA/cm² /14/. For optimizing the process in total, the current density level also has to be fitted. The restrictions are, firstly, overheating the target and, secondly, the blowing-up of the ion beam diameter due to space - charge effects /15/, described by the "Boersch-formula"/16/. The first limit has to be figured out experimentally. The second depends on the first:[3]

$$D_B = \kappa \cdot I \cdot L / (U_E \cdot v \cdot \alpha_0) \qquad (1)$$

where are

D_B =: beam-diameter

L =: travelling path of the ions,

I =: total beam current,

U_E =: potential difference,

v =: ions velocity,

α_0 =: half aperture angle and

κ = $9 \cdot 10^9$.

With the beam - target system given above, I = 1.0 mA, L = 0.1m, U_E = 500 V, $v = 5 \cdot 10^4$ m/s and $\alpha_0 = 5°$ one gets for D_B = 7.2mm. This number has to be added quadratically to the geometrical enlargement of the beam. Thus, a suitable transfer of an ion beam to a "plasma"-beam or an "atom"-beam, which has a zero ampere net current, would be helpful.

As outlined, we are already able to propose theoretically an optimistic answer as to the question of our title. Although there are still some imponderables, an experimental answer will be given at the conference.

---

[3] SI - units are used

References

/1/ R.A. Jones: Opt.Engng 25 (1986) 785

/2/ K. Becker and K. Beckstette: Proc. Int. Cong. for Ultraprecision Technology, Aachen, FRG, 1988

/3/ V. Giggel and A. Leitel: J. Optoelect.4 (1989) 525

/4/ S.R. Wilson, D.W. Reicher, and J.R. McNeil: SPIE 966 (1988) 74

/5/ A survey is given in: "Sputtering by Particle Bombardment", Vols.I and II, ed. R. Behrisch, Springer Verlag, Heidelberg, Berlin, New York, 1981 (Vol.I) and 1983 (Vol.II)

/6/ J.M.E. Harper, J.J. Cuomo, P.A. Leary, G.M. Summa, H.R. Kaufman, and F.J. Bresnock: J. Electrochem. Soc. 128 (1981) 1077

/7/ J. Lindhard, V. Nielsen, and M. Scharff: Kgl. Danske Videnskab. Selskab, Mat.-Fys.Medd 36 No.10 (1968)

/8/ J.F. Ziegler, J.P. Biersack, and U. Littmark in: "The Stopping and Ranges of Ions in Matter" (Vol.1); Pergamon Press, New York, 1985

/9/ L.N. Allen, and R.E. Keim: SPIE 1168 (1989) 33

/10/ J.M.E. Kaufman, J.J. Cuomo, and H.R. Kaufman: J. Vac.Sci.Technol. 21 (1982) 737

/11/ R.M. Bradley and J.M.E. Harper: J. Vac. Sci. Technol. A6 (1988) 2390

/12/ G. Carter, M.J. Nobes, F. Paton, J.S. Williams, and J.L. Whitton: Rad. Effects 33 (1977) 65

/13/ P. Sigmund: J. Mat.Sci. 8 (1973) 1545

/14/ P.C. Zalm in: "Handbook of Ion Beam Processing Technology", eds: J.J Cuomo, S.M. Rossnagel, and H.R. Kaufman, Noyes Publications, Park Ridge, NJ, (1989)

/15/ E. Goto, T. Soma, M. Idesawa, and T. Sasaki in: Proc. Symp. on "Elektron Ion Beams in Science and Technology", ed. R. Bakish, Princeton, 1978

/16/ H. Boersch: Z. Phys. 139 (1954) 115

Some Aspects of Ultra-precision Science and Engineering in Medicine and Biosciences

Dr. D. W. Robinson, B.Sc, D.Phil,
Fellow of Royal Society of Medicine
Nanotechnology Research Coordinator
Science and Engineering Research Council
Polaris House, North Star Avenue, Swindon, Wilts, England.

The understanding, diagnosis and treatment of human disease is becoming increasingly dependent upon interdisciplinary science, especially in the fields of biology, chemistry, physics and engineering.

Four decades have seen major advances in genetic engineering, molecular biology, medical sensors and instrumentation, joint replacement and minimally invasive surgical techniques to name but a few.

The impact of ultraprecision science and engineering on medicine and biosciences is a relatively recent event.

This paper discusses some state of the art ideas and developments which are now being considered or taking place and which directly involve techniques and structures on a size scale ranging from atomic to the micron levels - aptly described as nanotechnology (1) (2) (3).

Much of the work described is being undertaken in UK universities and research establishments with support from the Science & Engineering Research Council (SERC), other Research Councils, and government bodies such as the Department of Trade & Industry. SERC has been particularly active in encouraging interdisciplinary research through its specially promoted initiatives and programmes on medical engineering, biomaterials and nanotechnology.

Elsewhere in Europe, the USA and Japan there is growing awareness of active research in progress on the medical and biological implications of nanotechnology.

The next decade will see significant health care benefit, both therapeutic and cost effective, from this new application of ultraprecision science.

The Scope of Nanotechnology

Nanotechnology is generally concerned with processes and products on a scale ranging from 0.1 nanometres to 100 nanometres (1 nanometre = 10^{-9} metre) but capable of influencing, or being influenced by sizes outside this range, ie atomic and supramicron scale. An example of the latter is where large optical surfaces need to be processed to nanometric surface tolerances; the former could typically include nanometric devices for atomic manipulation.

Traditionally, nanotechnology developed by the 'top down' route, the processes involved included extending the precision levels of conventional machining techniques, the application of high energy particle beams (ions, electrons) for ablation, drilling and deposition, laser radiation and selective chemical etching. It was driven mainly by demands from the micro-electronics, optical and defence industries for smaller devices and finer surface structures.

The 'bottom up' route to nanotechnology received popular publicity from the ideas suggested by Drexler in the USA with his "engines of creation" (4) - molecular scale machines capable of self replication and construction from individual atoms and molecules. Initially this was viewed by the scientific community as intriguing science fiction with ideas such as nanoscale robots circulating in the human blood stream to scour unwanted blood clots and atheroma from the cardio-

vascular system. However recent developments may lend some credibility and challenge this scepticism.
At a more focussed, shorter term level the 'bottom up' approach is receiving attention in the field of ceramics where nanometre scale particles appear to play a vital role in the structure and properties of the bulk material and atomic engineering using scanning tip microscopy is becoming a reality.

In summary therefore, nanotechnology is now at a stage where it is possible to measure, manipulate and manufacture on a submicron to atomic scale and with corresponding fine tolerances:

(a) Surface coatings, shapes, holes, devices, particles (5).

(b) Materials ranging from metals, semiconductors, ceramics, and plastics both flexible and rigid.

(c) Structures with mechanical, optical and electronic features.

Applications to Medicine

(a) Orthopaedics

Applications for nanotechnology can be envisaged in two main areas.

The replacement of diseased joints by prosthetic devices based on metal/plastic articulations is a well established surgical technique giving medium to long term relief to many patients. Two problems, however, overshadow these otherwise impressive procedures.

Biocompatibility of the implant towards body tissue limits the effective life and eventually leads to loosening of the device and possibly irreversible infection. Recent research suggests that fine surface modification of a metal implant may provide a porous submicron matrix into which suitable host enhancing substances could be adsorbed and anti-inflammatory drugs incorporated. These matrices might also promote direct tissue ingrowth and regeneration (6).

The deposition of thin diamond like coatings onto various prosthetic devices may influence interfacial degradation and protein absorption. They could also reduce the tendency for thrombo-embolisms to develop when the devices are in direct contact with blood.

The second main problem with orthopaedic implants is frictional wear. Here the main contribution of nanotechnology is likely to come from specialised, wear resistant, thin coatings produced by, for example, ion implantation and reactive deposition techniques. However, as with the fixation of prostheses, the surface treatments should enhance rather than hinder biocompatability and host response.

(b) Tissue Regeneration

Over 20 years ago an interesting and unusual effect was observed whereby heterotopic bone cells formed in the matrix of a synthetic plastic sponge material implanted in a live animal (7). At about the same time human foetal membranes were being used experimentally to re-surface diseased bone with some success (8).

The ability of non-biological or biologically immune micro structures to promote new cell formation and growth provides an exciting new prospect for treating diseased and damaged tissues.

The influence of substrate topography on cell behaviour and morphology has been known for some time and recent interdisciplinary studies using microfabricated surface structures have thrown more light on the way in which direct cell movement takes place.

Nanostructure assisted tissue regeneration with perhaps the added feature of an electrical/electronic input (eg using a silicon based device) could find wider application to wound healing as well as bone disease.

(c) Nerve Repair

Repair of severed nerves, especially those with fine structure, demands great dexterity and skill by the surgeon. The use of an implant containing an array of sub-micron sized channels has been suggested as a guidance matrix to encourage re-growth.

(d) Filtration Systems

The presence of unwanted proteins in body fluids is usually associated with disease. An in vivo filtration system using an implanted nanostructure has been suggested to filter proteins from cerebro spinal fluid in the brain, thus alleviating encephalitis. Similarly, devices might be envisaged for selective removal of proteins thought to be responsible for the inflammatory condition in arthritic disease. Both of these are speculative ideas but they may be worth considering in the light of the high incidence of these crippling conditions.

(e) Microsurgery

Minimally invasive procedures for surgical intervention are very cost effective and lead to much less trauma for patients. Undoubtedly, these are the techniques of the future. Whilst at present many of the devices used are within the size scale of microns to millimetres, nevertheless there is likely to be a trend towards smaller devices made to more exacting tolerances within the nanometric range.

Using the enabling technology of the micro-electronics industry, it is now possible to make micron and sub-micron sized actuators, valves and electric motors which could be adapted to remotely operated surgical devices. A prototype micro-motor with cutter blades on the rotor is already being considered for use in ophthalmic surgery.

(f) Dental Surgery

Conventional techniques for treating dental decay involve removing considerable amounts of tooth tissue in excess of the diseased material in order to accommodate the mechanical limitations of the filling material. Pulsed excimer laser machining is currently being investigated and offers the prospect of faster, higher precision cavity preparation. In this way new filling techniques with precision fitting plugs may be considered.

(g) Imaging

Improved resolution ultrasonic imaging should be feasible if the transducer and detector elements could be reduced in size. Some of the ingenious nano-structures now being developed offer such a prospect and may be capable of incorporation within a catheter system thus accessing internal organs more effectively.

(h) Drug Delivery Systems

The micro-encapsulation of drugs and their controlled release through breakdown of the capsule wall by attack from internal body fluids is well known.

A further development of these ideas could include built in sensing and monitoring devices within the capsule to refine the delivery process - the so called "smart pill". Success of any such development will require significant use of nanotechnology techniques. Nano-structured microcapsules could also be devised to improve drug release when multi-component drug therapy is indicated.

Application to Biosciences

(a) Biosensors

Biosensors involving optical techniques such as ellipsometry and surface plasmon resonance depend critically upon nanometric thickness metal films. Future developments may be expected to involve more complex hybrid thin film structures, possibly incorporating bioactive molecules within inert substrates.

(b) Biomolecular Architecture

Arrays of nanometric sized holes in a dielectric material are now being developed as discrete sites for the immobilisation of individual protein molecules whose shape and size can then be optically recognised. Together with scanning tip technologies discussed later, this provides important additional data on the functional behaviour of biomolecules.

(c) Haemorheology

The rheological properties of human blood cells provide important and often early diagnostic evidence for the onset of certain diseases. Flow cytometry techniques are now being developed using cells with submicron sized flow paths etched in silicon. By incorporating electronic/electrical features into these devices and also depositing biological cells onto the surface, it should be possible to mimic blood flow in actual vascular systems. Such devices could be used for diagnostic and also pharmacological studies.

(d) Scanning Probe Technologies

This is perhaps the most exciting new development in the

nanotechnology field and one which is increasingly being used and studied for biological applications.

Scanning tunnelling microscopy (STM) relies upon an electron transfer mechanism occurring between a conducting tip and substrate separated by a few angstroms. It was originally considered inappropriate for biological systems but provided they are immobilised onto a conducting substrate such as graphite, it is possible to obtain some remarkable images of protein molecules, viruses and antibodies.

Atomic force microscopy (AFM) drags a fine tip across a surface containing atomic/molecular features and notes the minute flexural displacement of the probe as a measure of the size of the feature. It can be employed directly with biological systems but some doubt has been cast on the method in that it can distort the 'soft' biomolecules, thus giving a misleading image.

Recent research has shown that STM can be used to "pick and place" individual atoms on a surface (9). Such a manipulation technique opens up exciting prospects for carrying out changes to DNA at a molecular/atomic level - genetic engineering in a real sense.

(e) Cell Studies

Nanoscale devices combining both fine structural features and an electrical monitoring capability have been used to study electrical activity in cells and also directed cell movement, a phenomenon generally recognised as vital to the development of multi cellular organisms (10). Topographical response of cells plays a role in wound healing and regeneration.

Conclusions

Ultraprecision science and engineering is now poised to make a significant impact on medicine and biosciences. The essential background research and development is required now in order to ensure that practical benefit to human health care is realised within the next 10 years. Such a timescale may be considered somewhat long but it is nevertheless realistic for this type of development.

Acknowledgements

The Science & Engineering Research Initiative has supported the Nanotechnology Research Initiative and encouraged research and investigation into new interdisciplinary areas, some of which are described here. Their help is gratefully acknowledged.

In addition, the author wishes to thank his many friends and colleagues in medical schools, universities and research establishments in the UK who have contributed stimulating ideas and provided research information, some unpublished for this review paper.

References

(1) Taniguchi, N Annals of CIRP 1982 23(2)

(2) McKeown PA 'High Precision Manufacturing and the British Economy
1986 Clayton Lecture. Proc. Inst. Mech. Engrs.
1986 Vol. 200 No. 76

(3) Taniguchi, N 3rd International Precision Engineering Seminar, Interlaken May 1985
Precision Engineering July 1985, 7, (3)

(4) K E Drexler "Engines of Creation - the Coming Era of Nanotechnology. Fourth Estate 1990.

(5) Bullough T J, Humphreys C J & Devenish R W
Materials Res. Soc. Symp. Pro. Vol 157 1990
also Phil Mag Letters 1990 Vol 61 No. 4 p 181.

(6) Private Communication from Professor D. Williams

(7) Winter G D & Simpson B J Nature Vol 223 July 5 1990

(8) Coombs R, Gristina A & Hungerford D "Joint Replacement - State of the Art" p 31 Orthotext, London 1990.

(9) D Eigler & E Schweizer Nature Vol 344 5 April 1990

(10) Clark P, Connolly P. Curtis A S G, Dow J A T & Wilkinson C D W - Biosensors & Bioelectronics 1990 Elsevier Science Publishers Ltd.

Microelectronic and Nanoelectronic Interfacing Techniques for Biological Systems

P. Connolly*, S. Britland *, I. Hussain ,W. Monaghan* , G.R. Moores+, J Shen.

*Dept. of Electronics and Electrical Engineering, +Dept of Cell Biology
University of Glasgow
Glasgow, G12 8QQ
Scotland ,UK.

Extended Abstract

A new field of research and development has emerged in recent years which relies on the availability of ultra-high precision technology and cross-disciplinary collaboration between electronic engineers , biologists , biochemists and chemists. By exploiting the techniques of photolithography and electron beam lithography devices can be created to interface with biological cells or molecules such as enzymes . This type of work is increasingly referred to under the heading of 'bioelectronics' and touches on topics such as biosensors and cell-electrode interfacing.
Although such systems have enormous potential for both industrial and medical applications, there are a variety of practical problems to be overcome. Currently much of the effort is being directed at the microelectronic interface and usuable devices and techniques are already emerging at this level.

The device described here has been developed for a range of bioelectronic measurements , particularly biosensors and cell-electrode interfacing , which incorporates microelectrodes and various insulating surface topographies particularly suitable for cell culture/physiological saline environments. The microelectrodes can be used to record voltages or impedances *in vitro* or can be adopted as sensors / chemical release points by suitably immobilising molecules on the electrode in conducting polymers. Fabrication of the devices employs standard semiconductor techniques and materials such as gold , silicon nitride and polyimide and the general techniques of photolithography employed can also be adapted for the microelectronic patterning of proteins.

Applications

The bioelectronic applications of the device described in this paper relate particularly to cell-electrode interfacing for in vitro applications or for the fabrication of multi-analyte biosensors.

The term biosensor refers to a transducer which incorporates a biological sensing element in which the response of the biological component to a particular environment is converted to (usually) an electrical or optical signal and used to analyse that environment . In one type of biosensor device an enzyme , which is essentially a protein catalyst , could be immobilised on an electrode. Ideally , for electrical measurement , the enzyme should breakdown a specific molecule and create more ions in the vicinity of the electrode. This could be monitored electrically by amperometric or conductimetric measurements and related to concentration of the molecule of interest. Many other types of devices have been suggested or are being developed for commercial application (for an introduction to the subject see Turner et al , 1987) . Many of the available sensors are of the dimensions of millimetres rather than micrometres and test for only one substance.The development of miniaturised , multi - analyte biosensors for diagnostics or *in vivo* monitoring is therefore an important aspect of current biosensor development .In particular, if the techniques employed by the semiconductor industry for mass fabrication of miniaturised devices can be properly harnessed , a cheap multi-analyte biosensor could become a reality. However, a variety of problems have to be overcome in adapting semiconductor device fabrication techniques for biosensors. Biocompatibilty, device encapsulation , immobilisation of proteins , controlled deposition of materials and electrical characterisation of devices all require further development for biosensor applications.

A second important aspect of bioelectronics concerns interfacing microelectronic technology to cell cultures allowing the possibility of long term *in vitro* monitoring for biological research [see for example Thomas et al,1972, Edell, 1986, Connolly et al , 1990]. Particular interest centres on detection of electrical signalling in neural cell cultures for the investigation of neural processing . Nerve cells in culture tend to form a multiplicity of connections and the 'circuits ' so formed are rather difficult to unravel for the analysis of individual cell-to cell signalling. We have previously shown that it is possible to use surface topographical features at microelectronic dimensions to guide the growth of cells in culture (Clark et al 1987 , Clark et al 1990) . It is possible to use non-invasive , extracellular microelectrodes for detection of cell electrical activity and , if combined with surface topographical features , a device to monitor small , recognisable networks of neural cells in culture is created [Dow et al, 1987 , Connolly et al,1989]. There is also an increasing interest in the bioelectronics field in the development of long-term , implantable devices for prosthetic interfacing or nerve repair . Some of the advantages of miniaturisation and semiconductor techniques for these applications include ease of packaging, reduction of rejection risks, 'on chip' clinical testing and good control of device characteristics.

The device described here serves two purposes. The first application is as an extracellular microelectrode array for use *in vitro* with neurons or other electrogenic cells. The microelectrodes are used to record electrical signals (known as action potentials) from a cell in contact with a planar microelectrode, and to ensure good recording, the insulation layer should be opened only at the microelectrode. To form networks of cells , and record signal transmission from cell to cell , it is necessary to provide some topographical or chemical guidance cues for cell growth. To provide this the insulating layers of the device are patterned with an interconnecting groove structure which guides neurite outgrowth in culture [Clark et al ,1987] . For devices used in physiological saline, the insulation must not be prone to ion penetration either during recording or under applied voltage on the electrodes. This means that silicon oxide or nitride layers , which are readily available as insulators in the electronics industry , are not likely to have the best long term insulating properties. In principle the use of polymer insulating layers , such as PMMA and polyimide , should solve this problem. However ,if the polymer is used as the only insulating layer the insulation to substrate adhesion is very poor in solution and the device is easily damaged by handling or cleaning. A combination of silicon nitride / silicon oxide and polymer insulating layers (where the polymer is protected by a top layer of oxide or nitride) provides the best solution to these problems. The fabrication process developed to create microelectrode arrays insulated by a triple layer of silicon nitride -polyimide -silicon nitride is described herein. The insulation on the device is patterned so that circles of 10μm diameter are opened through the insulation to the electrodes with interconnecting grooves for cell guidance .

The second application of the device is to provide an array of seperately addressable electrodes for the deposition /encapsulation of biological molecules . One of the most promising means of achieving this , by a technique which might be adapted for mass production , is to use the conducting polymers which can be deposited electrolytically on the electrodes. For example , it is possible to encapsulate glucose oxidase in polypyrrole to form a biosensor [Foulds and Lowe, 1986] or to use the polymer as a means of trapping a molecule during electrolysis which can later be released by voltage control of the electrode [Zinger and Miller , 1984]. In the experiments described here two different enzymes , glucose oxidase and urease were encapsulated by electrolytic deposition in the presence of conducting polymer.

Molecular Patterning

Ultra- high precision technology can also be applied in the control of molecular deposition on surfaces at micrometre [Murakami et al ,1986, Kleinfeld et al,1988], and possibly nanometre , dimensions. Semiconductor processes can be used to pattern molecules directly on surfaces by patterning surface chemistry so as to provide suitable molecular attachment sites . A further possibly useful refinement allowed by the introduction of semiconductor technology is to allow the patterning of proteins at microelectronic dimensions. To illustrate this we have taken the method described by Kleinfeld et al which allows the patterning of hydrophobic and hydrophylic areas to the dimensions of micrometres on a planar surface by employing 'lift -off' . We have somewhat refined Kleinfeld's method which is rather lengthy (12 hours refluxing in hexane required) and developed silanisation techniques which can be performed in relatively short times (a few minutes) under cleanroom conditions [Britland et al, 1989 , Britland et al, 1991]. Some applications for this technology lie in biosensor fabrication , biocompatibility treatments and surface preparation for microscopy. The patterned surfaces can also be employed to guide cell growth in culture and an example of this is given here.

Scanning Tunneling Microscopy

The manipulation of molecules at micrometre and sub-micrometre levels was, until recently , hampered by the inability to visualise the molecules on surfaces *in situ.* Labelling with metals and electron microscopy allowed detection of binding but removed the molecule from its natural, hydrated state.

The advent of the Scanning Tunnelling Microscope introduced by Binnig and Rohrer (Binnig et al, 1982) has allowed the imaging of surfaces in air with atomic resolution and biological molecules in air and water (see Hansma et al , 1988 ,for general introduction). There are however many possible misinterpretations of biological images in STM and confusion with artefacts in the image. Several groups are now working on the problems of imaging of biological molecules on surfaces including ourselves . Three examples of images obtained by us in air are given here to illustrate the potential of this type of microscopy for surface biological imaging.

References

Binnig , G. , Rohrer, H. , Gerber , Ch. ,Weibel, E. (1982). Phys Rev Lett ,49 ,57.

Britland , S.T. , G. R. Moores, G. R. , Clark , P. , Connolly , P. (1989) Patterning cell adhesion and movement on artificial substrata , a simple method. Proc. Anat. Soc. Great Britain and Ireland , 18 .

Britland , S.T. , G. R. Moores, G. R. , McGinn , B. , Clark , P. , Connolly , P. (1991). A method for immobilising proteins and synthesiing peptides in micro-patterns on planar substrata using microelectronics fabrication technology.Submitted for publication February,1991.

Clark ,P., Connolly ,P., Dow,J.A.T. , Curtis, A.S.G , Wilkinson, C.D.W. (1987). Topographical control of cell behaviour: 1. Simple step cues.Develop. ,99,439-448.

Clark , P., Connolly , P. , Curtis , A.S.G. , Dow , J.A.T. , Wilkinson , C.D.W. (1990). Topographical control of cell behaviour : 2. Multiple grooved substrata " Development , 108, 635-644 .

Connolly ,P., Clark ,P.,Dow,J.A.T, Curtis, A.S.G , Lind , R. ,Wilkinson, C.D.W. (1989). Extracellular electrodes for monitoring cell cultures . Institute of Physics Short Meeting Series No. 21 (IOP Publishing , UK).

Connolly ,P., Clark ,P.,Dow,J.A.T, Curtis, A.S.G ,Wilkinson, C.D.W. (1990). An extracellular microelectrode array for monitoring electrogenic cells in culture. Biosensors and Bioelectronics, 5,3,223-234.

Dow,J.A.T , Clark ,P., Connolly ,P., Curtis, A.S.G , Wilkinson, C.D.W. (1987). Novel methods for the guidance and monitoring of single cells and simple networks in culture. J. Cell Sci. Suppl., 8,55-79.

Edell,D.J. (1986). A peripheral information transducer for amputees : long-term multichannel recordings from rabbit peripheral nerves. IEEE Trans. Biomed. Eng.33,203-214.

Foulds,N.C., Lowe, C.R. (1986).Enzyme entrapment in electrically conducting polymers. J. Chem. Soc. ,82 , 1259-1264.

Hansma , P.K. , Elings, V.B , Marti, O. , Bracker, C.E. (1988).Scanning Tunneling Microscopy and Atomic Force Microscopy : Application to Biology and Technology. Science, 242 , 209-216.

Kleinfeld, D. , Kahler, K.H. , Hockberger, P.E. (1988). Controlled outgrowth of dissociated neurons on patterned substrates. J. Neuoroscience. 8 , 4098-4120.
Murakami,T.,Nakamoto,S. , Kimura , J. , Kuriyama , T., Karube , I. (1986). A

microplanar amperometric glucose sensor using an isfet as a reference electrode. Anal.Lett. 19,1973-1986.

Thomas, C.A. , Springer, P.A., Loeb, G.E , Berwald-Netter, Y. , Okun, L.M . (1972). A miniature microelectrode array to monitor the bioelectric activity of cultured cells. Ex. Cell. Res. , 74 , 61-66.

Turner, A.P.F. , Karube , I. , Wilson , G. S. eds. (1987) Biosensors . Oxford University Press.

Zinger , B. , Miller,L.L .(1984). Timed release of chemicals from polypyrrole films. J. Am. Chem. Soc. ,106,6861-6863.

Cutting Experiments in a Computer using Atomic Models of a Copper Crystal and a Diamond Tool

Toyoshiro Inamura
Nagoya Institute of Technology
Gokiso-cho, Shouwa-ku, Nagoya-shi 466 Japan
Nobuhiro Takezawa
Kanazawa University
2-40-20 Kodatsuno, Kanazawa 920 Japan

ABSTRACT

In order to analyze the mechanism of nanometer cutting, a method of atomic-scale cutting in an experiment using a computer has been developed based on the nonlinear finite-element formulation which regards atoms and atomic interaction as nodes and elements, respectively. This method can handle discontinuous phenomena due to instantaneous propagation of dislocation in a workpiece during cutting. Experiments carried out using two kinds of assumed potential energy between tool and workpiece atoms have revealed that the process of chip formation as well as the stress distribution on the tool face during cutting is strongly dependent on the type of interaction energy between the tool and workpiece, while the size effect for the specific cutting coefficient and the discontinuity of cutting force variation during cutting are common in both types of potential energy. The experiments have also shown that the intermittent drop of potential energy accumulated in the workpiece during cutting results in heat generation associated with plastic deformation of the workpiece, while the heat generation repeatedly causes impulsive temperature rise on the tool face during cutting.

1. Introduction

Ultraprecision cutting using a diamond tool has reached a level such that cutting with an undeformed chip of a thickness of 1nm has been reported.[1] However, the occurrence at the tool edge is unclear because there is no effective method of observation, especially for cutting under practical cutting speed. On the other hand, the usual analytical methods such as the finite-element method cannot, of course, be used for analyzing this atomic-scale phenomenon. Furthermore, any assumption concerning the cutting process such as formation of a shear zone or determination of chip flow direction, is inadequate for analysis because the cutting phenomenon itself is unknown. Thus, the proper method to investigate is still experimental, but should be based on a new approach. A possible candidate in this direction is an experiment using a computer, where a workpiece and a tool are both modeled as crystals with interacting atoms being arranged. Cutting in this case is performed by a computer by letting the tool crystal enter the workpiece crystal to induce atomic motion at their stable positions at each moment. Because cutting is done under a completely controlled condition, any momentary event can be observed to offer a chance of finding new results which may be common even to macroscale cutting. In this study, we carry out cutting experiments using copper and diamond crystals as workpiece and tool, respectively, to show that the cutting process is susceptible to the affinity between the tool and workpiece atoms. We also show that the cutting force changes discontinuously during cutting due to the instantaneous propagation of dislocation in the workpiece. This instantaneous motion of dislocation generates atomic vibration, which results in repeated impulsive temperature rise on the tool face during cutting.

2.Cutting Experiments

2.1.Cutting model

Although real nanometer cutting will be performed using a tool and a workpiece with surface adsorbate, surface reconstruction of crystal structure and defects with and without contaminants, all of these are excluded for simplicity in this study. In Fig.1, crystal models of a copper workpiece and a diamond tool are shown with the indicated orientation and size, using a unit of length "a" of the lattice size of the fcc structure of copper. The tool is composed of carbon atoms arranged in the diamond structure and has the rake angle of 6.3 degrees and the relief angle of 5 degrees. The workpiece is fixed at the hatched area and the depth of cut in set at 0.9025nm (9.025 Å). The interatomic potential energy is assumed to be

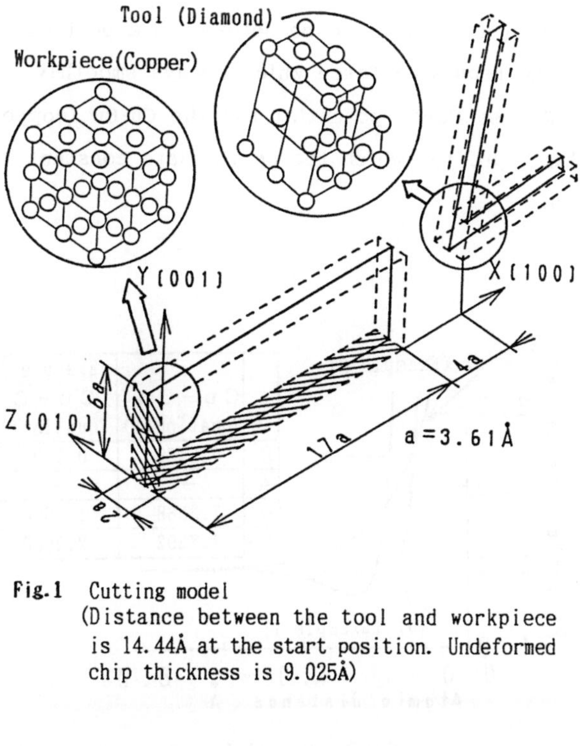

Fig.1 Cutting model
(Distance between the tool and workpiece is 14.44Å at the start position. Undeformed chip thickness is 9.025Å)

described as the sum of the pairwise empirical potentials which depend only on the distance of the atoms. The following two types of such potentials are used in this study. One is the "Morse potential" which has the form

$$\phi(r_{ij})=D[exp\{-2\alpha(r_{ij}-r_0)\}-2exp\{-\alpha(r_{ij}-r_0)\}] , \quad (1)$$

where r_{ij} is the distance between atoms "i" and "j", and D, α and r_0 are constants to be determined based on the physical properties of the corresponding material. This potential produces repulsive force in a short range, attractive force in a medium range and decays smoothly to zero in a long range. The other potential we use is the "Born-Meyer potential", which has the form

$$\phi(r_{ij})=A\,exp\{-\alpha(r_{ij}-r_0)\} \quad , \tag{2}$$

where A is also a constant to be determined experimentally. This potential produces only repulsive force which decays smoothly as the distance r_{ij} increases. We use Morse potential for the Cu-Cu (copper-copper) pair, as shown in Fig.2 (a), for which the appropriate constant values are given in the literature[2].

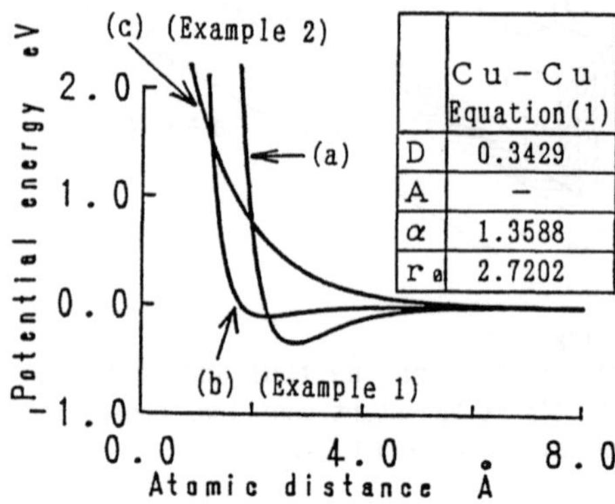

	Cu−Cu Equation(1)	Example 1 Cu−C Equation(1)	Example 2 Cu−C Equation(2)
D	0.3429	0.1000	−
A	−	−	0.4597
α	1.3588	1.7000	0.9545
r_0	2.7202	2.2000	2.5000

Fig.2 Potential energy curves used in the experiment

Because the potential between Cu-C (copper-carbon) atoms is still unclear particularly between surface atoms, we use both the Morse and Born-Meyer potentials for Cu-C with assumed constants, as shown in Figs.2 (b) and (c), to examine how the resultant cutting process differs between the

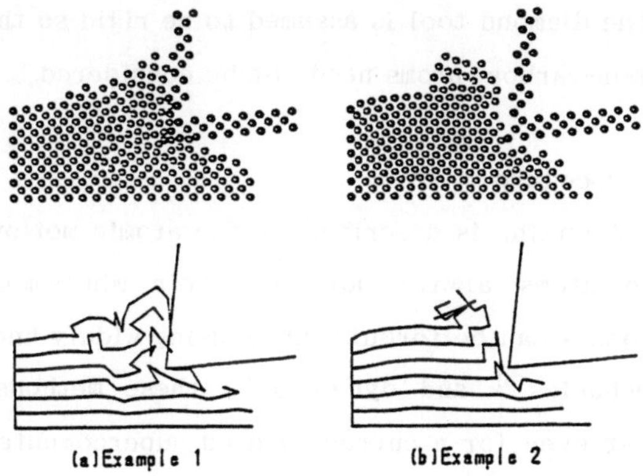

Fig.3 Process of chip formation observed in the experiments
(upper: particle representation
lower: schematic representation)

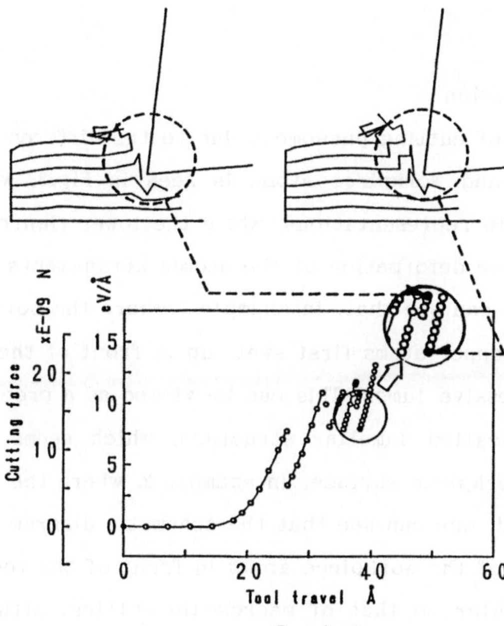

Fig.4 Cutting force variation due to instantaneous propagation of dislocation

two potentials. The diamond tool is assumed to be rigid so that the potential between C-C (carbon-carbon) atoms need not be considered.

2.2.Mathematical procedure

The process of cutting is described by the atomic motion of a tool and a workpiece, where atoms always move to their minimum-energy position. Although this process can be formulated by using widely known methods such as molecular mechanics[3] and dynamics[4], these methods demand a high computational cost even for a currently used supercomputer. Thus, we have developed a new and computationally more efficient procedure based on the nonlinear finite-element formulation[5], where atoms and pairwise atomic potentials are regarded as nodes and elements, respectively. The detailed description of the formulation will be skipped here due to limitation of space.

3.Results and Discussion

The difference of cutting phenomena due to the difference of potentials between the tool and workpiece atoms is seen in Fig.3, where the upper figures are particle representations, while the lower figures are schematic diagrams to show the deformation of the atomic arrangments during cutting. In this figure, one can see that, in example 1 where the potential curve (b) in Fig.2 is used, copper atoms first swell up in front of the tool, and then seem to form successive lumps. This can be viewed as a process producing a chip with the so-called lamellar structure, which seems to be made by buckling of the workpiece surface. In example 2, where the potential curve (c) in Fig.2 is used, one can see that the schematic diagram clearly shows a shear deformation of the workpiece atoms in front of the tool. This type of deformation is similar to that of macroscale cutting, although the shear angle is about 45 degrees and is considerably larger than that of macroscale cutting. This large shear angle could be attributed to the direction of the workpiece crystal. The variation of cutting force during cutting and how it relates to the motion of dislocation in the workpiece is illustrated in Fig.4

in the case of example 2. The lower graph of this figure shows that the cutting force increases gradually as the tool touches and begins to push the workpiece. However, the cutting force then suddenly drops down to a certain value as the tool progresses and similar changes are repeated afterwards. The reason for this sudden change can be understood by looking at the upper figures which show the workpiece deformations before and after the force drop. These figures clearly show that the sudden drop of the force is the result of the sudden motion of dislocation generated by the tool. This type of sudden change in cutting force is also seen in the case of example 1. The specific cutting force around the tool travel of 40 Å in Fig.4 is about 43GPa, which may be one magnitude larger than that of macroscale cutting. This "size effect" will be attributed to the defectless crystal used in this study because the result is common to both of the examples and is obtained using a tool with an extremely sharp edge. The magnitude and direction of the force acting on each atom of the tool exhibits a clear difference between examples 1 and 2, as shown in Figs.5 (a) and (b), respectively. In Fig.5 (a), a tensile force distribution is formed on the rake face of the tool and the corresponding tensile stress normal to the (1,1,1) plane is about 40GPa, which exceeds the measured critical stress of diamond[7]. This implies

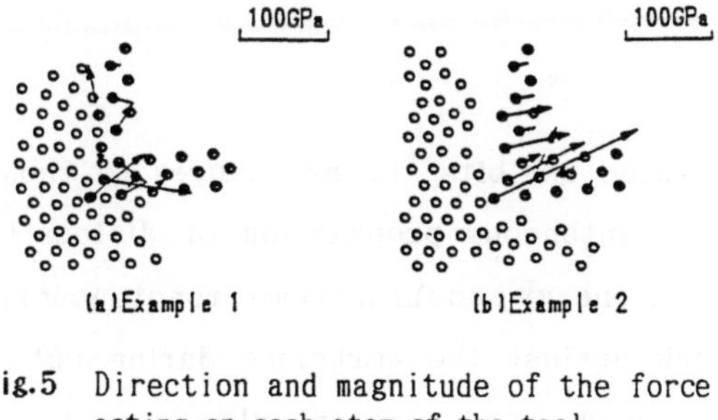

Fig.5 Direction and magnitude of the force acting on each atom of the tool (the scale indicates the equivalent stress)

that the diamond tool cannot maintain its sharp edge during cutting. On the other hand, in Fig.5 (b), force is compressive against the tool edge and thus will never break the tool edge. At any rate, we cannot say which case is true until the interaction between tool and workpiece atoms is clarified by future study. The total potential energy accumulated in the workpiece is plotted against the tool travel in Fig.6 in the case of example 1. This figure

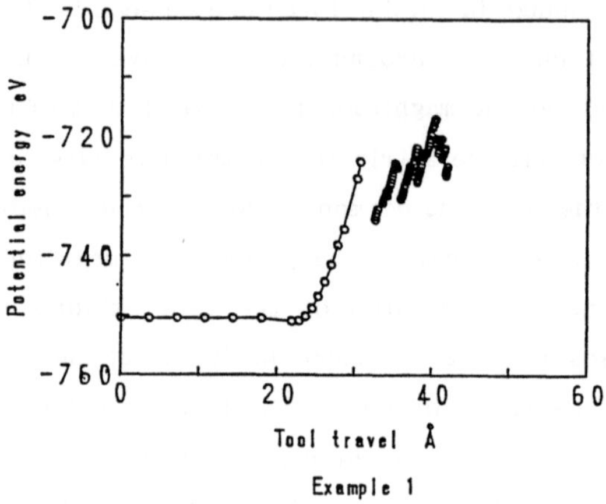

Fig.6 Change of potential energy during cutting

shows that the value exhibits similar changes to those in cutting force, reflecting the instantaneous propagation of dislocation. Here we should note that the total energy should increase monotonously because the tool is always doing work against the workpiece during cutting. Thus, when the potential energy drops, the corresponding amount of energy should be converted to the other kinds of energy such as kinetic, phonon and/or light emission. Note that the energy for surface generation is included in the potential energy. Here we neglect the energy of phonon and light emission, and examine the increase of the kinetic energy of each atom in the workpiece by computing the corresponding drop of the potential energy. This increase of kinetic energy can be interpreted as heat generation associated with plastic deformation of the workpiece. The relation between the kinetic energy increment ΔE and the resulting temperature rise $\Delta\theta$ can be given, using the Boltzman constant k, by

$$\Delta\theta=\left(\frac{2}{3k}\right)\Delta E \quad . \tag{3}$$

We set $\Delta\theta=0$ where the potential energy increases. Although ΔE in equation (3) should be the mean value of many atoms for $\Delta\theta$ to be a "true temperature" rise, we apply this equation to each atom because we are going to discuss, not the true temperature, but the propagation of the kinetic energy thus generated, for instance, to the rake face of the tool. The most direct description of the kinetic (=heat) energy propagation will be performed by using molecular dynamics. However, to save computational cost, we take an indirect approach based on the heat equation as follows: the solution of the heat equation with a point heat source is a time-decaying normal distribution function which starts from a delta function. Then we assume that the physical phenomena start at the time when the foot of the distribution spreads to the atomic size. We determine the corresponding height by using equation (3) for each atom and sum up each decaying distribution afterwards. The time origin in this case will be renewed each time the kinetic energy increment is observed. The temperature $\theta_t(x, y)$ at time "t" of the point (x, y) is given, using $\theta_0(j)$ as the temperature of the atom "j" at (x_j, y_j) and at t=0, $\Delta\theta_0(j)$ as the temperature increment of the same atom and D as the characteristic coefficient of heat diffusion, as follows

$$\theta_t(x,y)=\frac{1}{1+(t/t_0)}\sum_j\{\theta_0(j)+\Delta\theta_0(j)\}\cdot exp\left[-\frac{(x-x_j)^2+(y-y_j)^2}{4D(t+t_0)}\right] \quad , \tag{4}$$

where $t_0=2.8532\times10^{-17}$sec is determined so that $3\sqrt{2Dt_0}=2.4199$ (the size of the Wigner-Seitz cell) and $D=1.1403\times10^{-14}m^2/s$. The results of computation by equation (4) are illustrated in Fig.7 for a carbon atom on the rake face of the tool. The cutting speed in this case is set at 192m/min.

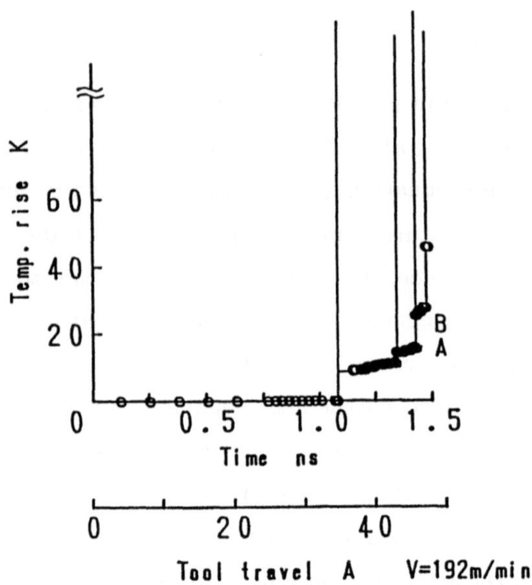

Fig.7 Resulting temperature variation on the rake face of the tool at the starting stage of cutting (example 1)

In this figure, we should note that there are many "impulses" in the temperature. Although it is difficult to observe these impulses in actual cutting because their time width is about 0.03ps, there is a possibility that they cause tool wear and/or tool breakage. Furthermore, it can be said that macroscale cutting must include the phenomena observed in the above in a more complex manner.

4. Conclusions

(1) A method of experiment using a computer has been developed to analyze the mechanism of nanometer cutting. This method can handle discrete phenomena due to instantaneous propagation of dislocation in a workpiece during cutting.

(2) Cutting experiments show that the process of chip formation is susceptible to the affinity between tool and workpiece atoms. An active interaction between them seems to result in chip formation without a clear shear zone in the workpiece, but with a lamellar structure produced by buckling on the workpiece surface. The nonactive interaction between them seems to result in chip formation with a clear shear zone in the workpiece. The corresponding force distribution on the rake face of a tool is tensile in the former case and compressive in the latter case.

(3) The specific cutting coefficient in nanometer cutting of a defectless crystal with a tool having an extremely sharp edge is about ten times larger than that in macroscale cutting. Cutting force in nanometer cutting repeatedly exhibits discontinuous changes during cutting, reflecting the instantaneous propagation of dislocation.

(4) The sudden motion of dislocation generates the kinetic energy of atoms, which appears as heat generation associated with plastic deformation of a workpiece, while the sudden heat generation repeatedly causes impulsive temperature rise on the tool face during cutting.

This study was supported by a Grant-in-Aid for Scientific Research from the Ministry of Education.

References

[1] N.Ikawa, S.Shimada, Y.Ohmori, H.Tanaka, C.K.Cry, J.S.Taylor and R.R.Donaldson, Chip formation process in nanometer level micro-cutting, Preprint of the Spring Annual Meeting of JSPE, 1989, 761 (in Japanese).

[2] M.Doyama and R.M.J.Cotterill, Stable and Metastable Tetravacancies in an fcc Metal, Phy. Rev., 137, 3A, 1965, A994.

[3] U.Burkert and N.L.Allinger, Molecular Mechanics, American Chemical Society, 1982.

[4] D.W.Heerman, Computer Simulation Methods, Second ed., Springer-Verlag, 1990.

[5] O.C.Zienkiewicz, The Finite Element Method, Third ed., McGraw-Hill, 1977.

[6] W.C.Rheinboldt and E.Ricks, Solution Techniques for Nonlinear Finite Element Equation, State-of-the-Art Surveys on Finite Element Technology, Chapter 7, ASME, 1983, 183.

[7] N.Ikawa and S.Shimada, Microfracture of Diamond as Fine Tool Material, Ann. CIRP, 31, 1, 1982, 71.

Precision Aspheric Generators for Off-axis X-ray Mirrors and Asymmetric Aspheric Lenses

Shigeo Moriyama

Central Research Laboratory, Hitachi Ltd.,
Kokubunji, Tokyo 185, Japan.

SUMMARY

This paper describes two types of newly developed precision aspheric generators. One is a fly cutting type machine for fabricating off-axis X-Ray metal mirrors. While the cutting spindle is rotating, the cutting depth can be dynamically controlled according to programmed data synchronized with the current rotation angle of the spindle. A figure accuracy of 0.1 μm and a surface roughness of 0.03 μm Rmax are achieved. Aspheric mirrors fabricated using this machine were built in an X-ray microscope, and demonstrated a 2 μm resolution. The other newly developed aspheric generator is for fabricating nonaxissymmetric aspheric glass lenses used in laser beam printers. Pairs of glass blanks are set around a turntable which rotates at about 4 rpm. The surfaces of the blanks are machined with a grinding wheel, where the grinding depth is dynamically controlled. The typical machining time for a lens is 30 minutes. A figure accuracy of 0.2 μm and a surface roughness of 0.05 μm Rmax are achieved.

1. INTRODUCTION

Aspheric mirrors and lenses have become widely used in optical systems because of their aberration reducing capabilities. Aberration-free characteristics are possible by using nonaxissymmetric aspheric optical components. However, conventional aspheric figures are restricted to rotationally symmetric figures because of difficulties in figure fabrication technologies. In order to overcome this problem, two types of aspheric generators have been developed for fabricating off-axis X-ray metal mirrors and modified toric like glass lenses. This paper describes the system architectures, and evaluates the two machines.

2. A FLY CUTTING MACHINE FOR OFF-AXIS X-RAY MIRRORS

2.1 Off-axis Mirrors for Soft X-ray Optics

Advances in diamond turning technology have made the use of aspheric mirrors in the visible light region practicable. However, it is difficult to apply the turning technique in fabricating off-axis mirrors, as shown in

Fig. 1, used in soft x-ray optics. This is because an unpracticably large turning machine is required for machining such long, nearly flat, grazing-incidence mirrors having a large radius. A new fly cutting machine[1] was developed to overcome this problem.

2. 2 Mechanical Structure

The mechanical structure of the newly developed machine is shown in Fig. 2. The work piece is set on the machine base, and its side face is machined with a single-point diamond tool fly cutter. The spindle is mounted on an X-slide and is continuously movable along the X-axis by as much as 600 mm. If the spindle axis is tilted a small angle about a Y-axis, cylindrical figures can be generated. Furthermore, numerically controlled, tool slide positioning along the Y-axis allows generating a generalized contour in an XY-plane. However, the combination of these figure-generating functions does not permit machining of a generalized figure. To achieve generation of a generalized figure, the tool cutting depth must be dynamically controlled while the spindle is rotating by utilizing a piezo-actuator fixed to the diamond tool.

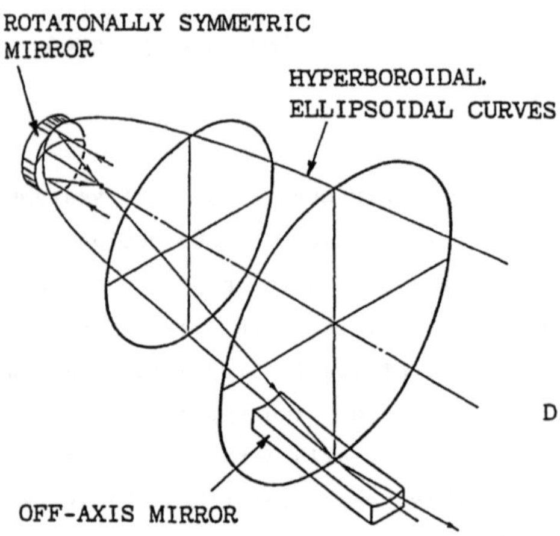

Fig. 1. Off-axis mirror for X-ray optics

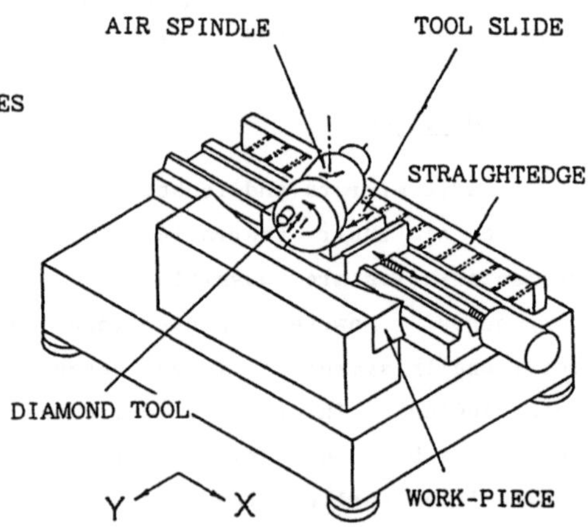

Fig. 2. Mechanical structure

2. 3 Cutting Depth Control

Figure 3 shows the cutting depth control concept for generating a generalized figure. A piezo-actuator is attached to the flange of the air-spindle, and a diamond tool is attached in turn to the piezo-actuator. The rotation angle of the spindle is detected by a rotary encoder, and the detected signal is used to read out the programmed cutting data from the

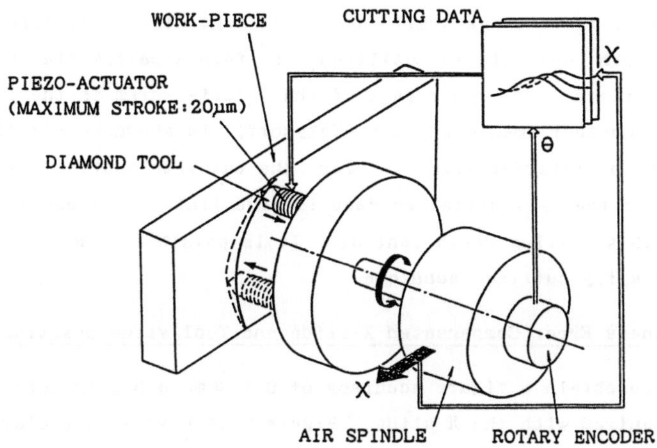

Fig. 3. Generallized surface machining by cutting depth control

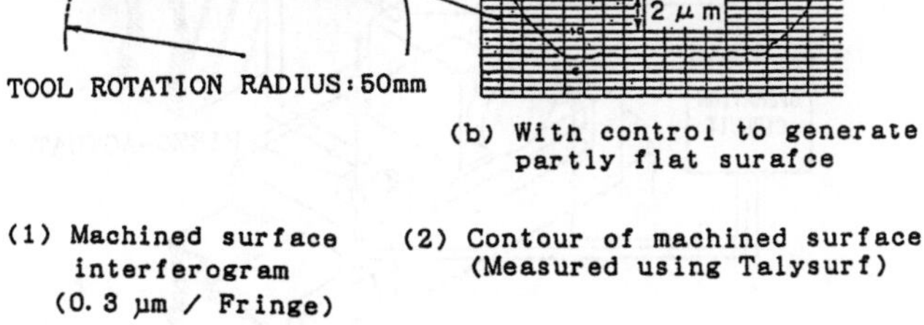

Fig. 4. Example of machined figure with cutting depth control

memory circuit. The read out cutting data signal drives the piezo-actuator through a slip-ring in contact with the spindle. In this system, the tool cutting depth is controlled according to the programmed data synchronized with the rotation angle of the spindle. Furthermore, by changing the cutting data according to the spindle X-position makes it possible to generate a generalized figure in the range of 20 μm, which is the maximum stroke of the piezo-actuator.

A machined example using the cutting depth control is shown in Fig. 4. The spindle axis is tilted a small amount to generate a cylindrical figure, and cutting depth control is utilized to form a partly flat area on the cylindrical surface. The upper part of the interferogram is the cut without control; its contour (measured with Talysurf) is shown in Fig. 4(a). The lower part of the interferogram is for the cut with control, so that the central part of the cylindrical surface is made flat, as shown in Fig. 4(b). By adopting this cutting depth control, it is possible to generate aspheric figure using a fly cutting technique.

2.4 Straightness Error Compensated X-slide and Tool slide positioning

In order to obtain a figure accuracy of 0.1 μm, a highly accurate linear motion is required with the X-slide. Figure 5 is a schematic diagram of the straightness error compensated X-slide. Fundamentally, the X-slide is guided with a V-flat sliding guide using Teflon sliders. Three optical gap sensors are mounted to the slide and face a straightedge that is mounted on the machine base. The straightedge is made of Zerodur, and its face is

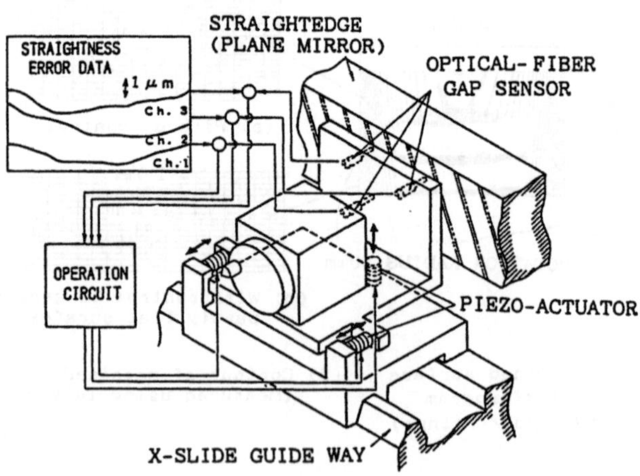

Fig. 5. Straightness error compensation by utilizing a straightedge

aluminized. The three piezo-actuators adjust the slide so that the three gap signals remain equal to the reference values. Straightness error data from the straightedge are stored in the memory circuit and are added to the servo reference signal to correct the error. With this error compensation system, straightness error, roll, and yaw can all be compensated. The measured straightness error was less than 0.04 μm_{p-v} for 280 mm travel.

To achieve a 0.01 μm positioning accuracy with the tool slide, a friction drive mechanism is adopted. The tool slide on which the spindle is mounted is linearly guided by using four air-bearings. A traction rod connected to the tool slide is forced into contact with the drive shaft of the torque motor by using a pre-loaded roller. Positioning control is carried out with a closed loop system utilizing a laser interferometer. The measuring resolution of the laser interferometer is $\lambda/120$ ($\simeq$0.005 μm). With this system, a positioning resolution of 0.01 μm can be achieved.

2.5 Evaluation

The developed machine was placed in an environmental chamber in which the air temperature was controlled within ±0.1°C. The workpiece chuck, air-spindle, and straightedge holder are cooled with silicon oil whose temperature is stabilized within ±0.005°C. Figure 6 shows an experimentally machined toroidal mirror with a short radius of 800 mm and a long radius of 100 m, and an ellipsoidal mirror with dimensions of 600 mm × 100 mm. A surface roughness of 0.05 μm is obtained for an OFHC surface. Figure 7 shows an interferogram of a machined flat surface to evaluate a straightness accuracy of the X-slide. As grazing incidence i iterferometer was used, one

Fig. 6. Experimentally machined toroidal mirror and ellipsoidal mirror

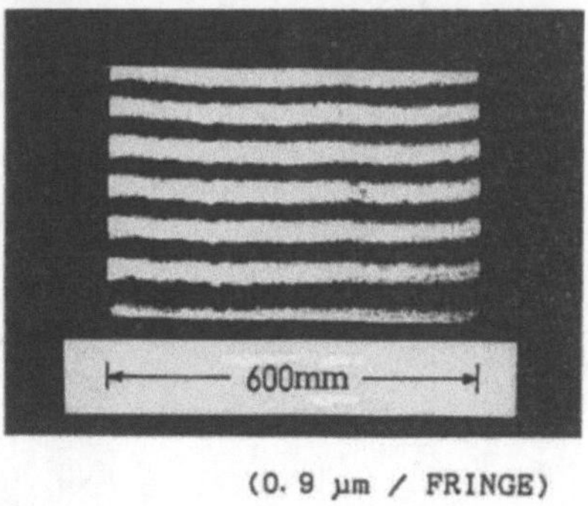

(0.9 μm / FRINGE)

Fig. 7. Interferogram of flat surface

interference fringe corresponds to 0.9 μm. The surface flatness is estimated at 0.2 μm for 600 mm × 100 mm.

A shallow spherical concave surface was machined to evaluate the aspheric figure machining accuracy. For the developed machine, sphere machining requires the same motions as for machining generalized figures. Namely,

(a) Cylindrical surface cutting, obtained by tilting the spindle axis accurately.

(b) Accurate tool slide numerical control for generating the spherical curvature.

(c) Cutting depth control for reducing the residual figure error

Figure 8 shows each shape and the interferogram obtained when each cutting motion is carried out independently. Machining with these three figure generating motions combined makes it possible to generate a spherical figure. The lower interferogram in Fig. 8 shows the interferogram of an experimentally machined spherical figure. Newton rings are shown, and 0.1 μm accuracy was confirmed by measuring the fringe deformation. Based on this experimental result, a figure accuracy of 0.1 μm can be expected for machining other asymmetric aspheric mirrors.

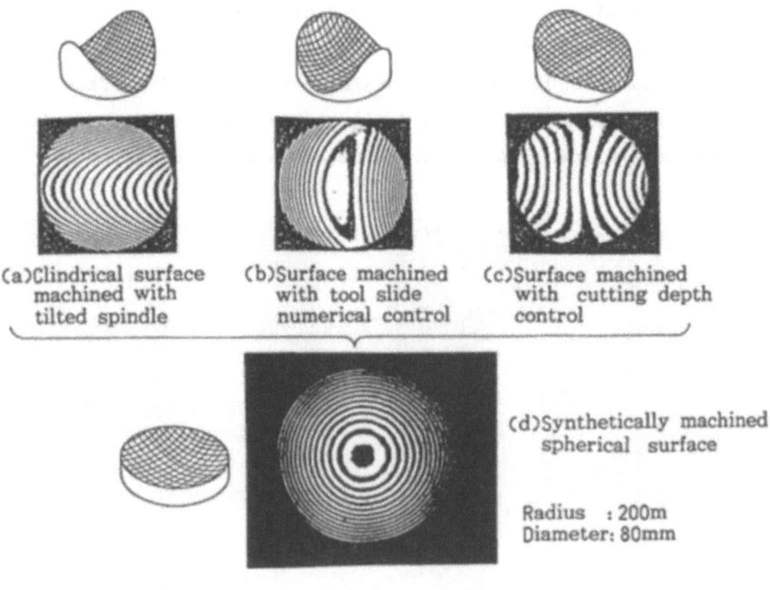

Fig. 8. Machine accuracy evaluation by cutting a spherical surface

2.6 Application for X-ray Microscope

Two pieces of ellipsoidal mirrors were fabricated using the developed machine. They were arranged in a crossed mirror geometry (Kirkpatrick-Baez) configuration, as shown in Figure 9, to form an X-ray source image. The mirror parameters are shown in Table 1. A focusing property with this configuration was evaluated by Y. Suzuki and F. Uchida[2] for a scanning X-ray microscope. Synchrotron radiation was used as the X-ray source. A transparent X-ray mask was set on the forcusing plane, and the transmitted X-ray beam intensity was detected by an X-ray sensor. Figure 10 shows the obtained image at a wavelength of 2.0 Å. A 2 μm line and space pattern is clearly observed.

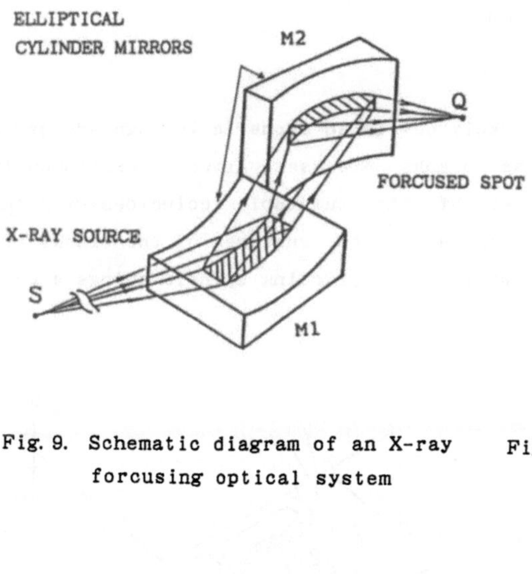

Fig. 9. Schematic diagram of an X-ray forcusing optical system

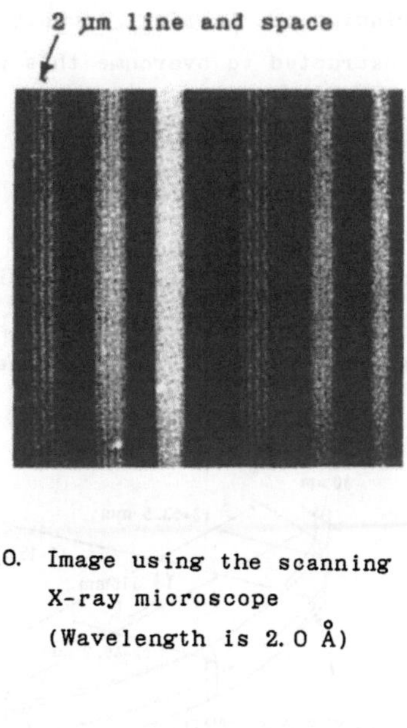

Fig. 10. Image using the scanning X-ray microscope (Wavelength is 2.0 Å)

Table I. Mirror parameters

	First mirror (M_1)	Second mirror (M_2)
Major axis of ellipse	28045 mm	28045 mm
Minor axis of ellipse	7.5 mm	5.3 mm
Eccentricity	$1-3.6\times10^{-8}$	$1-1.8\times10^{-8}$
Mirror length	30 mm	30 mm
Average grazing angle	3 mrad	3 mrad
Nominal demagnification (a/b)	1/466	1/934

3. AN ASPHERIC GENERATOR FOR GLASS LENSES

3.1 Nonaxissymetric Aspheric Lenses for Laser Printers

In a laser beam printer, an F-θ conversion lens system is used to eliminate the scan pitch error caused by the polygon mirror facets. A conventional F-θ lens system consisting of two or three spherical lenses and a long cylindrical lens has a large aberration characteristic in its wide beam scanning. Aberration-free characteristics are possible[3] by using a nonaxissymmetric aspheric lens as shown in Fig. 11. The figure is similar to a toroidal, but not axissymmetric. The laser beam scanning direction radius R is different from sub scanning direction radius r, and radius r is gradually varied along beam scanning axis L. Furthermore, the variation rate is not symmetric to central axis Y. However, this type of nonaxissymmetric glass lens is difficult to fabricate with conventional lens grinding or lapping techniques. A new aspheric generator[4] has been constructed to overcome this problem.

3.2 Operation Principle

The operation principle of the newly developed aspheric lens generator is shown in Fig. 12. Pairs of glass blanks are set around a continuously rotating turntable. The diameter of the turntable coincides with the diameter 2R of the lens to be machined. The surfaces of the blanks are machined with a grinding wheel, while the grinding spindle swings around

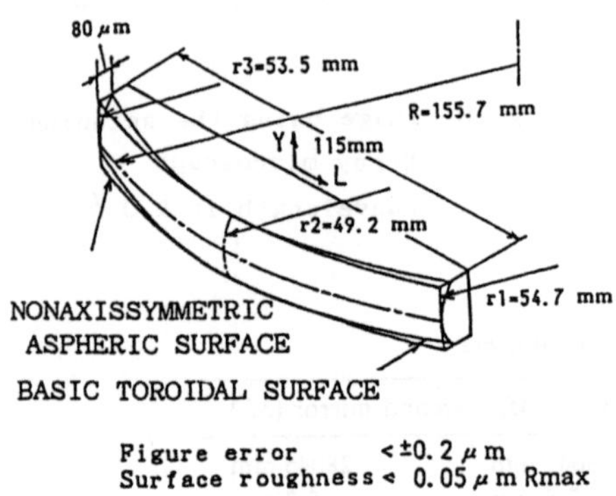

Fig. 11. Nonaxissymmetric aspheric lens

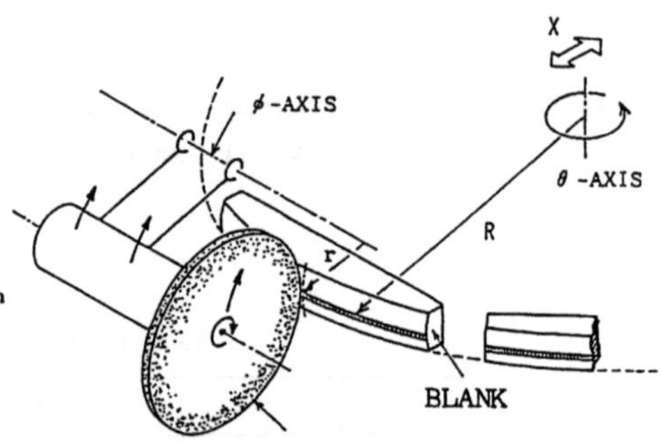

Fig. 12. The aspheric generation Principle

axis φ to generate short radius r. The spindle is swung step like after each revolution of the turntable. The grinding depth is dynamically controlled by actuating the turntable position X according to the aspheric figure data linkage with the rotation angle of the turntable. The figure data is exchanged through linkage with each step swing motion of the spindle. Using this figure generating principle, pairs of lenses with arbitrary surfaces can be machined at the same time.

3.3 Mechanical Structure

Figure 13 shows the mechanical syructure of this newly developed machine. The turntable, on which the six glass blanks are set, is supported by angular contact type ball bearings, and is driven by a DC servo motor to rotate at about 4 rpm. An air-bearing type grinding spindle with a built-in motor is used, which typically rotates at about 7500 rpm. A resin-bonded diamond wheel with a diameter of 100 mm is set on the end of the spindle shaft. The air-spindle housing is supported by two ball bearings and is swung by a pulse motor. One step angle of the pulse motor corresponds to about 50 μm on the blank surface. In this construction, the figure accuracy of the machined lens depends on the rotational accuracy of the swing bearings.

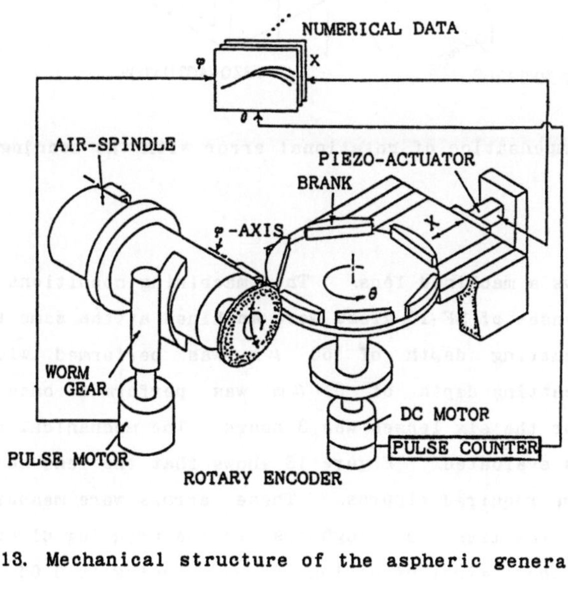

Fig. 13. Mechanical structure of the aspheric generator

To eliminate the rotational error of the swing bearings, which exceeds 0.1 μm, a new rotational error compensation system is used. This is shown in Fig. 14. A capacitance gauge senses the rotational error of the bearings as gap fluctuation between the gauge probe and the rotational shaft. A piezo actuator adjusts the bearing support frame to maintain a constant gap signal. As a result, the rotational error caused by the bearings can be compensated for. The turntable carriage slides on a linear roller guide, and its position is numerically controlled by a piezo actuator, according to the aspheric figure data. The piezo actuator can drive the carriage a maximum of 150 μm. Since the piezo actuator has a large hysteresis characteristic, a closed-loop control system utilizing a capacitance gauge is used. The positioning resolution of the system is 0.01 μm.

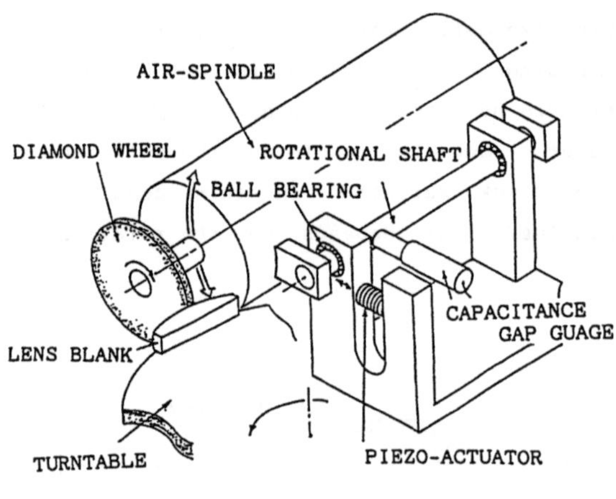

Fig. 14. Compensation of rotational error with the bearing

3.4 Evaluation

Figure 15 shows a machined lens. The machining conditions are shown in Table 2. Six lenses of SF-10 glass were machined at the same time. Rough grinding with a cutting depth of 50 μm was performed twice, and fine grinding with a cutting depth of 5 μm was performed once. The total machining time for the six lenses was 3 hours. The mechanical accuracy of a machined lens was evaluated. Figure 16 shows that the lens achieved errors of 0.2 μm for the required figures. These errors were measured using the Form Taly Surf. The measured roughness in the grinding direction is 0.05 μm Rmax, and in the nominal direction to the grinding is 0.05 μm Rmax.

An F-θ lens setup was used to evaluate the optical properties of the machined lens. Figure 17(a) shows the focusing spots with the F-θ lens system. Laser printing requires an ellipsoidal figure spot. Within the field angle of +29 and -29 degrees, 60 μm × 100 μm stable laser spots were obtained. In contrast, focusing spots with the F-θ lens system using a conventional toroidal lens without the aspheric compensation become enlarged at an angle of ± 15 degrees because of astigmatic aberration. Furthermore, the spot figure at an angle of +15 degrees is different from the one at -15 degrees. This shows the superior results of the nonaxissymmetric aspheric compensation.

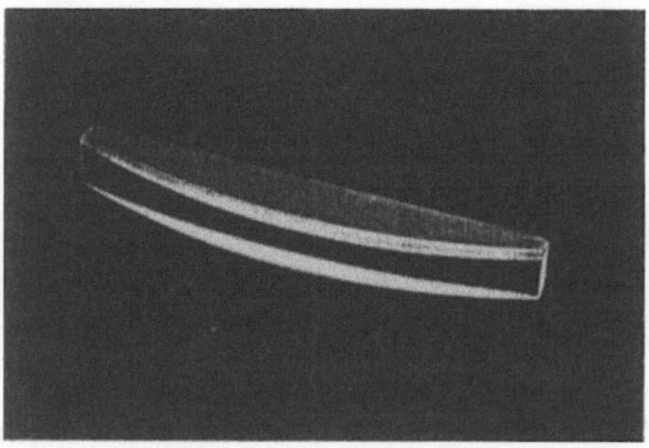

Fig. 15. The machined lens

Table 2 Grinding conditions

Material	SF-10 (glass)
Grinding wheel	Resin bonded diamond #800 (rough), #3000
Size of wheel	ϕ 100 mm × 4 mmt
Wheel speed	7500 rpm
Depth of cut	50 μm (rough), 5 μm
Feed	50 μm
Grinding fluid	CIMCOOL

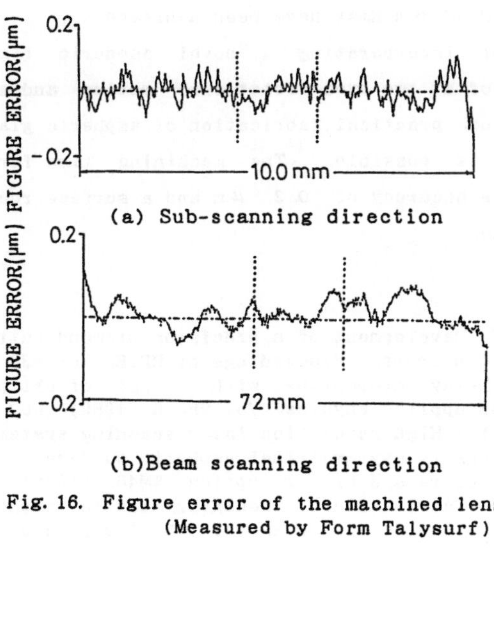

Fig. 16. Figure error of the machined lens (Measured by Form Talysurf)

(a) Beam spots by the machined lens

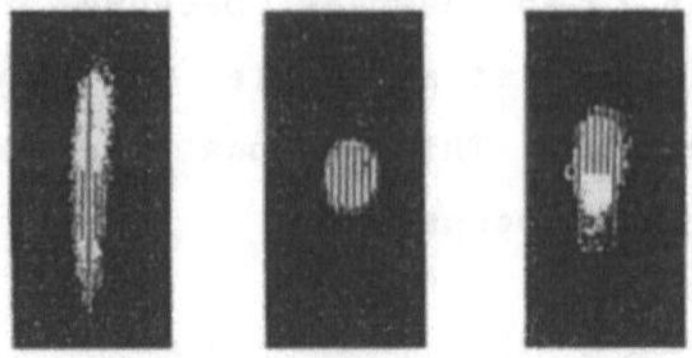

(b) Beam spots by a conventional torlic lens

Fig. 17. Beam spot properties

4. CONCLUSION

(1) As a result of incorporating cutting depth control synchronized with the tool rotation angle, the generation of free figures by diamond cutting has been made practical. The developed machine accepts a work-piece with maximum dimensions of 600 mm × 100 mm. A figure accuracy of 0.1 μm and surface roughness of 0.03 μm Rmax have been achieved.

(2) As a result of incorporating a novel aspheric figure-generating mechanism consisting of a two-axes rotational motion and a one-axis fast servo controlled motion, practical fabrication of aspheric glass lenses with a generalized figure is possible. The machining time for a lens was 30 minutes, and a figure accuracy of 0.2 μm and a surface roughness of 0.05 μm Rmax were achieved.

5. REFERENCES

1. S. Moriyama et al. ; Development of a precision diamond cutting machine of asymmetric aspheric mirrors. Proceedings of SPIE, Vol 830 (1987) 98.
2. Y. Suzuki et al. ; X-ray microprobe with a pair of elliptical mirrors. Japanese Journal of applied Physics, Vol 28, 9 (1989) 1660.
3. Akira Arimoto et al. ; High resolution laser scanning system using an F-θ lens with a rotationally asymmmetrical aspheric surface. Tech. Digest of 1988 Conf. on Lasers and Electro-optics, WM48, (1988).
4. S. Moriyama et al. ; Fabrication of nonaxissymmetric aspheric lenses for laser printers. Proccedings of SPIE, Vol 1167 (1989) 219.

Design and Modeling of a Linear Microactuator

P. Cheung, R. Horowitz, and R. Howe
Department of Mechanical Engineering
University of California at Berkeley
Berkeley, California 94720, USA

Abstract

This paper describes the design and modeling of a linear microactuator of polysilicon made by Silicon surface micromachining technology. The microactuator, two micrometers thick and less than one millimeters on the side, is built on a silicon wafer and is actuated by electrostatic forces on its interdigitated fingers. By means of a simple electro-mechanical model, dynamic equations are derived and values of force, mass, drag, and spring constant applying to the microactuator are estimated; and from these values, a transfer function is obtained and an acceleration as much as 470g is found possible. These values are then used to predict dynamic performance of the microactuator, and furthermore to study its potential use as a feedback motion control device. While the fabrication and control of these microactuators are still in the research stage, the immediate goals are to develop position detection schemes and to incorporate detection and driving circuits with the microactuator on the same wafer.

1 Introduction

For the past decade, researchers have investigated micro-devices of polysilicon, which are mechanical structures typically less than a millimeter on the side and are made from the existing technology used to fabricate integrated circuits. In the early eighties, polysilicon microsensors were developed to detect the viscosity of surrounding fluids by measuring the sensors' resonant frequencies [1]. Several years later, linear microactuators driven by electrostatic forces were developed using silicon surface micromachining and were able to travel within a range of 10 um(micrometers) or more [4]. These microactuators now start to gain the attention of control engineers because, with the integration of sensing circuits and driving circuits, these microactuators will become very useful as feedback motion control devices. To introduce these microactuators to researchers in the field of motion control, this paper describes a recent design of polysilicon microactuator, analyzes its features, characterizes its dynamic behaviors using a simple electro–mechanical model, and attempts to predict its performance and formulate a transfer function for the microactuator.

An application of the microactuators seems to be in future magnetic rigid drives, which may feature a conventional servo motor and a second fine-positioning (micro) actuator on the slider to improve tracking performance. An important source of vibration in a rigid drive comes from the pendular motion of the slider and suspension: because this vibration comes from somewhere between the servo motor and the read-write head, its effect on tracking cannot be corrected properly by a single actuator. A fast microactuator mounted between the slider and the head can be used to correct tracking errors due to this kind of vibration as well as other errors that the servo motor has failed to correct. Furthermore, with a second actuator carrying each head, all the disk surfaces inside the same drive can be tracked simultaneously and together they can form a disk array, which boosts data transfer rate but without the costs and reliability problems that come with several rigid drives.

In addition, microactuators may also be used as micro-valves, micro-pumps, mechanical memory-cells, moving light-reflectors, as the carrier of a scanning tunnel microscope, or in some other emerging applications.

2 Design of a Linear Microactuator

There are many ways to design a microactuator, possibly most are not developed yet. The design here follows from the scheme of [4] which was developed at the University of California at Berkeley.

This linear microactuator is a planar structure built on the surface of a silicon wafer. It is approximately 900 um long, 450 um wide, and only 2 um thick, and it comprises five essential parts: The body, driving fingers, suspension beams, anchors, and position-detection fingers (see Figure 1). At present, the body is only a small rectangular piece of polysilicon connecting the other parts.

Four branches of driving fingers extend from the corners of the body; and fifteen fingers protract from either side of each branch and interleave with stationary polysilicon fingers to form an interdigitated pattern. When a voltage is applied to the stationary fingers (with the microactuator remains as electrical ground), electrostatic attraction

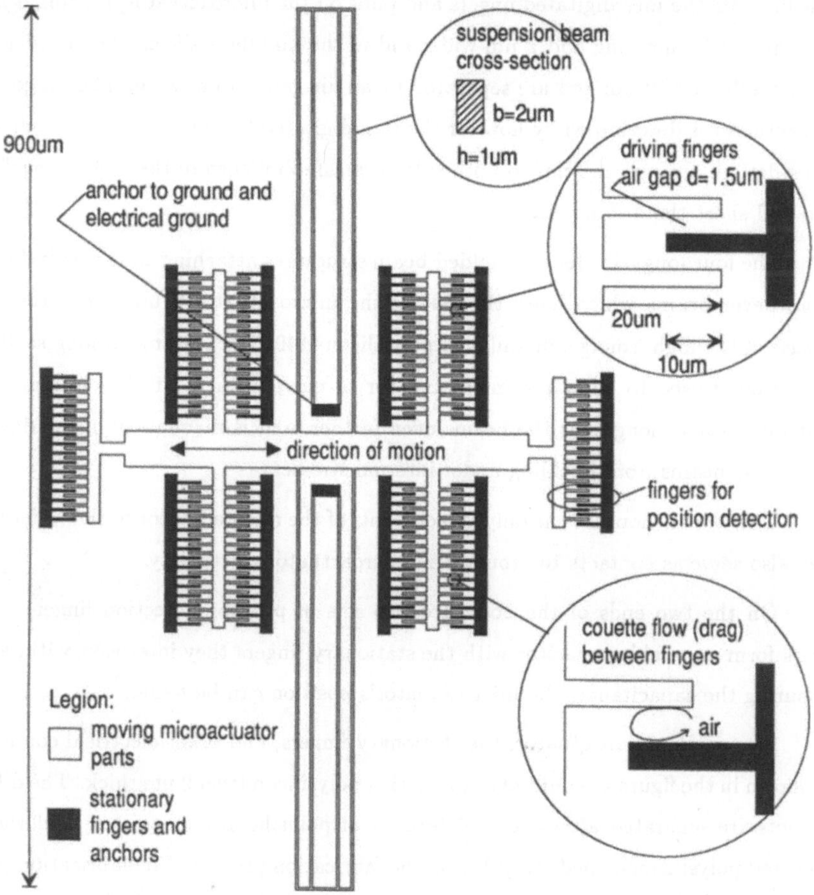

Figure 1: Top view of a linear microactuator.

forms between the interdigitated fingers and pulls on the microactuator, making it move. Each finger is 20 um long and 2 um wide; and in the middle position, the interdigitated fingers overlap by 10 um and are separated by an air gap 1.5 um wide. The range of the microactuator is determined by how far the interdigitated fingers can go into each other. In this design, a range of 8 um to both sides is a safe limit before the fingers touch each other and short the driving circuit.

The four long, slender, and folded beam structures attaching to the body function as cantilever beams which hang the rest of the microactuator 2 um above the wafer. Because of the high Young's modulus of polysilicon (140 GPa), beams as long as 400 um per section is used to allow the microactuator to move linearly within the range. The folding of sections lengthens the beams even further, and is mandatory in the design to prevent the beams from buckling under internal stresses.

The two anchors are the only attachments of the microactuator to the wafer below. They also serve as contacts to ground the microactuator electrically.

On the two ends of the body are two sets of position-detection fingers. These fingers form a variable capacitor with the stationary fingers they interleave with, and by measuring the capacitance, the microactuator's position can be found.

The entire microactuator, the stationary fingers, and other electrical connections not shown in the figure are built of a conducting polysilicon layer 2 um thick. The different elements are separated after a complete layer of polysilicon is deposited, and then the unwanted polysilicon etched away during the fabrication process. The fabrication process employed to build this microactuator is called Silicon Surface Micromachining, which is a chemical process that modifies the surface of a silicon wafer to build tiny planar mechanical structures, down to feature sizes of 1 um or less [2]. Even though the process flow in silicon surface micromachining is an important part of making the microactuator, in the spirit of this paper it will not be discussed further.

Polysilicon is a form of silicon also known as polycrystalline silicon and is distinguishable from single-crystal silicon by having small grains of crystal lumped one over another. Because it has mechanical properties equal to or exceeding steel, polysilicon is chosen as the structural material for this microactuator. Another reason for choosing polysilicon which is not obvious at this point is that polysilicon microactuator can be fabricated next to CMOS circuits which will enable sensing and driving circuits to be placed

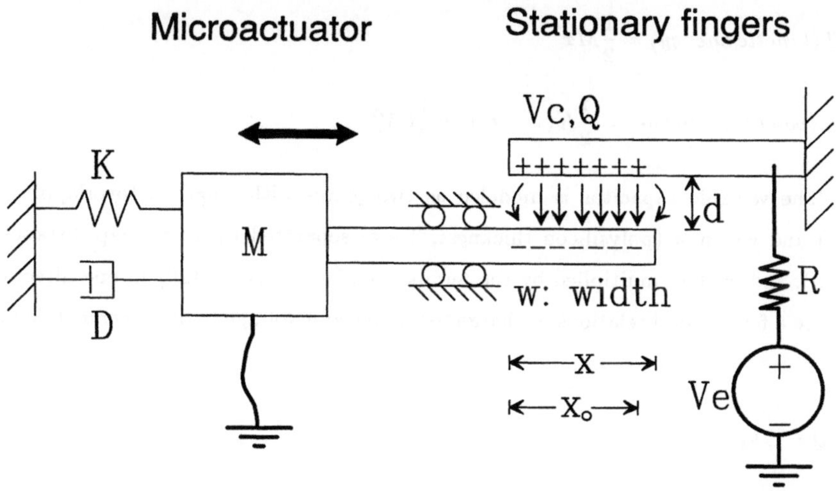

Figure 2: Electromechanical model of the linear microactuator

close to the microactuator to reduce noise interference. The combined microactuator-circuit fabrication will also ensure a low cost and high volume production.

3 Electro-mechanical Modeling

The mechanical behavior of the linear microactuator is analyzed using a simple electro-mechanical model (see Figure 2). The microactuator is modeled as a mass (M) effected by spring (K) and damping (D) representing the suspension by the cantilever beams and the air drag respectively. The microactuator is constrained to move without friction in the direction of the fingers.

The electrical counterpart is modeled by a driving voltage source (Ve), a variable capacitor (C) with voltage (Vc) and charge (Q), which represents the capacitive interaction between the moving and the fixed fingers, and a resistor (R) in series, which represent finite resistance of polysilicon. In addition, the microactuator is also ground electrically via the anchors.

To obtain the dynamic equations of the model, the Lagrange's Equation is operated on (T-V), with T and V defined as follows:

$$T(kinetic\, energy) = \frac{1}{2}M\dot{x}^2$$

$$V(potential\, energy) = \frac{1}{2}K(x - x_o)^2 + \frac{1}{2}CV_c^2$$

The variable capacitor is modeled as two plates with length x (overlapping of fingers) and width w (polysilicon thickness, 2 um) separated by an air gap distance d (1.5 um), and is then multiplied by the number of air gaps (N = 120) in one direction. There are a few general relations of charge to voltage and voltage to displacement defined below:

$$Q = CV_c \tag{1}$$

$$C(= \frac{\epsilon_o A}{d}) = N\frac{\epsilon_o w}{d}x \tag{2}$$

$$\dot{C}(= \frac{dC}{dt}) = N\frac{\epsilon_o w}{d}\dot{x} \tag{3}$$

From the above equations, together with Kirchoff's loop equation ($V_e = V_c + R(dQ/dt)$), three differential equations with the variables: x, $\dot{x}$, and V_c are obtained. However, with typical dimensions and resistance value V_c will approach V_e in the order of several pico-seconds, so as a fast time constant variable, V_c can be separated from x and $\dot{x}$. Subsequently, the "slower" mechanical system behaves as a second order differential equation:

$$M\ddot{x} = F - K(x - x_o) - D\dot{x}, \tag{4}$$

with F being the external force in the Lagrange's Equation and also being the electrostatic force inside the variable capacitor in the model.

3.1 Finding the Electrostatic Force

From the Lagrange's Equation (not shown here),

$$F = N(\frac{\epsilon_o w}{2d})V_e^2 \tag{5}$$

Assuming a maximum driving voltage of 30 volts is applied to the interdigitated fingers, an electrostatic force $F = 637nN$ which remains constant with position will be generated to move the microactuator.

To vary the amount of electrostatic driving force, V_e can be controlled by two methods: First, Pulse Width Modulation (PWM) can be applied to the stationary fingers in both directions so that when the PWM is at 50 percent duty cycle the net pulling force vanishes. The use of PWM simplifies the driver circuits and also it applies a constant levitation force(a electrostatic repelling force that lifts the microactuator away from the wafer) [3] to the microactuator.

As an alternative, the magnitude of the presumably digital command can be fed into two digital-to-analog convertors(DAC) each driving the microactuator in one direction. The sign signal can then be used to enable the DAC to drive in the desired direction while disabling the other DAC.

3.2 Finding the Mass of Microactuator

Mass of a linear microactuator decreases in the cubic power to its dimension, hence it drops off tremendously when the dimension of the actuator is in the tens of micrometers. Gravitation effect from the actuator is relatively insignificant as Mg is only approximately 1 nN, whereas the electrostatic forces are several hundred times greater. In the following calculation, only half of the mass of the suspension beam structure is included in the microactuator mass because on average, the beams only move half as much distance as the microactuator does:

$$\begin{aligned}
\text{Given } A &= 30046\mu m^2 \text{ (total surface of microactuator)} \\
w &= 2\mu m \text{ (thickness of microactuator)} \\
\rho &= 2300Kg/m^3 \text{ (density of polysilicon)}
\end{aligned}$$

$$\text{Thus } M = 1.38 * 10^{-10} Kg (= Aw\rho) \text{ (microactuator mass)}$$

3.3 Finding the air drag

The drag around the microactuator is obtained in the following equation by summing the Couette flow effects of air contributed from between the microactuator and the wafer and between the interdigitated fingers under ambient conditions:

$$D = \Sigma \frac{\mu_{air} A}{d} = 3.42 * 10^{-7} N/(m/s), \tag{6}$$

where μ_{air} is the Coefficient of viscosity of air, 0.000018 Kg/m.s; the area (A) between the microactuator and the wafer is $30046 \mu m^2$, and the corresponding air gap is 2 um wide; the total area between interdigitated fingers, including all the driving fingers and position-detection fingers, is $6000 \mu m^2$, with an air gap distance, 1.5 um, equal to the separation of fingers. The drag on the frontal area of the microactuator, however, is neglected to simplify calculation.

3.4 Finding the Spring Constant of the Suspension

There are four individual folded-beam structures suspending the microactuator, each structure comprises two 400 um long polysilicon beams that can be modeled as fixed-guided beams. Hence, the total spring constant (K) for the eight-beam system is twice that of a single beam, or $K = 24EI/l^3$. The Young's modulus (E) of polysilicon has earlier been measured to be approximately 140 GPa (under different process condition the modulus has found to vary drastically), and the area moment of inertia of a beam is calculated to be $I = 0.167 \mu m^4$ (see beam cross-section in Figure 1). Therefore the spring constant is found to be:

$$K = 0.00875 N/m.$$

4 Predicted Performance

The F, M, D, and K values obtained in the last section can be use to predict some dynamical performances of the microactuator. If two of the three factors M, D, or K are ignored, the maximum acceleration, terminal speed, and maximum spring deflection can be obtained, which, together with a given range of travel, can be used to predict bang-bang performance. Lastly, a transfer function and the natural frequency can also be estimated.

Hence, ignoring the other two factors,

Acceleration (F/M)	=	4612 m/s
or,	=	470g
		(ignores D and K)
Terminal speed (F/D)	=	1.86 m/s
		(ignores M and K)
Maximum spring deflection (F/K)	=	73 μm
		(ignores M and D)

The acceleration of the microactuator is almost ten times the typical maximum acceleration one can find in a magnetic rigid drive. Hence it seems that the tracking errors of the servo motor can be corrected by this microactuator if it is applied inside a rigid drive as a fine-positioning actuator. However, once some load like a read-write head is mounted on, the total mass will slow down the microactuator drastically. This concern prompts researchers to develop ways to increase the actuation force, such as by thickening the microactuator.

Nevertheless, in the present design, this acceleration is maintained within 10 percent of 470g in the range of $\pm 8\mu m$. It is because the maximum spring deflection is almost ten times greater than the range, which means that within the range, the opposing force from the spring will be less than one tenth of the electrostatic force; and besides, the spring constant should be fairly linear within the range.

Analyzing the bang-bang performance of a moving system can always tells us how fast an object can move from one spot to another. Here, if we use the same acceleration as before (4612 m/s) and assume a 10 um distance to travel, then the total bang-bang time of travel is ($t = \sqrt{L/A}$) = $93.1\mu sec$ and the peak velocity reached is ($V_{peak} = \sqrt{AL}$) = $0.21 m/s$.

The peak velocity reached by a bang-bang operation is only eleven percent of the terminal speed, therefore the drag factor do not seem to affect acceleration significantly; adding to this the square root factors in the bang-bang equations, the accuracy of the predicted bang-bang performance should be within ten percent.

Combining the values obtained from above and following from equation (4), we arrived at a transfer function of the microactuator:

$$G(s) = \frac{X(s)}{F(s)} = \frac{1}{M}\frac{1}{s^2 + 2\zeta\omega_n + \omega_n^2}, \tag{7}$$

$$\begin{aligned} \text{where M} &= 1.38 * 10^{-10} Kg \\ \omega_n &= 7950 \text{ radians/sec } (2\pi Freq_n) \\ \zeta &= 0.156 \text{ (damping ratio, or quality factor Q = 3.2)} \end{aligned}$$

Moreover, the natural frequency of the microactuator can be found as $Freq_n = \omega_n/2\pi = 1.26KHz$. This natural frequency may be too high for application in magnetic rigid drive because it is too close to the typical open loop crossover frequency found in most high performance rigid drives. Lower natural frequencies can be obtained if the beams are folded again or the beam width is reduced further.

5 Control, Position Detection, and Future Work

The microactuator is essentially a second order system, so if both position and velocity of the microactuator are obtained, state feedback techniques can be used to control the microactuator to any desired behavior. The foremost task of this microactuator research is to develop a practical method to detect position and velocity. The microactuator's position can be measured accurately by measuring the difference of the complementary horizontal capacitances between interdigitated fingers of the two position-detection structures. However, because this change in capacitance is in the order of femto-Farads, on-chip detection circuits must be used to ensure accuracy.

Detecting the microactuator's speed by means of measuring current coming out of a set of interdigitated fingers has been found to be more difficult than detecting position, as the signal under practical speeds are too small to be detected. Hence, velocity sensors are not included in this microactuator design. Nevertheless, the position signal can be differentiated to obtain velocity.

Research is being done now to incorporate CMOS sensing and driving circuits with microactuators on the same wafer [6]. To increase the actuator force, the microactuator can be made thicker perhaps by a technology different from Silicon surface micromachining. Moreover, to keep the suspension beams from being too long, other methods to suspend the microactuator is being studied.

Furthermore, the control of a dual-actuator system using the microactuator as one of the actuator are also being studied [5]. Once a sample of this microactuator is fabricated, mechanical behaviors such as higher modes of resonance can further be characterized to obtain a better understanding of this new kind of actuator.

6 Conclusion

Microactuators less than 1 mm on the side, with feature sizes down to 1 micrometers, can now be built of polysilicon by Silicon surface micromachining. These microactuators have potentials to become accurate feedback motion control devices so researchers have started to study their characteristics. The device presented in this paper is a planar structure with a thickness of 2 um. It is driven electrostatically by applying a voltage on a stationary set of fingers which interleave with the microactuator's fingers and move the latter. Long cantilever beams suspend the entire microactuator 2 um above a silicon wafer, where the microactuator is fabricated on.

The dynamic behavior of the microactuator is identified by using a simple electromechanical model. The actuation force, mass, drag, and spring constant applied onto the microactuator are estimated. These figures are then used to predict some dynamic performance such as the maximum acceleration, bang-bang travel time, and the transfer function. Acceleration up to 470g seems possible with this "fast" microactuator; it can traverse 10 um in 93.1 usec in a bang-bang operation; its natural frequency is estimated at 1.26KHz and it is lightly damped with a damping ratio of 0.156.

References

[1] Roger T. Howe. *Integrated Silicon Electromechanical Vapor Sensor.* PhD thesis, University of California at Berkeley, 1984.

[2] Roger T. Howe. Surface micromachining for microsensors and micoractuators. *J. of Vacuum Science and Technology*, Nov/Dec 1988.

[3] William C. Tang, Martin G. Lim, and Roger T. Howe. Electrostatically balanced comb drive for controlled levitation. BSAC, EECS, U of California at Berkeley, 1990.

[4] William C. Tang, Tu-Cuong H. Nguyen, and R.T. Howe. Laterally driven polysilicon resonant microstructures. *Sensors and Actuators*, 20:25–32, 1989.

[5] J.Y. Yen, K. Hallamasek, and R. Horowitz. Track following controller design for a compound disk drive actuator. *ASME Journal of Dynamic Systems, Measurement and Controls*, Sept 1990.

[6] Weijie Yun. Cmos metallization for integration with micromachining processes. Master's thesis, U of California at Berkeley, 1989.

Photon Scanning Tunneling Microscopy and Reflection Scanning Microscopy

J.P. GOUDONNET, L. SALOMON, N. CERRE, F. DE FORNEL

Université de Bourgogne, Laboratoire Physique du Solide, Section Photoélectricité, UA 785, Faculté des Sciences Mirande, BP 138, 21004 DIJON - FRANCE

and

T.L. FERRELL

Oak Ridge National Laboratory, Health and Safety Research Division, Oak Ridge, Tennessee 37831

Summary

The Photon Scanning Tunneling Microscope (PSTM) is the photon analogue to the Electron Scanning Tunneling Microscope (ESTM). It uses the evanescent field due to the total internal reflection (TIR) of a light beam in a prism modulated by a sample attached to the prism. The exponential decay of the evanescent field is characterized by the penetration depth d_p and depends on the angle of incidence θ, the wavelength and polarization of the incident beam. Changes in intensity are monitored by a probe tip scanned over the surface, and the data are processed to generate an image of the sample. Images produced by a prototype instrument are shown to have a vertical resolution of about 3 Å and a lateral resolution of about 100 Å.

To image non transparent samples, we have developed the Reflection Scanning Tunneling Microscopy (RSM), a method for producing a point light reflected by the sample. The microscope which uses an optical fiber coupler is easily installed on an existing PSTM.

I. PHOTON SCANNING TUNNELING MICROSCOPE

1) Introduction

As indicated before by Pohl and al [1] and generally speaking, Scanning Near Field Optical Microscopy belongs to the family of SXM Microscopies which exploit a distant dependent interaction for guidance of a probe tip and local information about this interaction.

The foundation for the operation of the Photon Scanning Tunneling Microscope (PSTM) [2] is the Total Internal Reflection (TIR) of light incident on an interface between materials of different refractive indices (n_1 and n_2) when the incident beam lies in the medium of higher index (n_2).

TIR takes place when the angle of incidence θi (relative to the normal) exceeds the critical angle given by :

$$\theta_c = \arc\sin\left(\frac{n_2}{n_1}\right)$$

For $\theta_i > \theta_c$ an evanescent field is produced in the medium of lesser index. The intensity of this field decreases exponentially with increasing distance from the interface according to :

$$I \sim \exp\left\{ - 2kz\left[\sin^2\theta_i - \left(\frac{n_2}{n_1}\right)^2\right]\right\} \qquad (1)$$

where k is the magnitude of the wave vector of the incident light and z the distance from the interface [3].

If another medium of index greater than n_2 is brought within the evanescent field, such as the tip of a sharpened optical fiber, a situation arises analogous to that of a finite tunneling barrrier in quantum mechanics. That is, a potential barrier for particles of energy below the barrier height is perfectly analogous to the present case, although the wave functions vary in character depending upon the type of particle [2].

2) Experimental set up.

The experimental set up has been already described [2, 4, 5, 6)

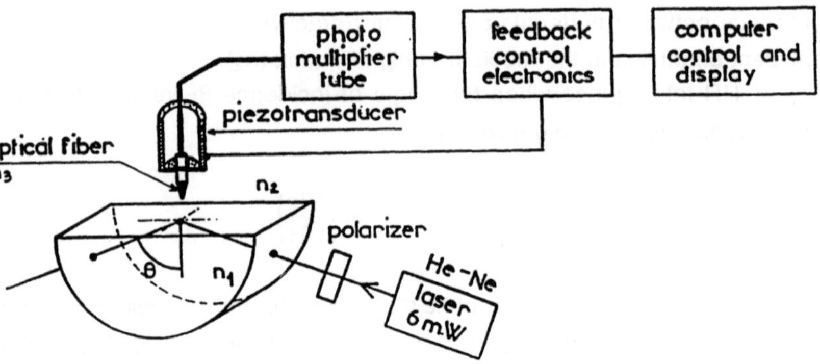

Fig. 1 Schematic of the experimental set-up. The evanescent field near the sample is probed by a fiber optic which is scanned by a piezotransducer.

The PSTM consists of a probe fiber optic tip mounted on a piezoelectric tube (Fig. 1) similar to those used in scanning tunneling microscopy [7]. Voltages applied to the tube allow detailed control of the motion of the tip in three dimensions. Depending on the length and diameter of the tubes different scanning area are possible ranging from 1μm x 1μm up to 10 μm x 10 μm tangential to the sample. z motion range is 3.3 μm.

As the tip is scanned, feedback electronics is used to maintain a constant photomultiplier tube current by lowering or raising the tip. The control voltages can be processed on a computer with suitable software to provide the three dimensional data in various forms.

The light source is a 2 mW He-Ne laser (λ_0 = 6328 Å). A laser is not essential. A white light source has been successfully used with a monochromator to yield a continuous range of wavelengths. Photons from the incident beam are totally reflected on the base of a glass semicylindrical prism and tunnel through the region between the tip and sample. In most cases the sample is placed on a microscope slide coupled to the base with an index-matching fluid. The presence of the sample on the TIR surface changes the evanescent field intensity and provide topographical information about the sample surface as well as information about the optical properties of the sample.

3) Resolution and images

To make a PSTM sensitive to dimensions smaller than those detected by the other optical microscopes, it is important that the distance across which the signal changes is on the same scale as the required resolution.

The decay length d_p for the evanescent field intensity as given by Eq.1 is :

$$d_p = \frac{\lambda_0}{2\pi} \frac{1}{\sqrt{n_1^2 \sin^2 \theta_i - n_2^2}} \qquad (2)$$

and depends on the wavelength λ_0 and the indices of refraction n_1 and n_2.

Figure 2 shows the penetration depth d_p calculated as a function of θ for two values λ_1 = 6328 Å and λ_2 = 2500 Å of the wavelength and n_1 = 1,458.

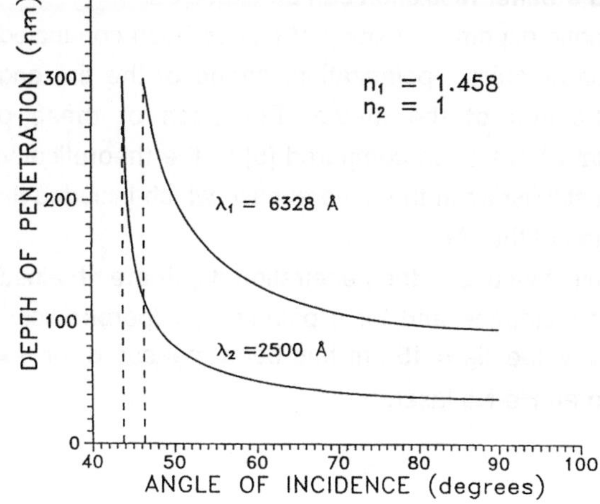

It can be seen that d_p is smaller when λ decreases. For a given wavelength $\lambda_0 = 6328$ Å, d_p varies when n_1 is changed as it can be seen on Fig. 3.

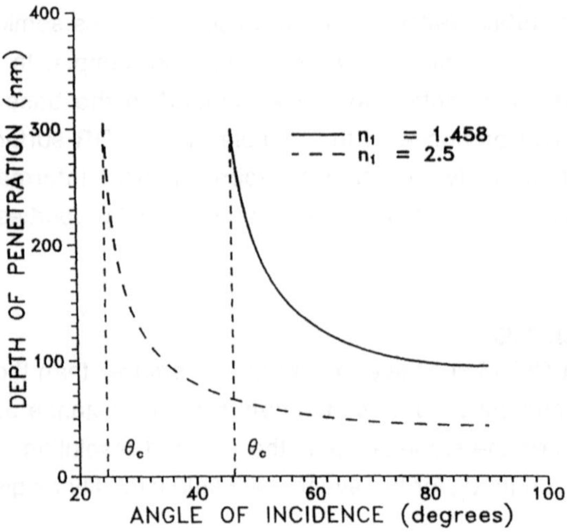

Fig. 3 Penetration depth calculated as a function of the angle of incidence for two values of the index of refraction $n_1 = 1.458$ (solid line) and $n_1 = 2.5$ (dashed line) of the prism.

From Fig. 2 and Fig. 3 it can be seen that for $\lambda_0 = 6328$ Å a penetration depth of 100 nm can be obtained with a glass prism (n1 = 1.458) illuminated at near grazing incidence (85°). A shorter decay length is obtained with a shorter wavelength and a better resolution can be expected.

A systematic exprimental study of d_p has been conducted as a function of the following parameters : polarization, shape of the tip, angle of incidence, indices of refraction of the media. For each of these parameters, the experimental data have been compared [5] to the theoretical values calculated from equation established in the general case which includes the value of n_3 the index of refraction of the fiber.

The smallest values of the penetration depth are obtained for large values of the angle of incidence and for s polarization (perpendical to the phase of incidence). The value $d_p = 45$ nm has been measured for $\theta = 85°$ by using a glass prism and an He-Ne laser.

By using X-rays near grazing incidence and a scintillator mounted on the fiber, a molecular scale resolution could be obtained if the samples were mounted on a metal prism. There are many possibilities for surface science research in this case.

The PSTM was calibrated in operation by looking at a holographic grating sample with a known line spacing of 0.6 μm. The grating structure was formed in a layer of transparent photoresist on a quartz substrate. A PSTM image of this sample is shown on Fig. 4.

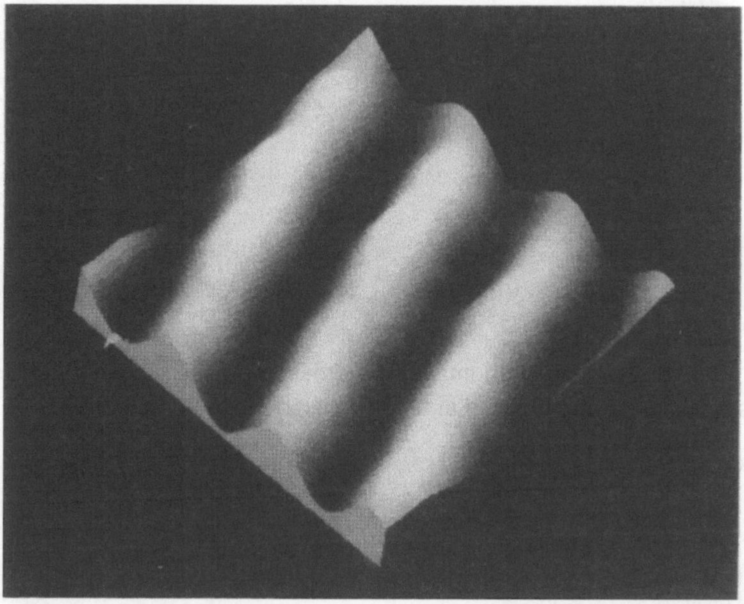

We also used the PSTM to image the surface of a clean microscope slide. Fig. 5 shows a 2 x 2 μm region of the surface. The subwavelength - sized roughness is seen with corregations on the order of 70 Å deep.

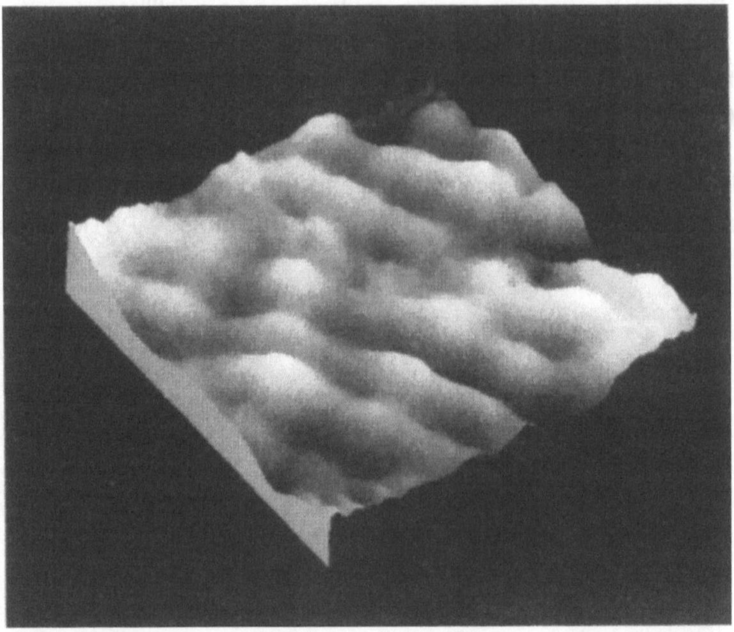

On Fig. 6 we show an image of a crossed grating etched into glass. The interference lines were crossed at 20°. The scan range is 3 x 6 μm.

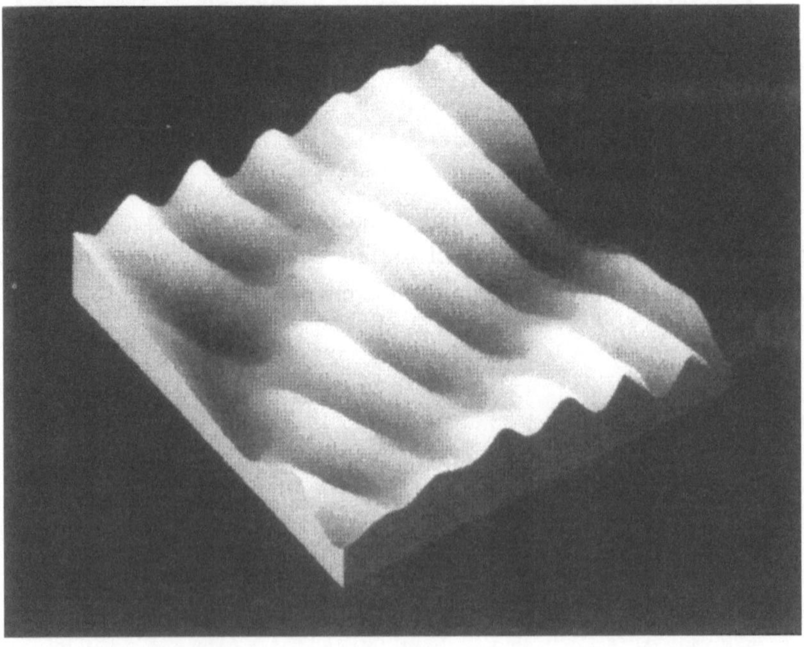

4) PSTM Spectroscopy

Suppose that a sample to be observed by Optical Scanning Tunneling Microscopy has a spatially varying chemical composition on a scale of tens or hundreds of nanometers. If spectroscopy can be carried out simultaneously with the microscopy, then a chemical map can be superimposed on the topology. The specific compound in a given region can be identified and distinguished in the image, say by absorption or Raman Spectroscopy. To prevent confusion between an absorbing site and a site at which there is a valley in the sample, the tip can be dipped at the site to determine the exact form of the intensity curve.

In the case of absorption spectroscopy, a sample mounted on a microscope slide and coupled to a hemicylindrical or prism base is examined by scanning and dipping the probe tip at each scan point at a given wavelength. The wavelength of the light is then changed and the process is repeated. The wavelength absorbed at each site provides an identification of the chemical compound at that site, although for infrared wavelengths there are some alterations that are utilised. By assigning a different color to each compound and superimposing this information on the height - shaded topographical image, the chemical map is made readily apparent.

II. REFLECTION SCANNING MICROSCOPE (8)

The experimental set-up is shown in Fig. 7

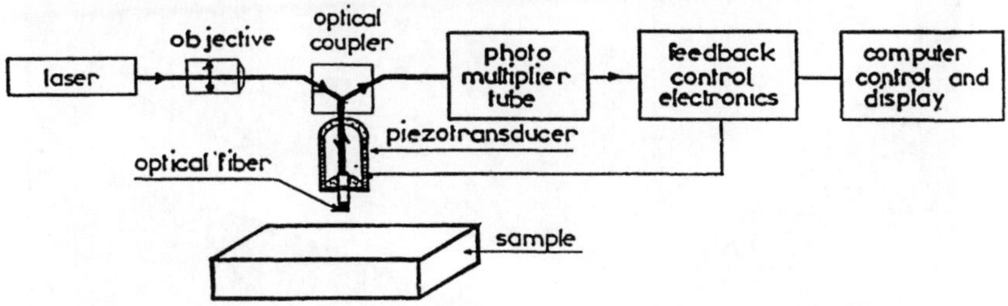

Light from a 5mW He-Ne laser (λ = 632.5 nm) is coupled to the end of an optical fiber by a microscope objective. A commercial coupler, made by fusion of two monomode optical fibers with 3 μ cores and 125 μ cladding diameters, is used to separate the incident light on the sample and the reflected light guided to a photomultiplier tube.

The optical fiber is mounted in a piezoelectric tube which allows controlled movements on a subnanometer scale.

The piezotube itself is mounted on an inchworm type micropositionner for a conventional and accurate corse approach. The rest of electronics, scanning system, feedback control, computer control, image processor is similar to the STM and PSTM electronics.

We have studied two different samples : a metallic mirror and a regular metallic grating. The mirror was used to compare the experimental data with the theoretical calculations and to give an estimation of the decay length. Images of the already well characterized grating were supposed to provide enough information for evaluating the interaction of the tip with the sample in order to avoid the artifacts or miss interpretations of images that could occur in most of this new generation scanning microscopes.

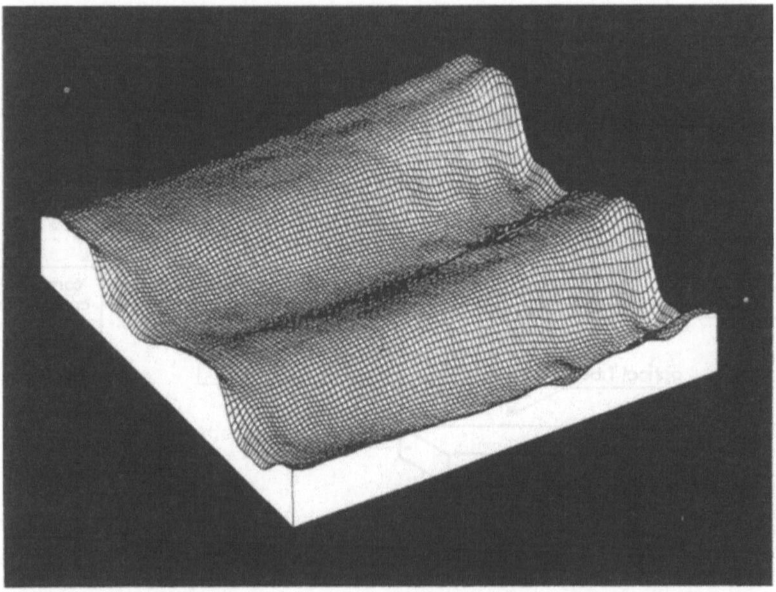

Fig. 8 is an image of a metallized silicon wafer with a scan range of 3 μm x 3 μm

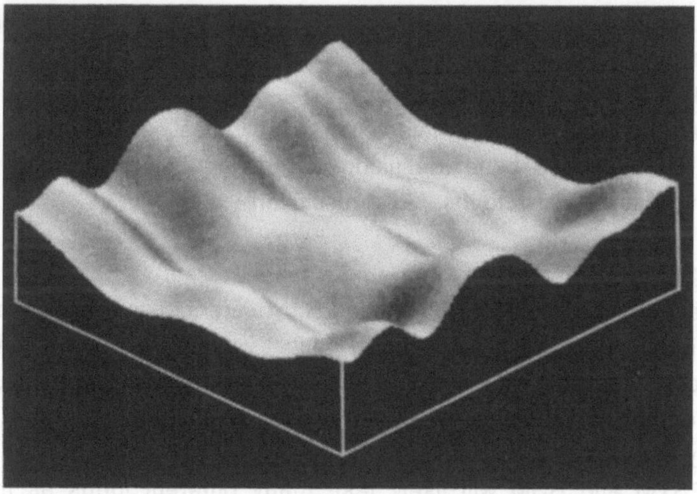

Fig. 9 shows an image (4 μm x 4 μm of a nickel grating on silicon.

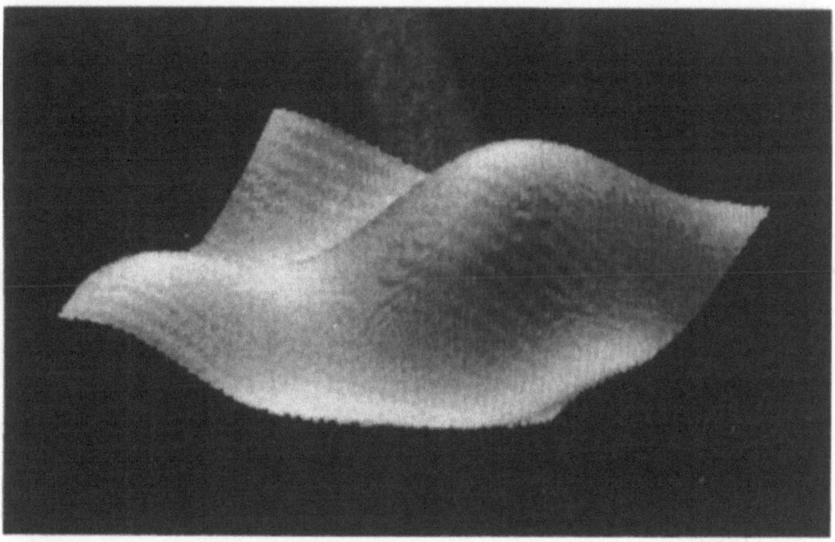

Fig. 10 SiO_2 photoresists posts on Si

The RSM images we have obtained, so far, show that a resolution better than 0,3 μm can be reached making this microscope an excellent tool for studying thick or non transparent samples.

A better understanding of the interaction of the optical fields in and out the optical fiber, as well as a better definition of the parameters will lead to a resolution which can be close to the PSTM resolution.

III. CONCLUSION

A variety of applications of PSTM's and RSM's are forthcoming : there are possibilities in microlithography, probing of optical waveguides, and testing of optical surfaces. Biological work could benefit considerably from high resolution chemical mapping of samples in a more natural environment than that demanded by other probes.

The PSTM and RSM will likely take many different forms as these applications evolve, but all will use the basic principle of tunneling as provided by fundamental physics.

REFERENCES

1. D.W. POHL, U.C. FISHER and U.T. DURIG, J. of Microscopy, 152, 853, (1986)
2. R.C. REDDICK, R.J. WARMACK and T.L. FERRELL, Phys. Rev. B, 39, 767, (1989)
3. N.J. HARRICK, Internal Reflection Spectroscopy, J. Wiley and Sons (eds), 1967
4. F. DE FORNEL, J.P. GOUDONNET, L. SALOMON, E. LESNIEWSKA, Conf. SPIE, Paris, April 1989, SPIE Proc. Vol. 1139
5. L. SALOMON, F. DE FORNEL and J.P. GOUDONNET proposed to J. of Opt. Soc. Am. (1991)
6. D. COURJON, K. SARAYEDDINE and M. SPAJER, Opt. Commun., 71, 23, (1989)
7. G. BINNIG and H. ROHRER, Surf. Sci., 126, 236, (1983)
8. N. CERRE, F. DE FORNEL and J.P. GOUDONNET, proposed to J. of Appl. Optics

Design and Performance Considerations of the Continuos Four-roll Bender: A Precision Machine for the Roller Bending of Plates

Dr. M. Hua
Data Products Components(HK)Ltd.
5/11 Floor, Blocks C&D
Sime Darby Industrial Est. Centre
420 Kwung Tong Rd., Kowloon
Hong Kong

Dr. K. Baines
Senior Lecturer
Department of Manufacturing Eng.
City Polytechnic of Hong Kong
Kowloon
Hong Kong

Professor D. H. Sansome
Managing Director
Technoform-Sonics Limited
9 Enterprise Trading Estate
Pedmore Road
Brierley Hill
West Midlands DY5 1TX

Summary

The production of high quality tubular sections by the roller bending of plates through a set of three- or four-rollers requires a thorough understanding of the mechanics of the process. Although some analyses have been conducted of the three-roll pyramid roller bender there is little information relating to the inherently more precise four-roll configuration. To simulate a prototype four-roll bender an extensive design study was undertaken from which evolved an experimental model bender. The comprehensive instrumentation, which included roll load, bending and torque transducers, allowed single and multi-pass continuous bending processes to be studied.

Notation

Symbols and subscripts used in this paper are either in standard usage or defined in the text.

Introduction

Cylindrical tube sections are widely used in most engineering structures, such as pressure vessels, heat exchanger shells, boiler chambers etc. They also form a major structural module of modern off-shore oil rigs, tunnels and industrial buildings.

Large diameter sections are more economically manufactured by welding together plate which has been bent to form cylindrical lengths. The existing plate bending techniques available are stamp-bending and stretch-bending etc. However, they require a relatively heavy tooling and labour investment.

The roller bender is a more versatile machine and is becoming more widely used for the production of tubular sections. The advantages of the roller bender over the plate-bending machine are that:

(i) It may be used to bend plate with precision to different sectional shapes.

(ii) It can bend plate to different radii, with a minimum radius just larger than that of the top roll of the bender.

(iii) It minimises the remaining straight section of a finished bent plate.

The advantage of the roller plate bending process has been fully realized and various types of roller benders developed to fulfil different working functions [1], [2], [3] and [4]. However there is little documentation of the mechanics of the roller bending of plate and it is for this reason that a study [5] of the more precise four-roller plate bending process was initiated. Such a bender, the Verrina, was installed at Head Wrightson Teesdale (HWT) and to realise its full potential a model bender was designed at Aston University to simulate and optimise the HWT bender.

Objectives

The objectives of this section of the research were to:

(i) produce a reliable theoretical prediction of the forces and torques on the bender rolls for various plate conditions, and

(ii) verify the theoretical analysis by carrying out the relevant experiments.

Model design

To produce a tested theory which would predict data for the roll bending process, it was decided to design and build a model bender which would:

(i) simulate the Verrina roll bender at HWT, and

(ii) embody various physical features of three- or four-roll benders.

Model making analysis

It is difficult to achieve all the geometrical configurations of the model bender as a particular scale of the Verrina bender. Consequently, various ratios of independent parameters relating the model to the prototype (the Verrina) were determined and subsequently used.

The following abbreviated analysis is an attempt to highlight the important stages in the model making process.

Consideration of the rolls

It was assumed that the rolls of the model would simulate the stiffness of the prototype, under its critical operation, if the deflection of the rolls was estimated on the basis of their being simply supported. Thus with the roll stiffness (λ) defined as δ/R it can be shown that:

$$\frac{\lambda_m}{\lambda_p} = \left(\frac{P_m}{P_p}\right)\left(\frac{L_m}{L_p}\right)^3\left(\frac{E_p}{E_m}\right)\left(\frac{d_p}{d_m}\right)^4\left(\frac{R_p}{R_m}\right) \qquad (1)$$

Equation (1) is the deflection ratio of the model and the Verrina bender respectively. Thus, the value of λ_m or λ_p will be obtained if all the ratios of the dimensionless quantities on the r.h.s. of either λ_m or λ_p are known or are assumed. The deflection analysis for the end-clamped roll also gave the same expression as (1).

Consideration of the bending force ratio of the model and prototype (P_m/P_p)

From plate bending theory, the elastic strain-stress relationship, the von Mises yield criterion for the plane strain problem and the assumption of a perfect plastic material, the yield strain at the interface (distance c from the neutral axis) can be expressed as:

$$ {}_{x}e_{o} = \frac{c}{R} = \frac{a}{R_y} = \frac{Y\,(1-\nu^2)}{E\sqrt{1-\nu+\nu^2}} \tag{2}$$

where Y = uni-axial tensile yield stress of the plate material

a = distance of the outer fibre from the plate neutral axis

R_y = radius of curvature as yield first occurs

The moment due to the internal resistance of the cross-sectional area of the bent plate can be expressed as the sum of the elastic and plastic components.

After integration, rearrangement and appropriate substitution for the external bending moment at the clamped edge:

$$ P = \frac{2Yb}{\sqrt{3}\,L}\left(a^2 - \frac{c^2}{3}\right) \tag{3}$$

These equations (2) and (3) gave an approximate estimation of the load (P) required to bend the plate to a radius R.

Consideration of the equivalent torque for plate bending (T_b)

This rudimentary analysis was made to estimate the torque on the top roll to bend the plate, assuming one end clamped and the other end loaded. By equating the total strain energy absorbed in the plate to the external work done by the equivalent torque it can be shown that:

$$ \frac{K_m}{K_p} = \left(\frac{Y_m}{Y_p}\right)^2 \left(\frac{A_m}{A_p}\right)^2 \left(\frac{L_m}{L_p}\right)\left(\frac{I_p}{I_m}\right)\left(\frac{E_p}{E_m}\right)\left(\frac{\theta_{dp}}{\theta_{dm}}\right)\left(\frac{R_p}{R_m}\right)\left(\frac{b_m}{b_p}\right)^2 \tag{4}$$

where $K = T_b/R$, and $A = a^2 - c^2/3$

Equation (4) is the equivalent torque ratio under the condition of maximum bending moment at the clamped edge.

A similar analysis of the forces and torques on the side and pinch rolls was carried out by taking account of the geometrical configuration of the roller bender for the pre-bending condition shown in Figures 1 and 2.

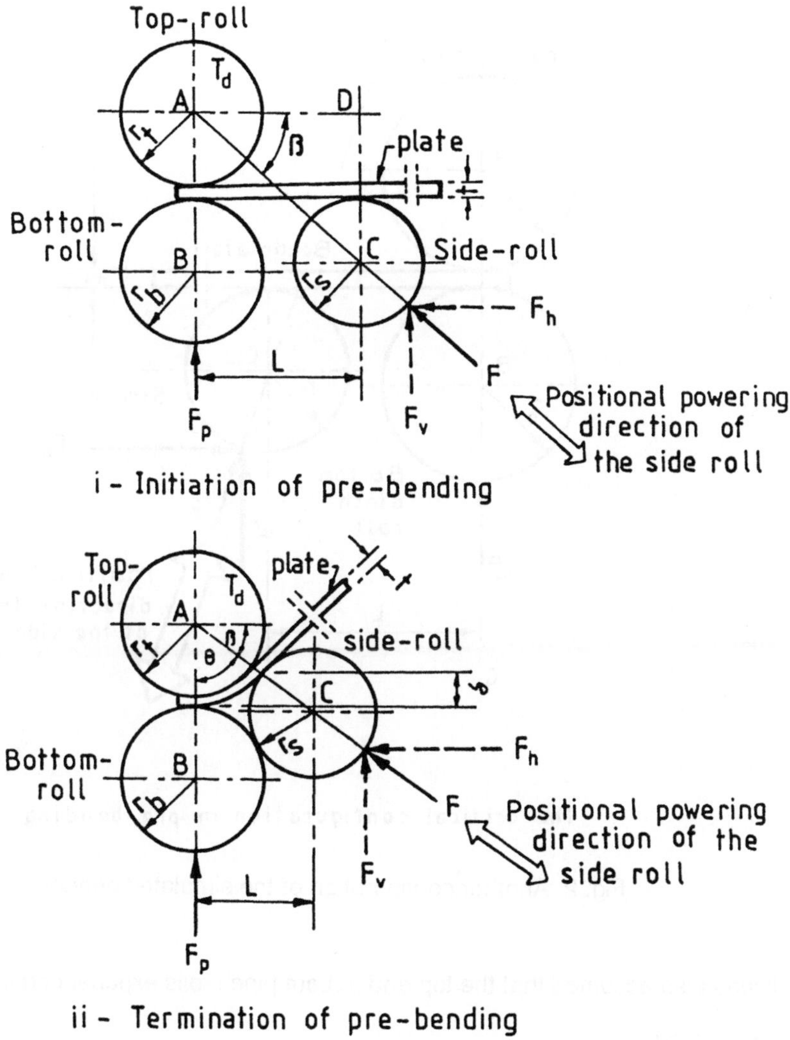

Fig. 1. Assumed configurations of the simulated bender

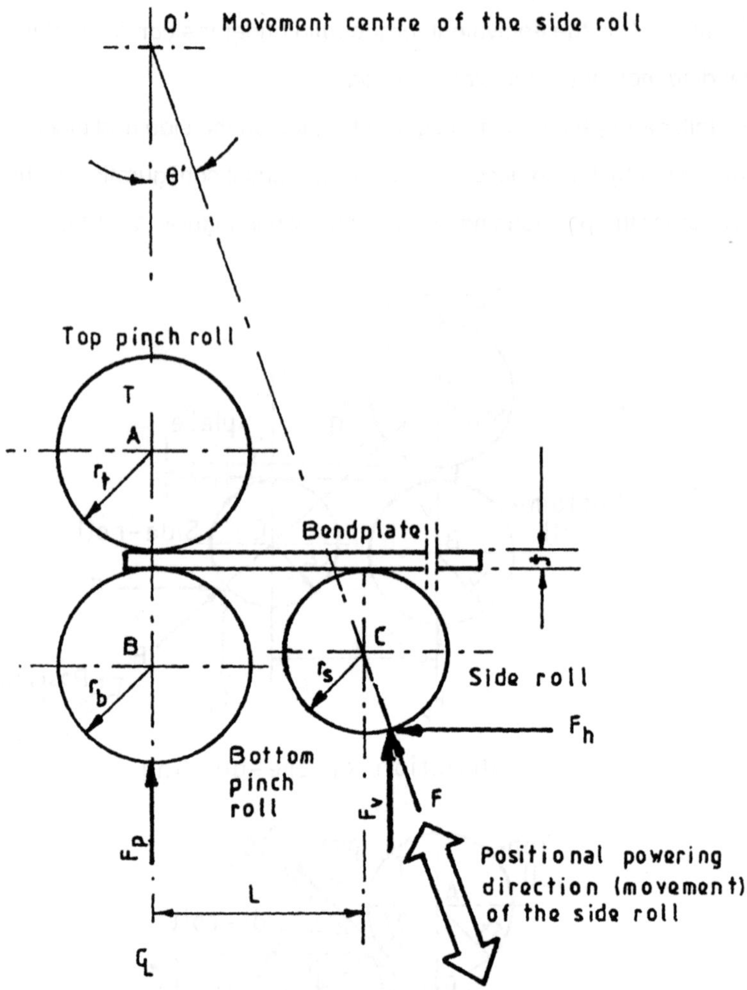

The critical configuration in pre-bending

Fig. 2. Another configuration of the simulated bender

It was also assumed that the top and bottom pinch rolls experienced the same force P_p.

Consideration of the acting force on the side roll (P) -

The Verrina configuration, Figure 2.

The substitution of equation (4) gives:

$$P = \frac{P_v}{\cos\theta^1} = \frac{2Yb\left[a^2 - \frac{(1-\nu^2)2Y^2R^2}{3E^2(1-\nu+\nu^2)}\right]}{\sqrt{3}\,L\cos\theta^1} \tag{5}$$

For a particular plate bend radius R and complete yielding:

$$P = \frac{2Yba^2}{\sqrt{3}\,L\cos\theta^1} \tag{6}$$

Consideration of the acting force on the pinch rolls (P_p) -

The Verrina configuration, Figure 2.

The estimated force on both pinch rolls can be expressed as:

$$P_p = \frac{P\sin\theta^1}{\mu} = \frac{2Yb\tan\theta^1}{\sqrt{3}\,L\,\mu}\left[a^2 - \frac{(1-\nu^2)^2\,Y^2R^2}{3E^2(1-\nu+\nu^2)}\right] \tag{7}$$

for a particular plate bend radius R, and complete yielding:

$$P_p = \frac{P\sin\theta^1}{\mu} = \frac{2\,Y\,b\,a^2\tan\theta^1}{\sqrt{3}\,L\,\mu} \tag{8}$$

Consideration of the drive torque (T) -

The Verrina configuration, Figure 2.

The least torque required for the drive has the following expression:

$$T = r_t P\sin\theta^1 = \frac{2Ybr_t\tan\theta^1\left[\left(a^2 - \frac{(1-\nu^2)^2\,Y^2R^2}{3E^2(1-\nu+\nu^2)}\right)\right]}{\sqrt{3}\,L} \tag{9}$$

for a particular plate bend radius R, and complete yielding:

$$T = r_t P\sin\theta^1 = \frac{2Ybr_t a^2\tan\theta^1}{\sqrt{3}\,L} \tag{10}$$

Numerical data analysis of the model

Data study

This considers the application of the numerical data to the model analysis to assess the relative design parameters for bending mild steel plate with $Y = 235\ N/mm^2$. With a model scale of 1/10, the application of expression (1), for $(\lambda_m/\lambda_p) = 1$, gives the forces on the model side roll to be 85 kN, and on the pinch roll to be 166 kN. The application of the equation (4) gives the bending torque to be 98.1 kNm.

A series of estimations were performed using the equations (5),(6),(7),(8), (9) and (10) for the Verrina bender when bending (see Figure 2) a full scale mild steel plate to the bend radius of 500 m. The results indicated that the data calculated with $c = 0$ and c precalculated with equation (2) were very close, within 2%. This data, as tabulated in Figure 3, for the former condition, gave good agreement with the original specification of the Verrina bender. Thus the model designed on the basis of the preliminary analysis can readily stimulate the original prototype at HWT.

The results shown in the table Figure 3 for the three critical geometrical layouts of the model (Figure 1 and Figure 2) with $c = 0$ showed that the geometrical layout in Figure 1.(ii) gave the highest values, within +50% of those for the layout in Figure 1.(i). The latter layout gave a safety factor of 2.5 for the case in Figure 2 and also could perform the bending tests for the case in Figure 1. A compromise for the drive torque requirements was made by selecting an existing Kopp variable speed drive which would generate a maximum torque of 2.5 kNm (at 4.5 rev/min) and operate at a speed of 4.5-40 rev/min.

Final model parameters

The important parameters for the model were:

(i) force on side roll to be 120 kN,

(ii) force on the top and bottom rolls to be 450 - 500 kN, and

(iii) torque of drive on top roll to be 2.5 kNm.

Data on the table was either extracted from the Verrina Bender at HWT or estimated from equations obtained from the preliminary rudimental analysis, for the pre-bending mode.

SPECIFICATIONS OF THE VERRINA FOUR ROLL BENDER AT HWT			
	Top Roll	Pinch Roll	Side Roll
Diameter (mm)	900	900	750
Maximum Roll Length (mm)	3050		
Daylight Between Pinch Roll (mm)	205		
Drive Torque (kN m)	981		
Drive Force (MN)		16.645	8.472

ESTIMATED DATA FROM THE APPROXIMATE ANALYSIS				
Layout	HWT Verrina Bender (Prototype)	HWT Verrina Simulation roll centre O', Fig.2.	Side Roll towards top roll centre A, Fig.1.(i).	Side Roll towards top roll centre A, Fig.1.(ii)
Pinch Roll Dia. (mm)	900	90	90	90
Side Roll Dia. (mm)	750	75	75	75
Max. Roll Length (mm)	3,050	305	305	305
Bend Radius (mm)	500	50	50	50
Plate Thickness (mm)	100	10	10	10
Side Roll Powering Angle, ψ^{o} (or ϕ^{o}).	20	20	47.06	49.43
Length of Roll (mm)	3100	310	310	310
Bending Arm (mm)	994	99.4	99.4	42.5
Top Roll Drive Torque (kN m)	136.16	1.362	4.017	5.556
Side Roll Drive Force (MN)	9.92	0.099	0.129	0.194
Pinch Roll Drive Force (MN)	16.93	0.169	0.445	0.615

Fig. 3. Tabulation of the design data analysis

The diameters of the top and bottom pinch roll and the side roll were 90 mm and 75 mm respectively with a full roll length of 310 mm.
The model was designed to bend a mild steel plate with a width of 305 mm, thickness of 10 mm, to a minimum bend-radius of 50 mm.

Discussion

Taking account of the model analysis and the required parameters and following consultations with HWT the design of the model bender was specified. In tabular form this appears in Figure 3.

The model bender comprised of the following five major parts:

(i) the framework of the model bender,

(ii) the mounting for the rolls,

(iii) the hydraulic actuators for the bottom pinch roll and the side roll,

(iv) the motor drives of the top roll and the side roll, and

(v) the load cells.

Most of the other design requirements are satisfied with this model. Thus the design of the model bender, shown in Figure 4, consisted of:

(i) Two 75mm diameter side rolls, supplying a maximum bending force of 120 kN through the hydraulic actuators at either end of each roll. The maximum deflection of the roll was 0.43mm at its full load condition.

(ii) A 90mm diameter pinch roll, supplying a maximum pinch force of 460 kN through hydraulic actuators at the ends of the roll. The roll gave a maximum deflection of 0.52mm at a full load of 500 kN. This roll was driven by a Kopp Variator which produced a torque of 2.50 kNm at 4.5 rev.min.-1.

(iii) A H.T.R.I. gear box in line with the top roll Kopp Variator for powering the top roll.

(iv) Two Kopp Variators, each generating a maximum power of 2kW, one for each of the two side rolls.

(v) Three prop-shafts; one connected to the top roll powering lines, and the other two connected to each of the two powered side rolls.

(vi) Two hydraulic hand pumps; one to actuate the side rolls, the other to actuate the bottom pinch roll.

(vii) A pair of vertical plate units to form the framework of the bending unit and to contain the components of the bender.

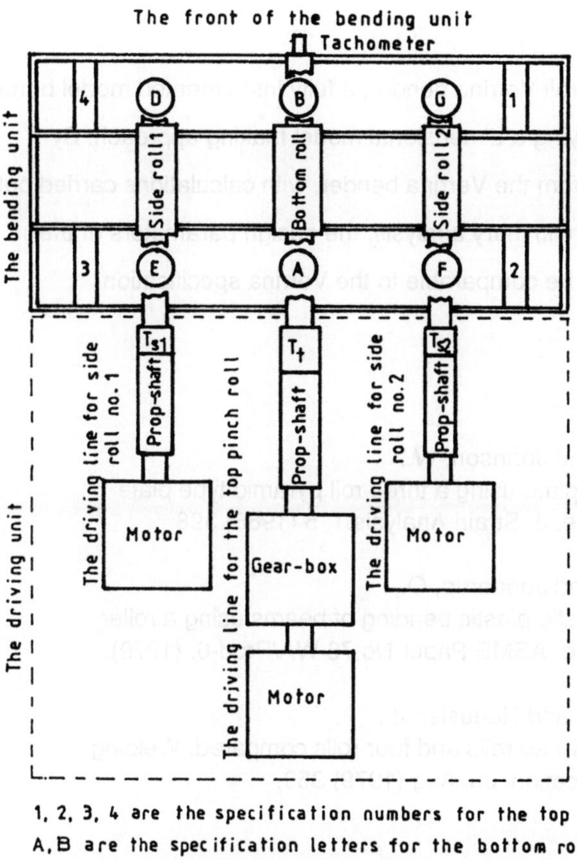

1, 2, 3, 4 are the specification numbers for the top load cells

A, B are the specification letters for the bottom roll load cells

D, E, F, G are the specification letters for the side roll load cells

T_{s1}, T_{s2} are the torquemeters for the side roll number 1 and 2 respectively

T_t is the top roll torquemeter

Fig. 4. Plan arrangement of the model bender

(viii) Six ram load cells to measure the axial and bending loads on the bottom pinch roll and each side roll, and four top load cells to measure the axial and bending loads on the top roll.

(ix) Three torquemeters, one to measure the top roll torque and the other two to measure the slip torque effect on each of the two side rolls.

(x) A base frame, consisting of a primary base-bed on which the bending unit was bolted, a secondary base-bed to which were attached three Kopp Variators and the H.T.R.I. gear box.

Conclusions

To simulate the four-roll Verrina bender, a fully instrumented model bender was designed employing a dimensional model making approach. By amalgamating data from the Verrina bender, with calculations carried out on the basis of the preliminary analysis, the design parameters of the model were found to be comparable to the Verrina specification.

References

I Bassett, M.B. and Johnson, W.,
The bending of plate using a three roll pyramid type plate bending machine. J. Strain Analysis.1, 5 (1966) 398.

2. Hanson, N.E. and Jannerup, O.,
Modelling of elastic-plastic bending of beams using a roller bending machine. ASME Paper No.78 WA/Prod-6. (1979).

3. Roggendorff, S. and Haeusler. J.,
Plate bending : three rolls and four rolls compared. Welding and Metal Fabrication. Jul/Aug (1979) 353.

4. Winship, J.T.,
Basics of roll bending. Amer. Mach. 127,2 Feb. (1983) 105.

5. Hua, M.
The mechanics of continuous roller bending of plates. Ph.D. thesis. Aston University. England. (1986).

Acknowledgements

The support of the Science and Engineering Research Council is gratefully acknowledged. The authors are also grateful for the invalued assistance of Head Wrightson Teesdale Limited (now ITM Head Wrightson Teesdale Ltd) and help from the technical staff of the Department of Mechanical and Production Engineering, Aston University, Birmingham, U.K.

Application of Variable Parametric Programming in Precision Machining Processes

Viet Phung
Senior Lecturer, Department of Mechanical and Production Engineering,
Nanyang Technological Institute, Singapore.

ABSTRACT

In applications that demand consistent and accurate finishing, the normal tool compensation technique which relies on pre-defined tool dimensions with fixed values is inadequate. This paper describes a technique of retro-fitting machines with modern CNC controllers so that tool wear can be computed and precisely compensated during the machining cycle. Application of this technique on a cylindrical grinding machine is discussed and results presented.

INADEQUACY OF CONVENTIONAL TOOL COMPENSATION

In most CNC (Computer Numerical Controlled) machines, tool-wear compensation is conventionally catered for by the use of preparation codes, typically G40 G41 G42, which allows user to store fixed values of tool dimensions (e.g. tool radius for lathe or tool diameter + tool length for milling machines). During the machining cycle, these fixed values are extracted by the controller to compute the required tool path motion via pre-programmed compensation equations (figure 1).

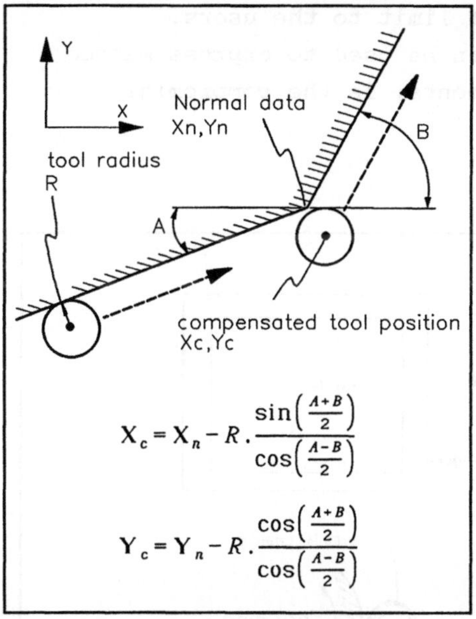

Figure 1
Conventional tool compensation

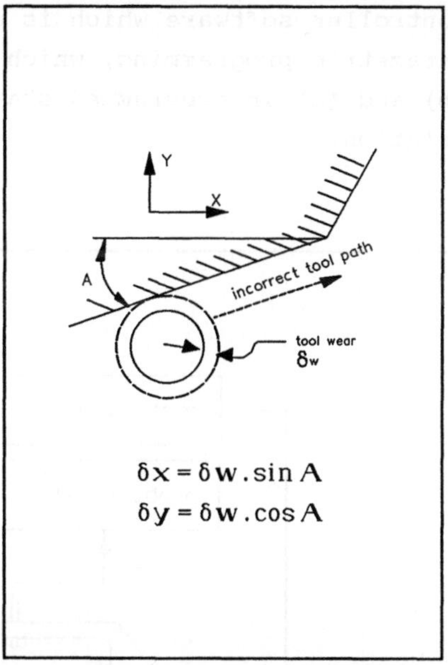

Figure 2
Error due to tool wear

When the tool dimensions change due to wear, the above mentioned computation is no longer valid and will lead to inconsistent and inaccurate finishing (figure 2) unless one of the following methods are attempted:

(a) Re-define tool dimensions.
(b) Offset machine axes by the same amount as tool wear.
(c) Modify compensation equations in such a way that rate of tool wear is taken into consideration.

Unfortunately, none of these methods are easy, if not impossible, to be executed in automatic mode on currently available CNC controllers. Method (a) can only be done manually. Attempts to apply methods (b) in automatic mode (figure 3) result in extra measuring devices , complicated interface to controller, and longer processing time due to pauses in the machining cycle for measurements to take place. Method (c) requires major re-writing of internal CNC controller software which is off limit to the users. Parametric programming, which can be used to express method (b) and (c) in programmed statements, is the compromise solution.

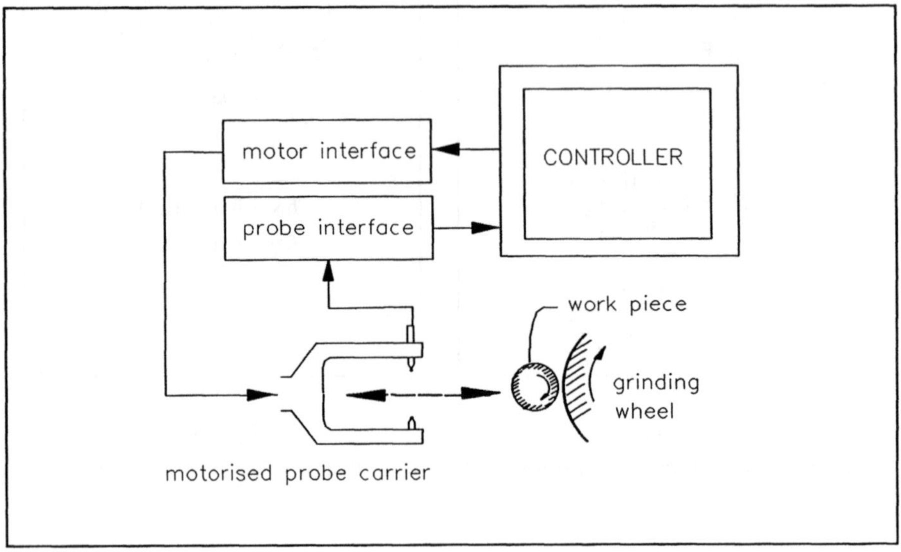

Figure 3 Adaptive control of work piece dimension

APPLICATION OF PARAMETRIC PROGRAMMING FOR TOOL WEAR COMPENSATION

Parametric programming involves the use of variables to represent values of measurements such as co-ordinates or feedrate. During program execution, the numerical values provided by parameter definitions or derived from parameter equations are used by the control system to affect tool path motion (figure 4).

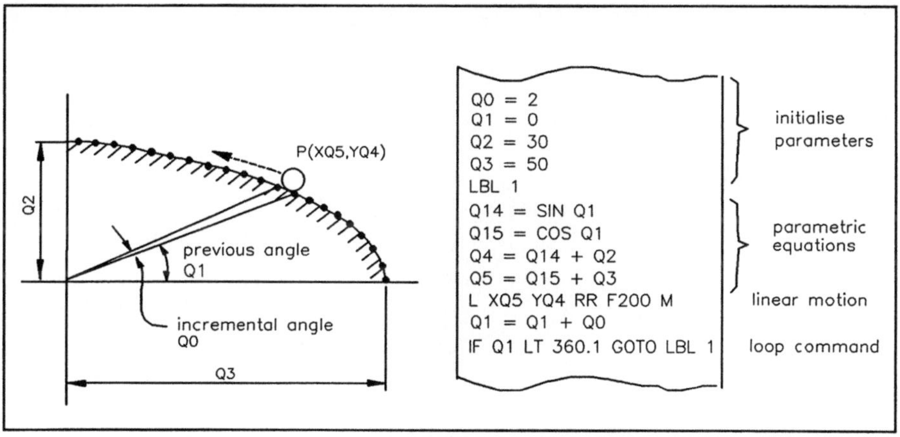

Figure 4 Parametric programming to generate ellipse

In precision processes, especially in processes involving high rate of tool wear such as grinding, the parametric programming can be used to express tool wear compensation equations so that required offset of co-ordinates can be executed inside the CNC program.

In cylindrical grinding process (figure 5), the work-piece (1) is clamped in, and rotated with, a chuck; the grinding wheel (2) rotates on its own axis and moves along axis Z at 45 ° to the work-piece axis; a dressing mechanism (3) cuts the required finishing profile on the face of the grinding wheel by having its diamond dressing tool traversed appropriately in X and Y directions. During the grinding cycle, the grinding wheel wears at high rate, and the diamond dressing tool also wears but at a much lower rate. Typically, the grinding wheel is required to be dressed a few times for each work-piece.

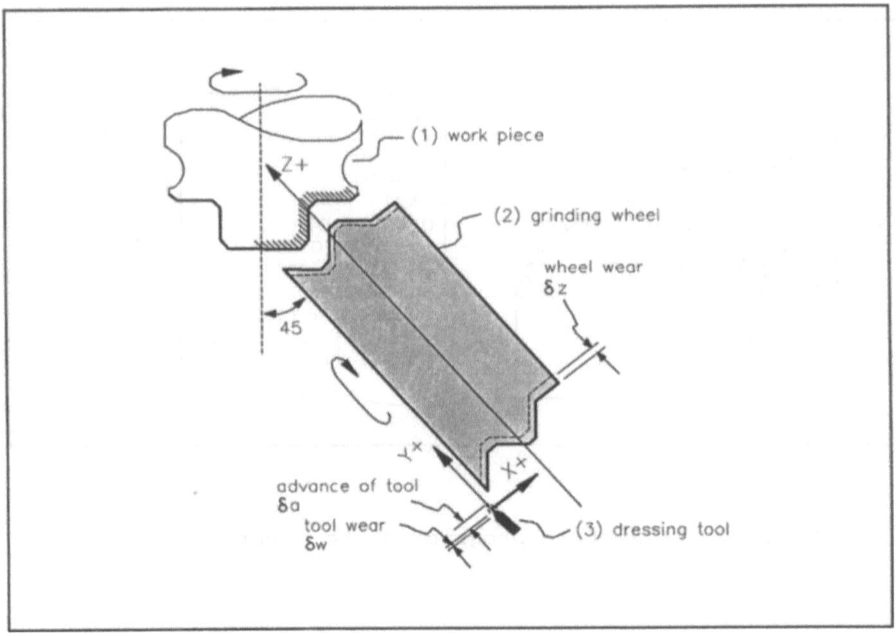

Figure 5 Cylindrical grinding process

In this application of grinding, tool wear compensation must be considered for two components - the dressing tool and the grinding wheel, i.e. Y axis and Z axes.

If n is the total number of dressing cycles, δa is the advance of the dressing tool per dressing cycle, and δw is the amount of wear on the diamond tool per dressing cycle[1], then the tool wear in the dressing head δy and the tool wear in the grinding wheel axis δz can be expressed by equations (1) and (2)

---

1 Tool wear δw per dressing cycle is best obtained from experiments prior to production runs. If feedrate, coolant and material properties can be maintained then tool wear is a function of volume of material removal which can be expressed in terms of tool advance δa, width of grinding wheel, and the effective radius of the grinding wheel ($R_{initial} - \delta z$). A good first approximation of δw is $\delta w = K * \delta a$ where K is a constant obtained from experiments.

$$\delta y = \sum^{n} \delta w \tag{1}$$

$$\delta z = \sum^{n} (\delta a - \delta w) \tag{2}$$

and the compensated positional values y_c and z_c due to tool wear are offset from the non-compensated values y_n and z_n by tool wear amount δy and δz:

$$y_c = y_n + \delta y \tag{3}$$

$$z_c = z_n + \delta z \tag{4}$$

Even though mathematically simple, equations (1) to (4) can result in rather complicated numerical-control programming[2]. It can also result in confusion to machine operators as to what is happening because values of desired end-block positions are different for each new work piece.

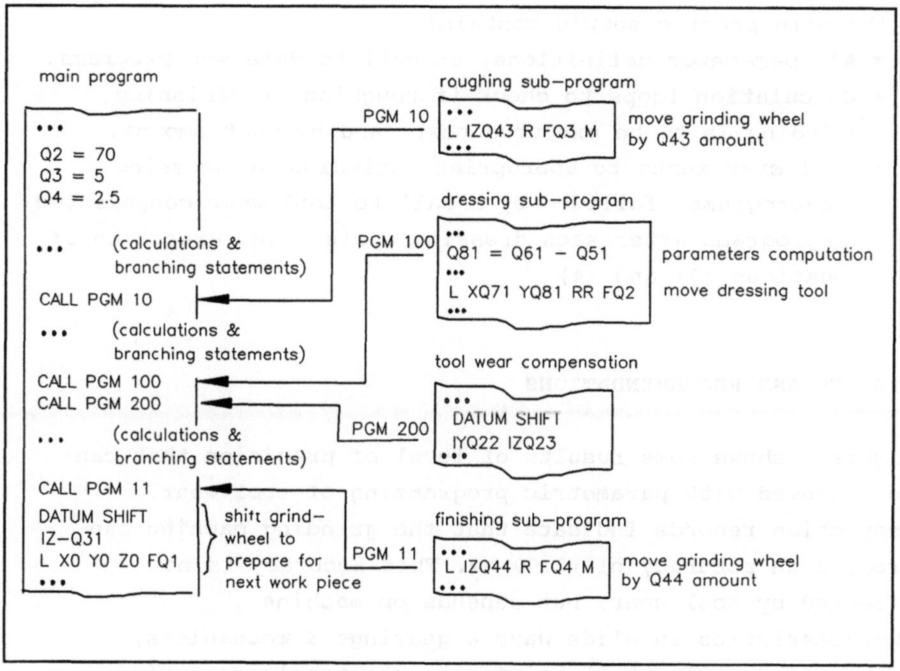

Figure 6 Structure of parametric program

2 Especially if δw changes for each new dressing cycle due to the fact that grinding wheel is getting smaller during grinding and after each dressing operation.

The best way to express the whole grinding operation in CNC programming is as shown in figure 6:

- Write subprograms for rough-grinding cycle, finishing-grinding cycle, dressing cycle.
- Write two subprograms to calculate tool wear amount δw and $(\delta a-\delta w)$ on Y and Z axis for each new dressing cycle.
- Write tool wear compensated subprograms to compute the required compensated offsets on Y and Z axis for each new dressing cycle, i.e. equations (1) and (2). The offsets δy and δz should be expressed using DATUM-SHIFT commands, so that during grinding operations end-block positions are displayed consistently from work piece to work piece and total tool wear can be checked by displaying datum values on the controller screen.
- The main program should contain:
  * All parameter definitions, or call to data sub-programs.
  * Calculation loops to check if roughing or finishing grinding is to be executed next, and by what amount.
  * Call statements to appropriate grinding or dressing sub-programs, followed by a call to tool wear compensated subprograms after each dressing cycle, i.e. execution of equations (3) and (4).

RESULTS AND RECOMMENDATIONS

Figure 7 shows some results of level of precision that can be achieved with parametric programming of tool wear. Inspection records indicate that the grinding machine can produce an accuracy of ±0.09 mm. This accuracy is not affected by tool wear, but depends on machine characteristics in slide ways & gearings & mechanisms, machine & dressing tool stiffness, electronic measuring encoders, and motion control system. And if tool wear is not compensated, error increases and results in an offset of the order of +0.25 mm. The error due to wear in the dressing tool is of the order of -0.03 mm, and the error due to tool wear in the grinding wheel is +0.28 mm, representing about 33% and 310% in comparison to machine accuracy.

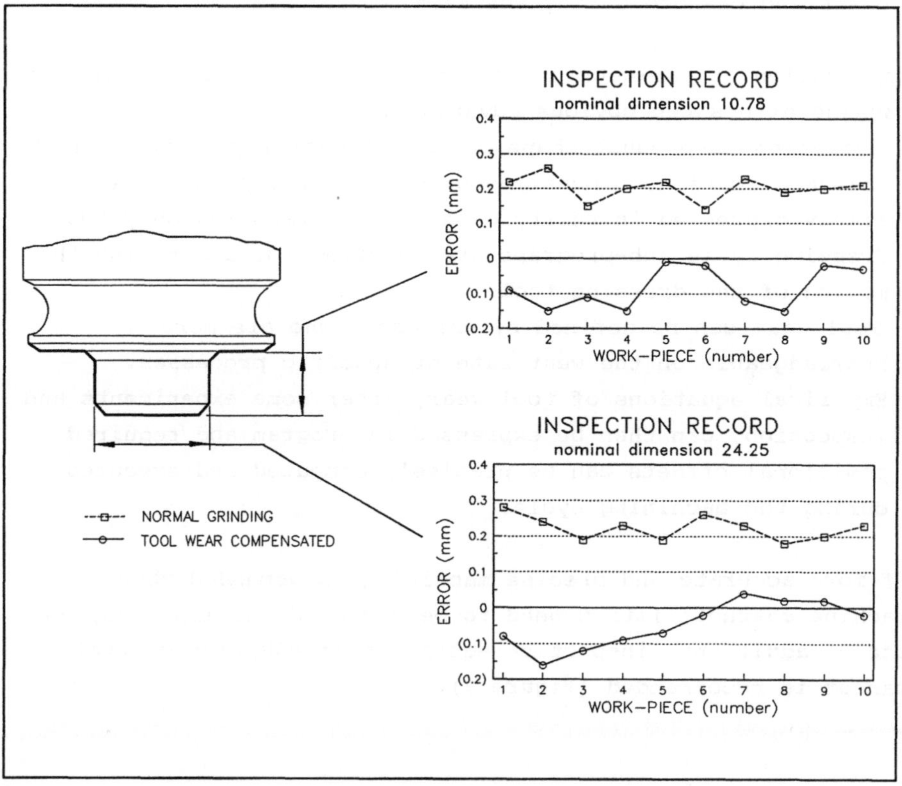

Figure 7 Inspection records of the grinding process

At this stage, drawbacks of parametric programming are due mainly to difficulties of programming and debugging, which will be minimised if future generation of CNC controllers have additional facilities:

- Allow display selection of CNC program statement in either original state, or in derived state where parameters are replaced by their current computed values.

- Allocate tool wear DATUM-SHIFT that can be used on top of other DATUM-SHIFTs, and allow for independent display of tool wear and positional datum.
- Extend existing tool compensation to cater for tool wear by allowing tool parameters to be expressed mathematically by users and changed during the machining cycle.

Application of parametric programming on cylindrical grinding machine offers two obvious advantages:
- Variation in product shape can be effectively programmed. In this application, dimensions of some thirty products were simply expressed in thirty data sub-programs and only two dressing cycle sub-programs are required for describing the motion of the dressing tool.
- Tool wear can be defined by end users who are more knowledgeable on the wear rate of specific processes. Empirical equations of tool wear, after some experiments and inspection, can then be expressed in program and required positional offsets can be precisely computed and executed during the machining cycle.

If more accurate and precise machining is demanded then machine characteristics need to be reviewed and improved, and use of additional inspection equipment in adaptive control manner is recommended (figure 3).

CONCLUSION

Parametric programming offers a method to compensate for tool wear in precision machining processes.

REFERENCE

Operating manual, Heidenhain TNC 355B/TNC 355Q contouring control,
Serial no. 23952421-10-5/88, published by Dr. Johanes-Heidenhain GmbH, Germany, 1988.

Micromachining of Hydraulic Elements

Janez KOPAC
Marjan ŽAKELJ

Faculty of Mechanical Engineering
University of Ljubljana
YU - Slovenia

Kladivar Žiri
YU - Slovenia

Summary

The current day requirements for high accuracy of shapes and surfaces on exacting functional components in the field of hydraulic control have urged manufacturers to start using high precision machining techniques. The paper presents how these requirements can be satisfied on a real example of a hydraulic control valve. The geometric prescriptions for this type of valve require an accuracy of +-0,001 mm while the technological procedure should achieve a roughness of R_a = 0,07 to 0,1 µm to ensure good servicing of the valve.

INTRODUCTION

The most vital components of a control valve are the housing and the control spool - Fig. 1. The housing is made of grey cast iron of quality GG 30. The control channels are cast since it is impossible to machine some of the passages by cutting - location X in Figure 1.

In this way the passages are smoother and the pressure resistances smaller. The control spool is made of steel C60. It is machined by cutting (turning) and hardened onto 56 to 62 HRc.

In the final phase fine grinding and lapping are applied. Both procedures are technically desribed. The test have shown that machining by superfinish is not appropriate as too much material is swept away on the spool edges - Figure 1 location Y. Theoretical research into the reason for superfinish defects has not yet been done. We have decided to solve the problem by using the lapping procedure which grants the achievement of high accuracy of geometric shapes and sufficient surface smoothness.

1. MACHINING OF CONTROL VALVE HOUSING

Housing of control valves and also other types of valves are castings of grey cast iron (HYL 30) or DIN GG 30. These housings are first machined by milling. In order to achieve the required shape and size and the parallelism of the faces simultaneous milling at both sides was applied. The housing premachined in this way is then machined on a CNC - turning machine. For the machining of the hole with control edges the turning procedure was used. The machining of these edges is important since they are very relevant for the correct functioning of the control valve, both their shape and dimensions playing an important role. Besides this, it should be mentioned that also the shape of the control channels is very important for subsequent machining. Sharp edges should be slightly chamfered already by turning otherwise a sharp bur is formed during honing which cannot be removed by deburring. The overmeasure which is necessary for honing is usually 0,05 mm. All control edges are made in one clamping. Thus besides centering also the positioning - centre distance - of control edges is achieved. The latter is important because a good match should be achieved between the spool and the housing. The next machining step is honing which is carried out on KADIA machine tool. This is a CNC machine tool for honing with three spindles and four jigs located on a swivel table. This enables an uninterrupted change of workpieces during operation. While the first spindle performs rough honing on one workpiece, the second spindle hones another workpiece to final measure and surface quality. The third spindle has a special brush made of artificial fibres containing silicon carbide or diamond grains. This brush does the deburring of sharp edges formed during honing. If the brush is fine enough then, besides deburring the sharp edges, a better surface roughness is achieved. The brush diameter is by 10 - 20 % larger than the diameter of the hole. Besides the correctly chosen honing parameters and cooling oil emulsion it is also good product design that plays an important role. The best results are obtained if the rings which become visible after honing are equally wide and uniformly distributed throughout the length. Usually adjustable honing thorns are used, however for the most demanding shapes the parameters required can be achieved only by using special honing tools - seising. Honing is mainly used for small hole diameters which is an additional problem. The diameters which can be honed range from 3 to 32 mm. The dimensional requirements which are satisfied fall within the tolerance limits +-0,001 mm. The ovality and conicity are achieved within the limits 0,001 mm. A bigger problem is how to achieve the required roughness on castings from grey cast iron where R_a ranges between 0,15 and 0,1 µm. Better results in terms of roughness are obtained on hardened parts where R_a ranges from 0,1 to 0,03 µm.

The control of dimensions and shape is carried out 100% on the machine itself with pneumatic measuring thorns enabling accurate readings and simple handling. Besides a 100% control carried out by the worker working at the machine, process control is performed by the controller, randomly choosing 3 - 5 housings in each round and making additional control measurements in the control room. Here on a high precision machine for measuring shape deviations, the shape, dimensions and surface roughness are controlled for the second time. The results are recorded on special measurement sheets and processed on PC using a statistical program. Some measurement results are presented in Figure 2. The results analysis show that the ovality and roughness range within the prescribed limits, taking into account that different control hydraulics manufacturers (Rexroth, Vickers, Duplomatik, Bosch) prescribe different tolerance limits depending on design requirements and machining applied. A statistical analysis of functional measures shows that, compared to other manufacturers, the accuracies achieved on our product fall in the medium range. Our design engineers do prescribe smaller roughness values, however they are not easy to achieve. Our recent efforts involve a change of the parameters of honing (higher splindle speeds - lower feed rates). The result is a roughness R_a in honing smaller than 0,1 µm.

2. MACHINING OF CONTROL SPOOL

Control spools are manufactured from different materials depending on the requirements and their function. With regard to the function, several spool version can be used which can all be built into the same housing. As an example we shall discuss the commonest kind of spool most frequently built into control valves. It is made of heat-treatable steel or cementation steel. In our factory all the spools are heat treated, with their hardness ranging from 56 to 62 HRc. Before heat treatment the spools are machined by turning and milling - Figure 3. Various grooves and cutouts on the spool circumference cause big problems in achieving the required accuracy of shape in grinding.

Since during servicing the spool slides against the housing it is subjected to wear. In order to reduce this wear and prolong the life time, the spools are specially designed and heat treated. After the heat treatment (hardening) the spools are cleaned and corrosion protected with phosphate coating. The next step is grinding. Depending on the shape and difficulty, the spools are ground on a coneless grinding machine or between the ends. Spools having complex shapes are ground between the cones. The tolerance limits for ovality and conicity are 0,001 mm while the dimensional tolerance limits are +-0,001 mm. To achieve these requirements in process measurements are made on MARPOSS with automatic switch off when the desired measure is reached. Grinding is divided into rough and fine grinding, the overmeasures ranging from 0,2 to 0,3 mm depending on the diameter and shape of the spool. At these strict requirements, besides the correct selection grinding parameters and the grinding wheel, an important role is played also by the cooling liquid and the planing of the grinding wheel. Apart from the correct choice of the cooling liquid in fine grinding, also the temperature of this liquid and the ambient temperature are important.

Surface roughness achieved by fine grinding R_a is below 0,1 µm, which correspond to N3 machining. The results of the measurements made on the spool are presented in Figure 4. If the roughness achieved by grinding does not meet the requirements it has to be improved by lapping between the rollers. This is especially frequently done if the components are hardened. By lapping the surface roughness of the spools is improved by one quality grade as can be seen from Figure 5. A third roller was added to the lapping device. It is connected to the pneumatic cylinder and this further to the time relais. Thus the lapping time and pressure of the roller on the workpiece can be exactly defined. This makes regular as well as experimental work a lot easier, since the two basic factors are exactly defined. A schematic of the lapping device is shown in Figure 6.

3. CONCLUSIONS

From the analysis of the results it can be noted that the adequate choice of the machining procedure (lapping - superfinish) is most important in achieving the required accuracy of shape and surface quality of machining. However high accuracy of shape and low roughness values of the machined surface are not the key issues of the leading European manufacturers of control hydraulics. The key issues representing the main guidelines are technological possibilities of machining at a still acceptable price achieving the upper yet allowable tolerances. The trend in today's control hydralics is to develop proportional hydraulics where the requirements concerning dimensional accuracy and quality of the machined surface are treated from a different aspect.

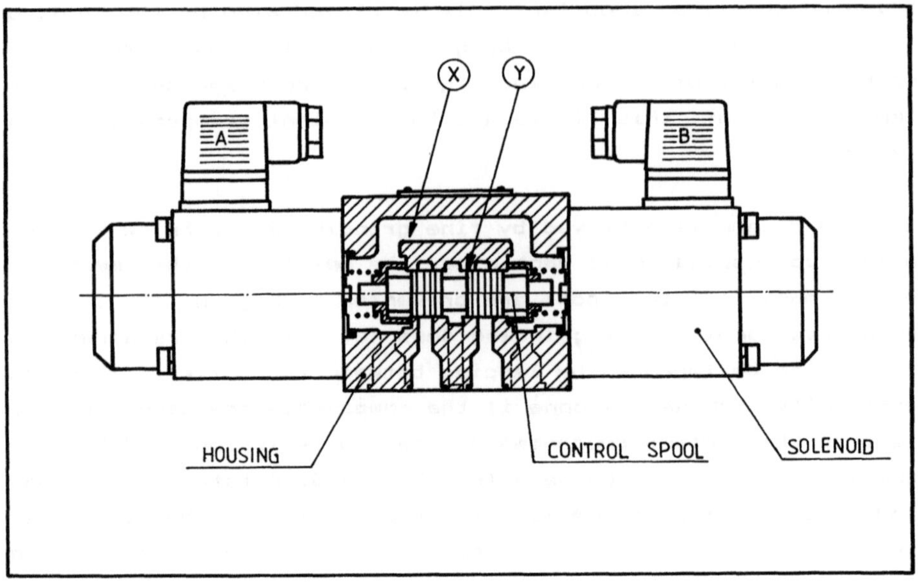

Fig. 1. Directional valves

Nr. of pieces	Ovality (µm)	Roughness (µm)
1	0,8	0,10
2	1,0	0,10
3	0,7	0,11
4	0,9	0,10
5	1,0	0,09
6	0,9	0,10
7	0,9	0,10
8	0,8	0,12
9	1,0	0,10
10	1,0	0,12
11	0,8	0,09

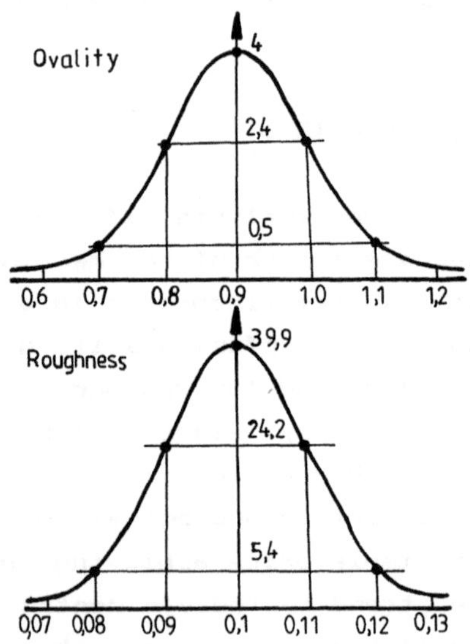

Fig. 2. Measurement results of roughness and ovality achieved by honing

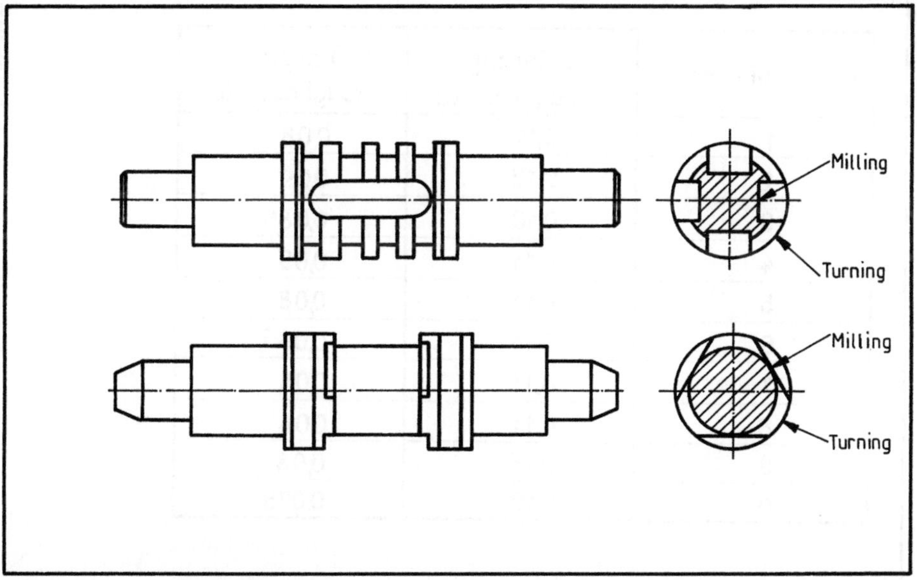

Fig. 3. Two control spools

Nr. of pieces	Ovality (μm)	Roughnes. (μm)
1	0,6	0,13
2	0,8	0,16
3	0,7	0,14
4	0,5	0,13
5	0,7	0,15
6	0,4	0,14
7	0,7	0,15
8	0,7	0,14
9	0,5	0,12
10	0,5	0,15
11	0,5	0,14

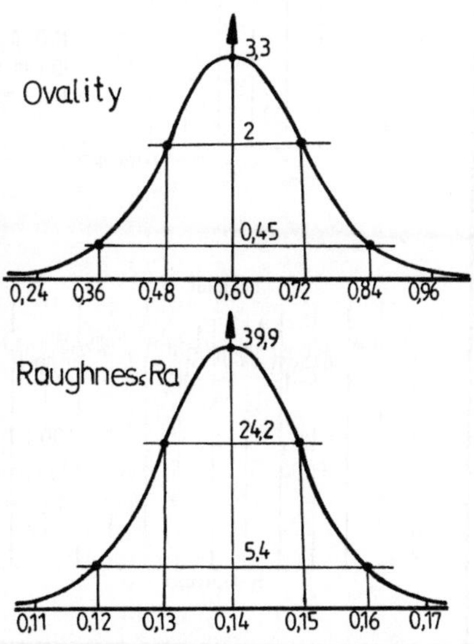

Fig. 4. Measurement results of roughness and ovality achieved by grinding

Nr. of pieces	Grinding Roughness Ra	Lapping Roughness Ra
1	0,12	0,08
2	0,12	0,09
3	0,10	0,085
4	0,11	0,09
5	0,11	0,08
6	0,11	0,085
7	0,11	0,08
8	0,11	0,08
9	0,15	0,08
10	0,12	0,075

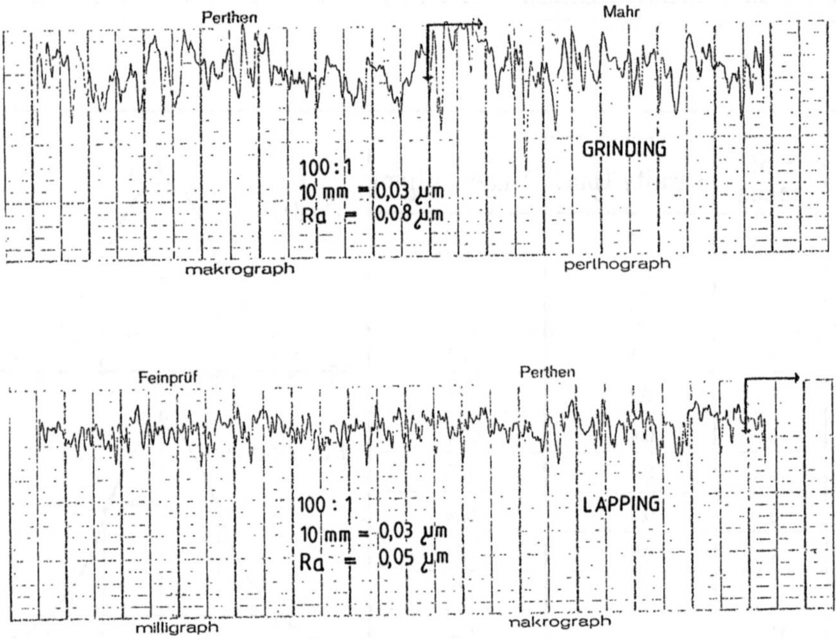

Fig. 5. Results – comparison of roughness between fine grinding and lapping

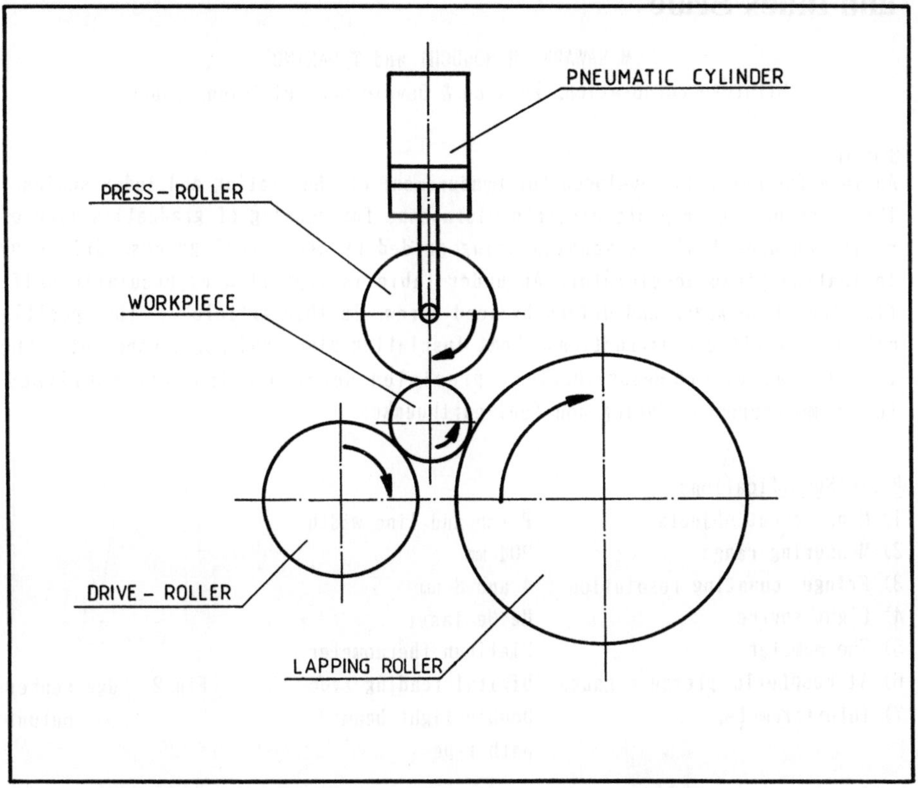

Fig. 6. Shematic presentation of lapping

References

1. M. Žakelj, J. Kopač: Quality assurance of pneumatic and hydraulic components, 1 st European Conference, Brussel 1990, B

2. M. Žakelj, J. Kopač: Quality assurance in manufacturing enginneering, Maribor 1990, YU

3. Kladivar: Catalogue; Hydraulics 7, Žiri 1988, YU

4. A.W. Stähli: Catalogue; Läpp technik

An Interferometer for Measurement of Grating and Index Scale

M. SAWABE, H. NOGUCHI and T. MAKINO
Mitutoyo corporation. Resarch & Development Division. Japan

Summary
An interferometer is developed for measurement of the grating and index scales. This machine uses a photo-electric microscope for sensing of graduation line of scale. An uper table is mounted being guided by cross roll guides, driven by lominating piezo accelerator. An under table is supported by preumatic half-floating slide ways, and driven by lead screw. In this article, the specifications, specific constructions, heat insulating plan and performance of the interferometer are describe. The precision deterioration are confirmed in the mmeasurement during non-feel earthquakes.

Basic Specifications

1) Measurement objects	Pitch and line width
2) Measuring range	204 mm
3) Fringe counting resolution	4 and 8 nm
4) Light source	He-Ne laser
5) Thermometer	Platinun thermometer
6) Atomospheric pressure gauge	Digital reading type
7) Interferometer	Double light beam path type

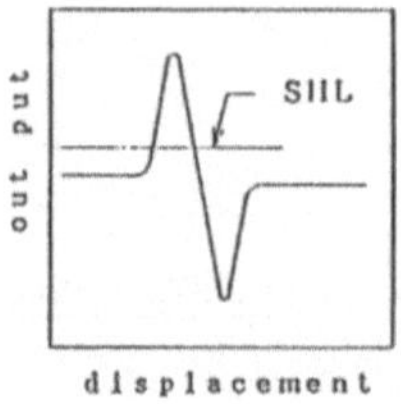

Fig. 2 Edge senser output

Fig. 1 Photogragh of the machine

Out-line of the interferometer

1. Edge senser

This machine uses a photo-electric microscope type edge senser for posision sensing of graduation line of scale. An output example when its light axis scans graduation line is shown in Fig. 2. When this output voltage reaches threshold level [SHL] provided in edge signal controller, interferece fringe counted value is latched so as the graduation line position is automatically detected.

2. Machine configration and heat insulating plans

Fig. 3 is a front view diagram of the machine. Referring to Fig. 3, An under table is supported by the pneumatic half-floating slide ways and driven by a lead screw. An uper table is driven by a laminating piezo-actuater. When under table moves till the commanded position, it stops, voltage is applied to the PZT and the uper table is fed with dead speed feed of 0.1 μm/s till graduation line detection pulse is generated. Therefore, under quient ambience where air fluctuation or vibration is few, the line detection is possible.

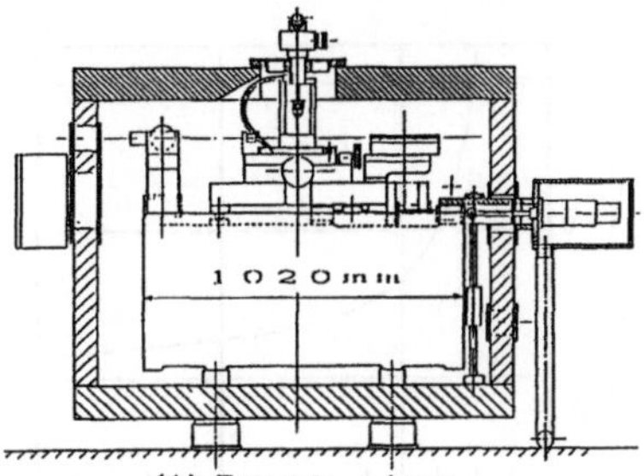

(A) Front view

Fig. 3 Machine configration

This machine uses 3 points support pneumatic half-floating slide-way in gravity loading surface of the under table, about 70% of moving mass is pneumatically supported by static pressure, remaining 30% is sustained by sliding face. Since coefficient of dynamic friction of the sliding face is about 0.1, apparent coefficient of dynamic friction becomes about 0.03. With this system, rate of air is as minim as 20 normal cm^3/min. that such flow rate causes thermal disturbance is few. The entire faces of the machine is covered by heat insulating enclosure, self-heat build-up parts, like as eye pieces of microsope and light source, laser head, AC servo motor etc. are mounted outside the enclosure. Under the room temperature fluctuation of ±0.05℃, temperature of objective scale placed inside the enclosre is stabilized to 0.003 ℃/h.

Calibration of the laser wavelength with iodine absorption stabilized He-Ne laser

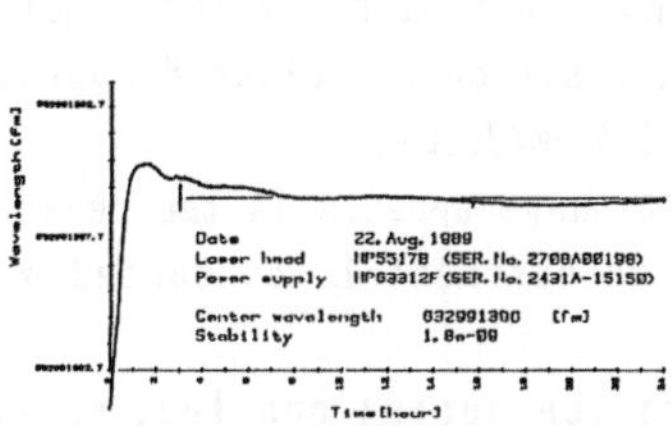

Fig. 4 Laser wavelength calibration

The wavelength is measured using our povinsly fabricated iodine absorption stabilized He-Ne laser[1], the mesured data is:

For No. 1 machine λ : 632.991 357 nm

$\Delta\lambda/\lambda$ 4.1×10^{-9}

For No. 2 machine λ : 632.991 385 nm

$\Delta\lambda/\lambda$ 1.8×10^{-9}

They are satisfactory quality for both.

Estimate of the errors due to the atomospheric pressure shift.

When atomospheric pressure is changing to upside or downside every moment, in the case of, we have cyclic erroes in the repetation measurement diagram, due to the delay for atmospheric pressure gauging using the digital barometer.
Fig. 5 shows the periodic phenomenuns.

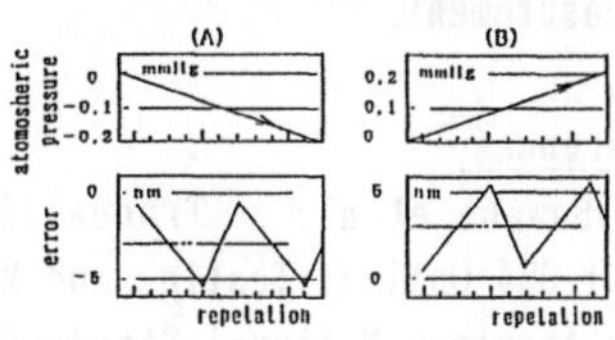

Fig. 5 Periodic errors due to the atomospheric pressure shift

Measurement repeatability

We performe repeat measuring test on the ordinary and non-feel earthquakes

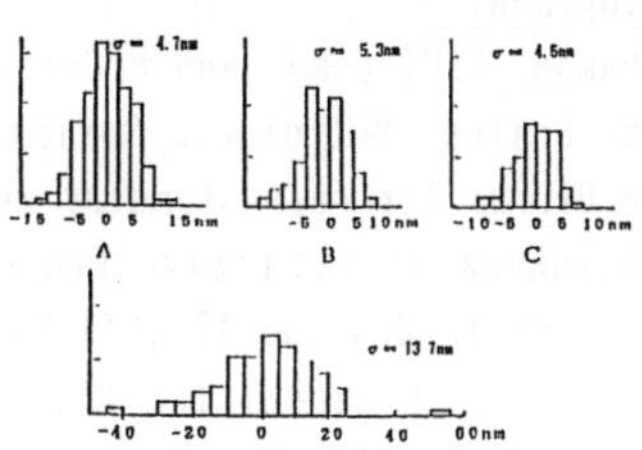

D. Precision deterioration during non-feel earthquakes.

Fig. 6 Errors distribution

occasions. (See Fig 6-D). We can confirm the precision deterioration during non-feel earthquakes.

Measurement uncertainty of the 8 nm resolution machine.

Applying ASME Measurement Uncertainty [2], we estimate the measurement uncertainty of the 8 nm resolution machine and get a result as shown in Fig. 7.

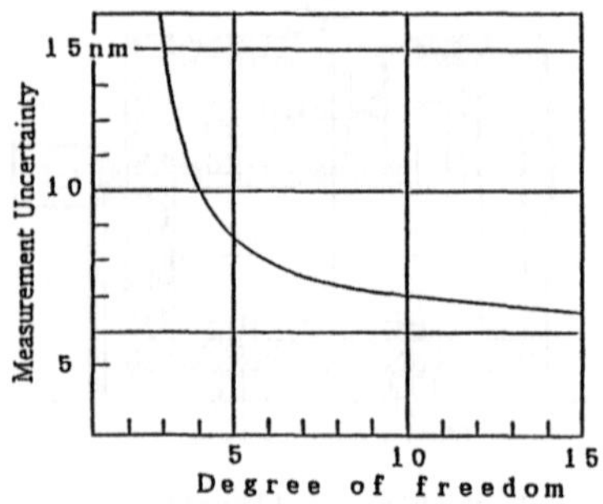

Fig. 7 Measurement Uncertainty

Proformance

1) A standard deviation of repetition measurement of 200 mm length became.
   a) On the ordinary occasion of 8 nm resolution machine 5.0 nm
   b) On the ordinary occasion of 4 nm resolution machine 4.5 nm
   c) On the non-feel earthquaks of 8 nm resolution machine 13.7 nm
2) In the measurement of 200 mm length with No. 1 and No. 2 machine
   a) Difference of coefficient of thermal expansion between them: 0.8 %
   b) Difference of measured values between them: 1.5 nm(0.75×10^{-8})
3) We can estimate in numerical values a standard deviation of the combined effects of optical fluctuation, graduation position detection dispersion, vibration, noise and other error factors to 3.6 nm/200mm.
4) We find that the periodicity of saw tooth shape appears in the measured values and the mean value is shifted when measurement is conducted with digital barometer.
5) We can confirm the precision deterioration during non-feel earthquakes. That is about 3 times of standard deviation in ordinary measurement.

Reforences

1) M.Sawabe et al: Traceability of optical length Measurements, SPIE vol,954 Optical Testing and Metrology II (1988), 448.
2) An American National Standard ANSI/ASME DTC19.1-1985.

Bibliographies

1) J.Hamon, P.Giacomo. and P.Carre': International Comparison of Measurement of Line Scales. Metrologia Springer Verlay 1987.
2) K.P.Birch: Air refrectometry and application to precision length measurement.
3) Y.Hurukawa et al: Design Procedure of Pneumatic Half-floting Slideway JSPE 47-2(1981), 204 and 47-10(1981), 1210 in Japanese.
4) H.Noguchi, Y.Saruki: The Development of an Interferometer for calibration Linear Scales(1), Mitutoyo Techinical Bulletin, 5(1989), 26.

Ultraprecision Machining - A Theoretical Analysis of the Mechanisms of Surface Formation

Professor G. Spur* and Dr. G. Byrne**

In ultraprecision machining the demands for a precise understanding of the exact mechanisms of chip removal are more considerable than they have ever been in the past. The inaccurcies due to unknown macro-gemetries when grinding and lapping cannot be tolerated and recourse is made to tooling of defined geometries to achieve tolerances of better than 10^{-5} mm. With such developments, it is evident that a clear understanding of the machining process and of the parameters which influence the surface integrity is imperative. A theoretical approach to the problem of engagement mechanics in ultraprecision turning is reported here. The fundamental concept behind the theory lies in the method of observation of the machining process. To date observation of the machining process has generally taken place either remote from the cutting tool/workpiece surface or from an imaginary zone within the cutting tool. In the latter case the point of observation is steadily moving away from the machined surface (Figure 1).

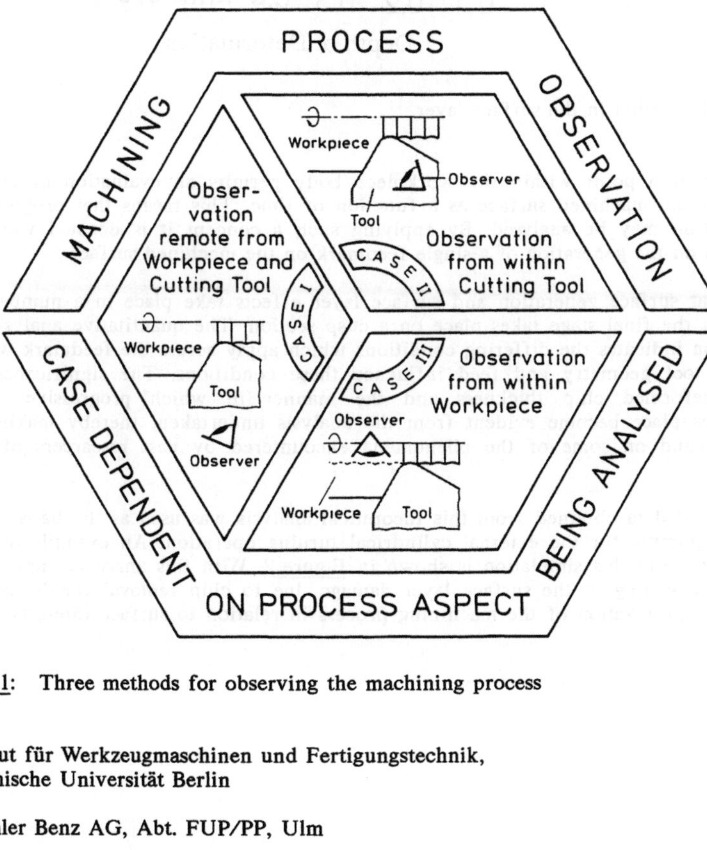

Figure 1: Three methods for observing the machining process

*Institut für Werkzeugmaschinen und Fertigungstechnik, Technische Universität Berlin

**Daimler Benz AG, Abt. FUP/PP, Ulm

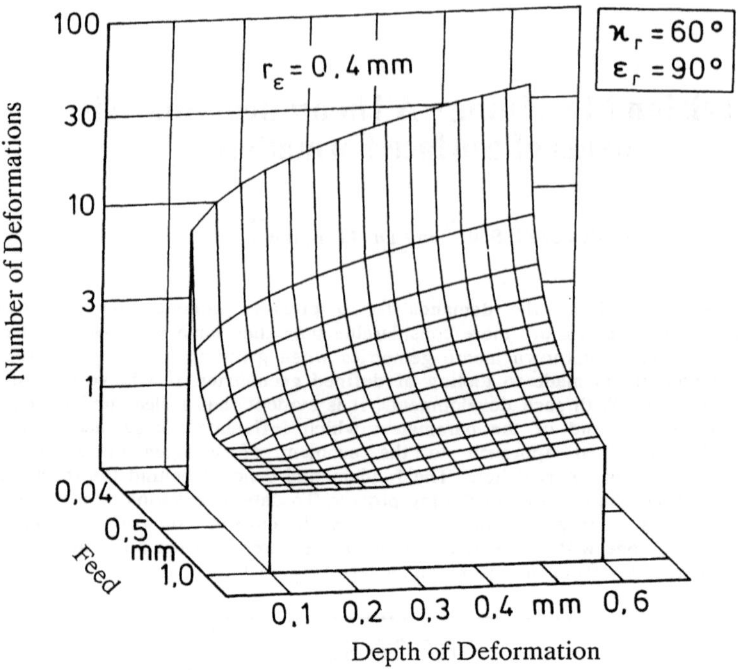

Figure 2: Deformation in the surface layer

Observation from a point within the workpiece body permits an evaluation of events taking place at the machined surface as a function of time. This means that progressive surface formation may be analysed. By applying such a concept it is deemed valid to focus attention on the generation of a single feedmark on the machined surface.

It is found that surface generation and surface layer effects take place in a number of stages and that the final stage takes place on a cusp section. The quantitative analysis of the cusp section indicates the differing conditions which apply across the feedmark width and how the tool geometry and feed influence these conditions. The significance of minimum undeformed chip thickness and the manner in which progressive work hardening takes place become evident from the analysis undertaken, thereby making it possible to expand on some of the phenomena encountered by past researcers of the subject.

The technological data obtained from this theoretical analysis was used as the basis for a simulation programme for the external cylindrical turning operation. An example of the results obtained from this simulation is shown in Figure 2. With this analytical approach the mechanisms leading to the surface layer damage due to chip removal can be better quantified and optimization of the machining process in relation to surface intergity may be achievable.

A Stylus Profilometer with one Nanometre Precision

E W PALMER

National Physical Laboratory
Teddington, UK

Summary

A stylus profilometer having a height range of 5 mm and a scan length of 100 mm has been developed for examining the form of optical and mechanical components. The specimen rests on a horizontally travelling carriage and is probed with a 2 mm diameter spherical stylus. The travel of the carriage and the height changes of the stylus are measured with two novel polarisation interferometers. The profilometer gives results on fused silica specimens which are repeatable to one nanometre.

The stylus profilometer (figure 1) gives results on the flatness of fused silica specimens which are repeatable to one nanometre (figure 2). The high stability of the machine is due to the use of low expansion materials and kinematic design.

The height changes of the stylus are measured with a double passed interferometer. One of the interferometer's retroreflectors is cemented on the stylus and the other is referenced to the carriage slideways. The output consists of two coaxial plane polarised beams whose phase difference changes by 360 degrees per 158 nanometres height change of the stylus.

The horizontal movement of the specimen carriage is measured with a variant of a folded Jamin interferometer. The sinusoidal signal from this interferometer is converted to pulses which trigger a computer to sample the signals from the stylus interferometer.

In order to reduce parasitic reflections and improve signal quality, each interferometer was designed to operate without retardation plates in the measuring arms. Roof prism retroreflectors are used instead of cube corners for the same reason.

Figure 1. Stylus profilometer having a height range of 5 mm and a scan length of 100 mm.

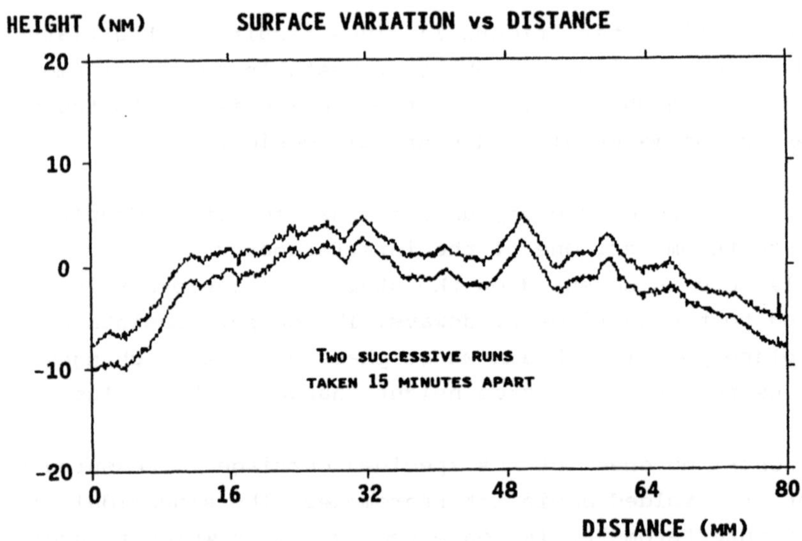

Figure 2. Two scans across the central 80 mm of a 100 mm diameter fused silica optical flat. The scanning velocity was 0.5 mm per second and the stylus load was 0.03 N. The repeatability is better than 1 nm.

An Infinite-stiffness Aerostatic Bearing with an Exhaust-control Restrictor

H. Mizumoto*, T. Matsubara**, H. Yamamoto*,
K. Okuno* and M. Yabuya***

* Faculty of Engineering, Tottori University, Tottori, Japan
** Nagoya Institute of Technology, Nagoya, Japan
*** Nachi-Fujikoshi Co., Ltd., Namerikawa, Japan

The aerostatic bearing is a crucial element for ultra-precision machine tools, and its stiffness should be as high as possible. To obtain an infinite bearing stiffness, we propose a new type of active restrictor. Differing from conventional inflow passive and active restrictors [1], the proposed restrictor controls the air volume exhausted from the bearing surface. Therefore, we have called this active restrictor an exhaust-control restrictor (ECR). Main advantage of the ECR is its stability for the occurrence of pneumatic hammer.

An aerostatic journal bearing with the ECR is shown in Fig. 1; the ECR consists of elements 5 to 10. A magnified view of the bearing surface and the ECR is shown in Fig. 2, where arrows show the air flow. Pressurized air supplied from the inlet is restricted by passive inflow restrictors, and makes an air film on the bearing surface. Then the air is divided into outer and inner flows. The inner-flow air is exhausted to the ECR, and restricted in the gap made by a diaphragm and a spacer. This gap is actively controlled by a piezoelectric actuator. Block diagram of control system is shown in Fig. 3. The displacement of the shaft is detected by a electric micrometer. Appropriate voltage is calculated by a micro computer, and supplied to a amplifier which drives the piezoelectric actuator. When the load for the bearing increases, the restriction gap is decreased by the piezoelectric actuator, and the pressure on the bearing surface can be increased without decreasing the air film thickness. Thus the bearing stiffness can be infinite.

Figure 4 shows the load carrying capacity of the bearing with the ECR and without. The supply voltage to the piezo-electric actuator is also shown. The displacement of the shaft is effectively suppressed by the ECR. Therefor, it is confirmed that the ECR can realize an infinite-stiffness aerostatic bearing. Pneumatic hammer does not occur during the whole measurements.

Reference

1. Mizumoto, H., et al.:Zero-compliance aerostatic bearing for an ultra-precision machine. Precision Engineering 12 (1990) 75-80.

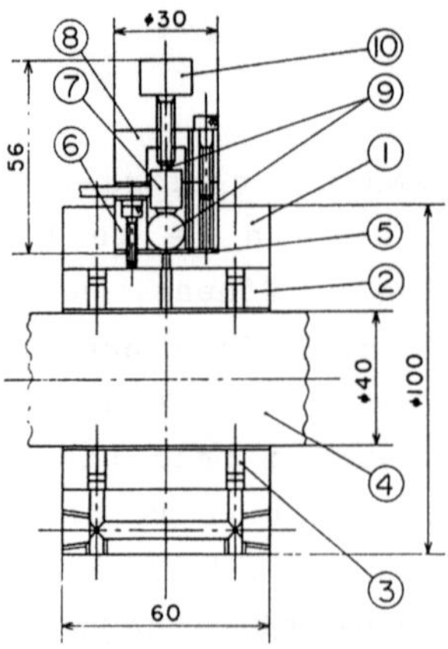

Fig. 1. An aerostatic journal bearing with an exhaust-control restrictor

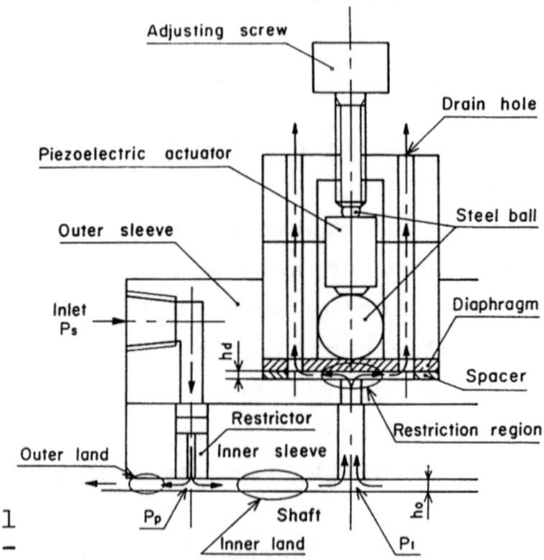

Fig. 2. Magnified view of the bearing with the ECR

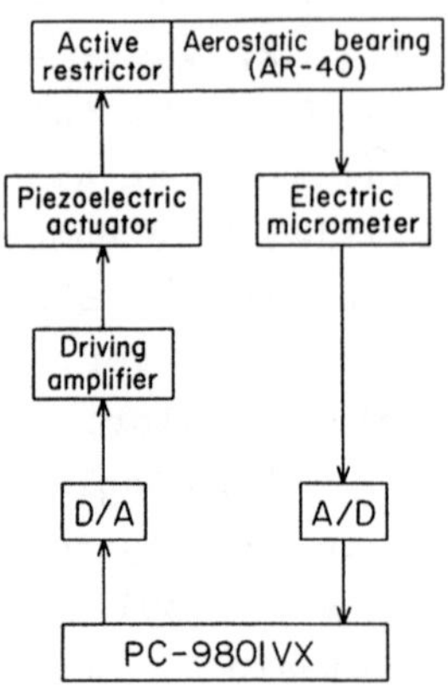

Fig. 3. Block diaglam of control system for the ECR

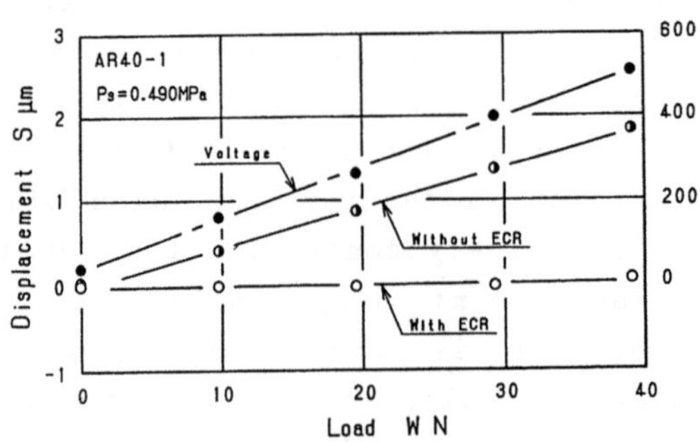

Fig. 4. Static load carrying property of the bearing with ECR

Analysis of Dynamic Coefficients for Multirecess Hydrostatic Journal Bearings with Journal Rotation by Finite Element Method

Y. TIAN and M. BONIS

Division systèmes mécaniques, Département de Génie Mécanique
Université de Technologie de Compiègne
B.P.649, 60206 Compiègne cédex, France

INTRODUCTION

Hydrostatic journal bearings are extensively used as supports of spindle in high precision machine tools because of their high rigidity and stable rotational accuracy. It is of great importance to predict the dynamic behaviour of hybrid journal bearings subjected to an external force with a given excitation frequency, because the presence of cross-coupling terms can have a significant effect on the dynamic response and stability of hydrostatic spindles, especially, owing to the fact that if the cross-coupling stiffness is different from zero, the hybrid journal bearing is not always stable. An analysis of dynamic characteristics of a capillary-compensated four-recesses hybrid journal bearing, taking into account compressibility of the lubricant in the bearing recesses and lubricant inertia in the bearing capillary resistor, is presented.

MATHEMATICAL MODEL

The pressure field over the bearing land area is governed by the Reynolds equation. For an incompressible lubricant, with the usual assumptions of lubrication problems, the Reynolds equation for non steady state can be written as

$$\frac{\partial}{R^2\partial\theta}\left[\frac{h^3}{12\mu}\frac{\partial p}{\partial\theta}\right] + \frac{\partial}{\partial z}\left[\frac{h^3}{12\mu}\frac{\partial p}{\partial z}\right] = \frac{1}{2}\,\omega\,\frac{\partial h}{\partial\theta} + \frac{\partial h}{\partial t} \tag{1}$$

where $h = C\,[1 - \varepsilon\cos(\theta - \phi)]$ (2)

The stiffness and damping coefficients of an hydrostatic multirecess bearing can be calculated by giving the journal a small displacement and a small velocity perturbation, respectively. Thus, equation(2) may be written as

$$h = h_o - \Delta h \tag{3}$$

where
$$h_o = C - x_o\cos(\theta) - y_o\sin(\theta) \tag{4}$$

$$\Delta h = \Delta x\cos(\theta) + \Delta y\sin(\theta) \tag{5}$$

$$\frac{\partial h}{\partial t} = -\Delta x\cos(\theta) - \Delta y\sin(\theta) \tag{6}$$

The perturbation in the film thickness gives rise to a similar perturbation in the film pressure:

$$p = p_o + \Delta p \tag{7}$$

where
$$\Delta p = p_x\,\Delta x + p_y\,\Delta y + p_x\,\Delta x + p_y\,\Delta y \tag{8}$$

Introducing equation (3) to (8) into Reynolds equation (1) and neglecting higher order terms, the perturbed equations can be obtained. After the finite element discretisation process and modification for appropriate boundary conditions, the final set of system equations is

$$[K]\ \{P_k\} = \{Q_k\} \tag{9}$$

where k= 0, x, y, x, y .

The pressure in the recess is determined by the continuity that the flow out of the recess equals the entry flow through the restrictor, for a capillary, it may be expressed as

$$\frac{k_c}{\mu(1+\rho r_c^2 \nu/8\mu)}(-\Delta p_r) = \Delta Q_r + \frac{\partial}{\partial t}(Ah_r) + V_o\beta\,\frac{\partial p_r}{\partial t} \tag{10}$$

then the recess pressure variations may be obtained with same perturbations as that previously considered.

STIFFNESS AND DAMPING CHARACTERISTICS

After etablishing the dynamic pressure solutions, the dynamic performance characteristics of hybrid journal bearing system can be computed. Figure 1 gives examples of stiffness and damping coefficients at concentric position.

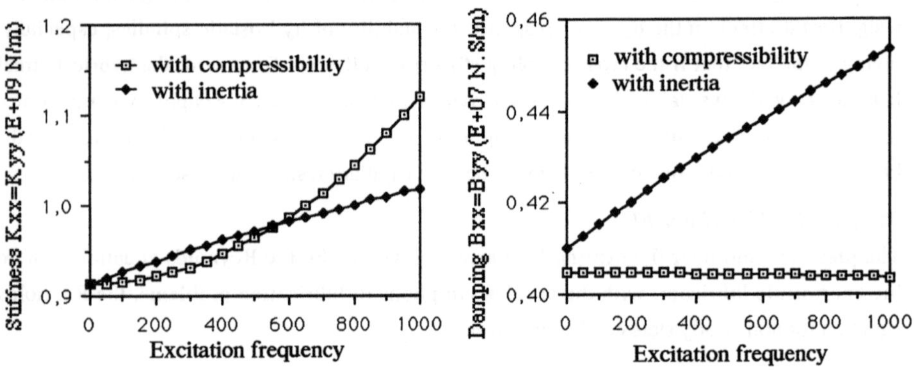

Fig. 1 Dynamic stiffness and damping versus excitation frequency

The results demonstrate that the lubricant inertia in the capillary increases the stiffness and damping capacities of the bearing. The effect of compressibility in pockets on damping, and, on cross-coupling stiffness coefficients remains small, but its influence on direct stiffness is the more high as frequency increases.

References

1. Ghosh M.K. and Viswanath N.S., Recess Volume Fluid Compressibility Effect on the Dynamic Characteristics of Multirecess Hydrostatic Journal Bearings with Journal Rotation. Journal of Tribology, vol.109, N°2, July 1987, PP.417-426

2. Lund J.W., Review of the concept of Dynamic Coefficients for Fluid Film Journal Bearings. Journal of Tribology, paper ASME n° 86-Tri-48, pp. 1-5

Measurement of Sub Surface Damage in Silicon Wafers

H.K. TÖNSHOFF, E. BRINKSMEIER, H. SEIBT

Institut für Fertigungstechnik
und Spanende Werkzeugmaschinen
Universität Hannover, Deutschland

Different machining processes during the manufacturing of semiconductor slices produce a damaged surface layer which influences the structural perfection of mono crystalline materials. The damaged layers which have a depth of up to 50 μm consist of microcracks, residual stresses and dislocations caused by the abrasion of the single crystal substrate. Therefore the surface layers have to be removed by etching of electrochemical methods. For these processes it is necessary to have information about the amount of subsurface damage and its depth. Different techniques for measuring these parameters are applicable. The X-ray diffraction technique is widely applied as a non-destructive method to measure lattice strains caused by internal stresses and to determine the half-width values of Bragg-reflection versus different steps of manufacturing and machining parameters. Fig. 1 shows the principle of X-ray diffraction of a mono crystal by the use of a four-axis diffractometer. The Bragg-reflection of the (620) lattice plane for two silicon wafers are shown. Different peak positions and half width values of Bragg-reflection can be observed. This might be the result of the surface damage caused by the machining process. Besides this X-ray measuring technique, a high resolution X-ray diffractometer with a multi-crystal monochromator, Fig. 2, is useful to describe the state of damaged surface layers. Fig. 2 gives an example of Rocking-curves of the (004) Bragg-reflection of ground and polished silicon-wafers. The ground wafers show a very high diffraction intensity in the wings of the Bragg-peak caused by a damaged surface. In opposition to this broadening the Rocking-curve of the polished, "perfect" wafer shows a small width at the half maximum of the Bragg-peak. It can be shown that the methods described above are sensitive to changes in the lattice of abraded silicon.

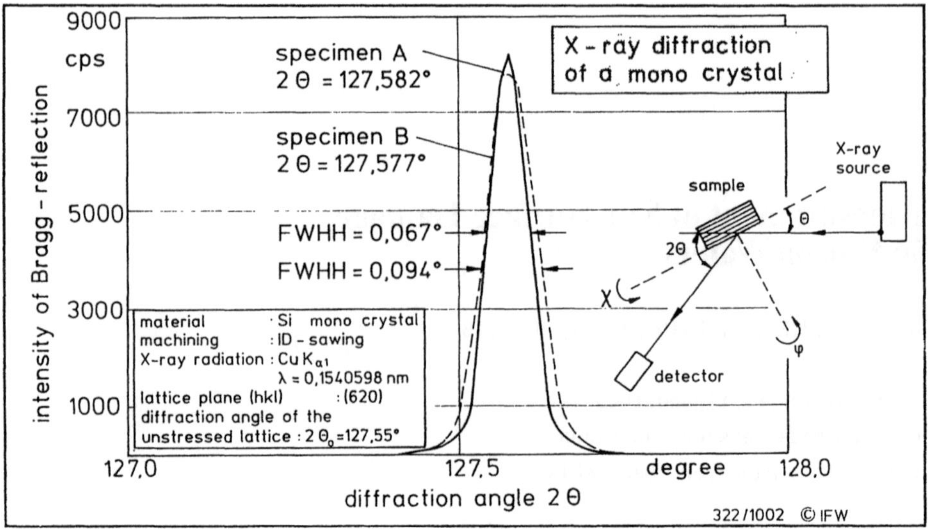

Fig. 1: X-ray diffraction of monocristalline silicon
(Four-axis-diffractometer)

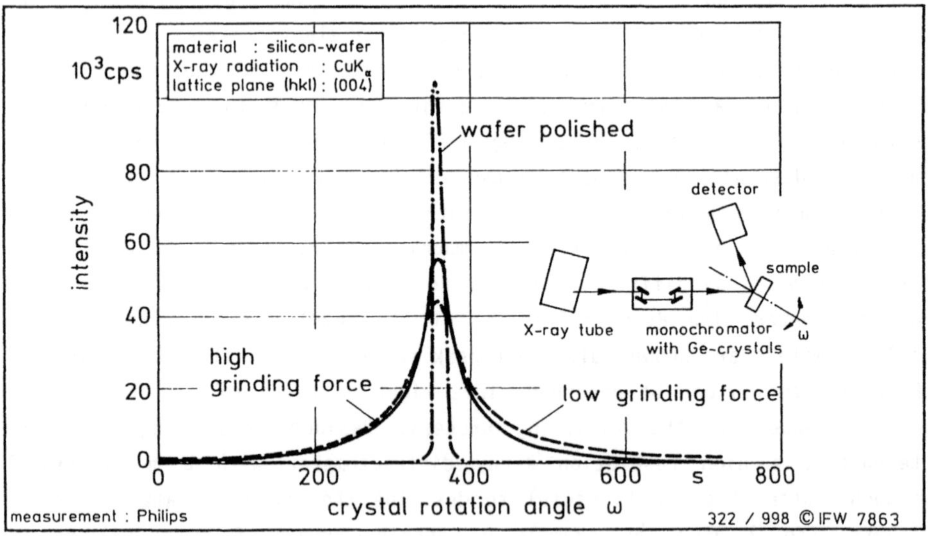

Fig. 2: Rocking-curves of machined silicon-wafers
(High resolution X-ray diffractometer)

Absolute Laserrefractometer

Dr. D. Frölich, H.-P. Meiser, D. Frerking
Spindler & Hoyer GmbH
Göttingen, Germany

Laser interferometers are widely used for accurate determination of length and for the calibration of precision lengthmeasuring equipment and tooling machines. The accuracy of an interferometer is determined by the uncertainty of the wavelength of the laser light, which is determined in part by the uncertainties of the laser frequency and the refractive index, n, of the air. In most applications the uncertainty of n is 1 - 2 orders of magnitude larger than the uncertainty of the frequency. Due to that fact a more precise determination of n will result in a significant increase in the accuracy of the interferometric measurement.

Up to now the following methods have been established for the determination of the refractive index of air:
1. The refractive index depends on temperature, pressure and the composition of the air. The so called Edlén formula expresses the dependence on temperature, pressure, and humidity. It is widely used for the calculation of n. This method has the disadvantage of a long time constant due to the temperature measurement and it does not take the generally unknown composition of the air into account.
2. A tracking refractometer determines the change of n during a given time. The initial value of n has to be measured separately. This method typically uses a highly stable interferometer whose measurement arm is floated with the air to be measured. A variation of n changes the optical pathlength in the interferometer which is used to calculate the change of n.

3. An absolute measuring refractometer is used for high precision measurement of n. It consists of a stable interferometer whose measurement arm is evacuated for each measurement. Because of its technical environment (vacuum pump) it is only used under laboratory conditions by calibration institutes.

We present a new absolute measuring refractometer which allows the determination of n whithout using a vacuum pump. The refractometer is a compact device which can be also used in the industrial environment.

The basic principle is outlined in Fig 1 and shows the vacuum cell with variable length. There are two interferometers (1 and 2) which are used simultaneously to measure the refractive index. The inner interferometer 1 measures the length variation of the vacuum cell using the vacuum wavelength of the laser. An actuator moves the window F_2 over a distance L while the window F_1 is held at a fixed position. The number of fringes measured by the interferometer 1 is :

$$N_1 = \frac{2L}{\lambda_0} \qquad (1)$$

λ_0 being the vacuum wavelength of the laser.

Moving the window F_2 produces a replacement of air by vacuum over the length L in the outer interferometer 2. This change in optical path length results in a fringe count :

$$N_2 = \frac{2nL}{\lambda_0} - \frac{2L}{\lambda_0} \qquad (2)$$

Using (1) and (2) n can be expressed as :

$$n = 1 + \frac{N_2}{N_1} \qquad (3)$$

This method has the following features :

- The refractive index is directly determined by the two fringe counts N_1 and N_2. Assuming that the value of n for air under normal conditions is appr. 1.00027, the relation N_2/N_1 takes on a value of appr. 1/3700.

- Using a simple counter for N_1 and precise zero crossing detection for N_2, this ratio is easily determined. With L = 6 mm an accuracy of 5×10^{-7} is easily achieved.

- The actuation of the vacuum cell allows measuring rates of 1 Hz. This large data rate can be used to analyze fluctuations in the refractive index. The averaging time can be adapted to the fluctuations.

- A very compact design is possible because no vacuum pump is needed. Thus the refractometer can be placed near the measurement volume of the interferometer and can also be used in an industrial environment.

Construction of the Refractometer

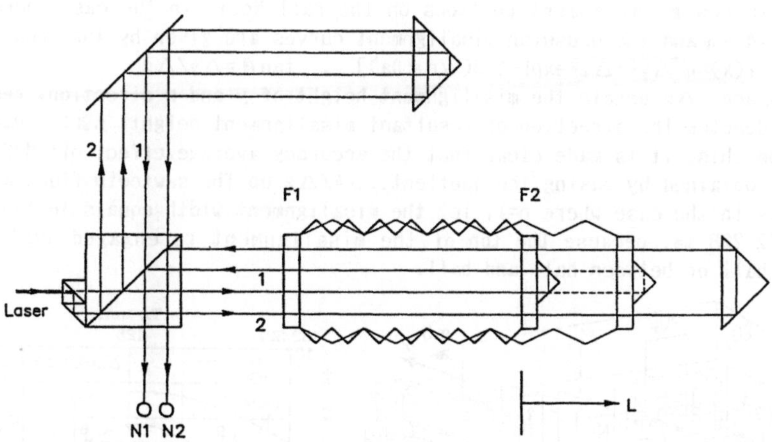

Fig. 1

Accuracy Average Effect of Linear Motion Ball Guides System for NC Machines – Theoretical Analysis

Shigeo SHIMIZU

Dep. of Precision Engineering
Meiji University, Kawasaki, Japan

Nobuyuki FURUYA

Dep. of Mechanical System Engineering
Yamanashi University, Yamanasi, Japan

Supposing the table of NC machine tool, the load distribution formula in LMBG system is derived under the combined loads and moments by considering the Gaussian shape misalignment between one of the rails and blocks. The theoretical analysis is carried out with various misalignments and preloads. When the misalignment width is less than the block length, the accuracy average effect, ie, the ratio between the table displacement and the maximum misalignment height, becomes 1/40 to 1/8. When the misalignment width exceeds the block length, it becomes 1/8 to 1/4. The effect of preloading on the accuracy, rigidity and load distribution of LMBG system is remarkable.

Fig.1 illustrates the X-table composed of LMBG system and its coordinate system. The load distribution formula of this system consists of 5 nonlinear simultaneous equations with the unknowns, y direction displacement $\alpha 1$, pitching $\alpha 2$, rolling $\alpha 3$, z direction displacement $\alpha 4$ and yawing $\alpha 5$.

Fig.2 shows the fluctuations of $\alpha 4$ and $\alpha 5$ of the table which has passed through various misalignment portions on the rail No.1, in the case where the preload $\lambda = 4\ \mu m$ and the Gaussian misalignment curves are given by the equation

$$f(x)=\sqrt{\Delta y^2+\Delta z^2}\exp[-4.5(x/n\kappa Da)] \quad , \quad \tan\theta=\Delta y/\Delta z$$

where Δy and Δz denote the misalignment height of y and z direction, respectively. θ denotes the direction of resultant misalignment height. $n \geqq 1$, κDa=ball pitch. From this, it is made clear that the accuracy average effect of LMBG system can be obtained by making the quotients, $\alpha 4/\Delta z$ up. The sawteeth fluctuations appear only in the case where n=1, ie, the misalignment width equals to two ball pitches, 12.738 mm, because the top of the misalignment is engaged just on a center of ball or between ball and ball.

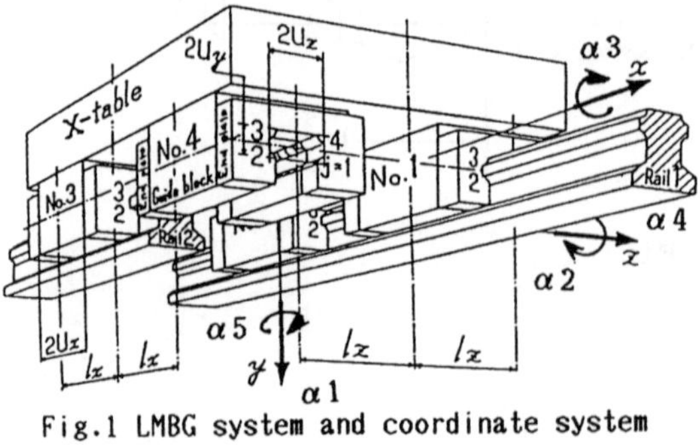

Fig.1 LMBG system and coordinate system

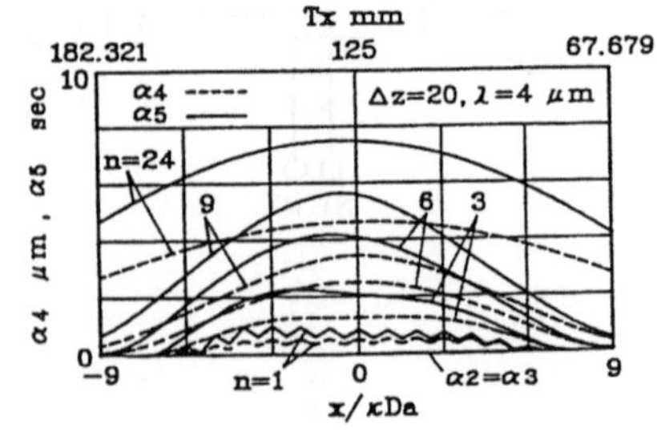

Fig.2 Fluctuations of $\alpha 4$ and $\alpha 5$

Accuracy Average Effect of Linear Motion Ball Guides System for NC Machines – Experimental Verification

Shigeo SHIMIZU

Dep. of Precision Engineering
Meiji University, Kawasaki, Japan

Hiroshi KOSHIISHI

THK CO.,LTD.
Tokyo, Japan

The accuracy average effect of LMBG system has been verified by the experiment using the X-table composed from 4 guide blocks and 2 rails. One of the rail is used as a reference rail and the other is bent by a wedge-shaped loading unit to forcedly apply a misalignment towards z axis. After the shape of misaligned rail is measured, the X-table is traveled over the misalignment portion. The displacements toward y and z axis, pitching, rolling and yawing are measured by means of 6 electric micrometer sensors setting up onto 2 straight edges beside the table. The measurements are carried out with 3 kinds of preloads and 4 kinds of misalignments. The results are as follows. When the preload is increased, the accuracy average effect is also increased. The good agreement is obtained between the experiment and the theoretical result, but the former is a little bit greater than the latter in the average effect. The authors think that this reason is mainly due to the LMBG system theory in which the Hertzian deformations of balls are only taken into account and the elastic deformations of guide block jaws, X-table etc., are neglected.

Fig.1 shows that the measuring result of z direction displacement and yawing in the case where the misalignment $\Delta z = 28\ \mu m$ and the preload $\lambda = 12\ \mu m$.

Table 1 also represents the accuracy average effect of this LMBG system. In which $\Delta\alpha 5$ is put such that the ratio between the misalignment height Δz and the half length of the pitch of guide blocks in x direction, lx=125 mm.

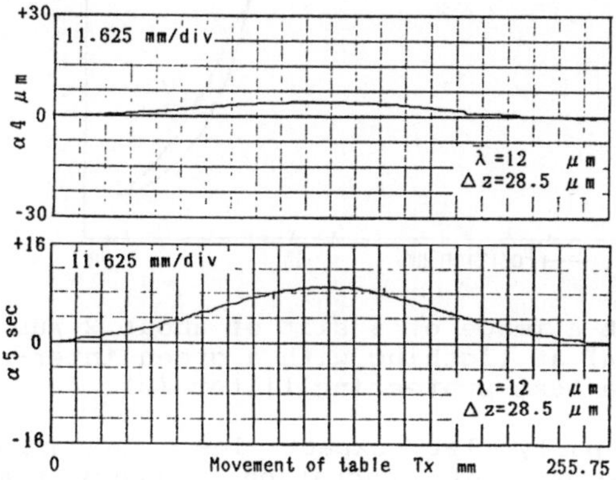

Fig.1 Measuring results for $\alpha 4$ and $\alpha 5$

Table 1 Accuracy average effect

Δz	λ	$\alpha_4/\Delta z$		$\alpha_5/\Delta\alpha_5$	
μm	μm	Theory	Ex.	Theory	Ex.
7.4	0	1/4.1	1/ 9.	1/4.2	1/4.0
	4	1/4.1	1/ 8.	1/4.2	1/6.9
	12	1/4.2	1/14.	1/4.3	1/8.8
15.9	0	1/4.4	1/5.5	1/4.5	1/3.6
	4	1/4.5	1/7.3	1/4.6	1/5.9
	12	1/4.5	1/9.7	1/4.7	1/6.4
21.1	0	1/4.2	1/4.9	1/4.4	1/3.4
	4	1/4.3	1/6.3	1/4.4	1/4.9
	12	1/4.4	1/7.7	1/4.5	1/5.2
28.0	0	1/4.2	1/4.5	1/4.2	1/6.9
	4	1/4.3	1/6.1	1/4.4	1/4.7
	12	1/4.4	1/7.1	1/4.5	1/5.3

Investigation of the Geometrical Properties of CCD Line Image Sensors

J. Tschirnich, F. Gehring, D. Hantke, H. Philipp, G. Sparrer

Physikalisch-Technische-Bundesanstalt, Germany

The use of CCD image sensors in measuring and positioning systems with highest precision increases. In order to fulfil the demands to resolution and accuracy (present up to the submicrometer range), the knowledge of the geometrical properties of the CCD detectors is basically important.
This poster describes measuring technique and evaluation to analyse the geometrical properties of CCD line image sensors and presents results of measurements. As known in optical investigations an optical edge or a very narrow optical slit are imaged as sharp as possible on the pixels of the detector to be analysed. The CCD sensor is relatively moved to the fixed image of the edge or image of the slit. Its positions are synchronously registered with the corresponding electrical output voltage of CCD sensor at a sufficiently dense sequence of measuring points. In this way the edge method yields the integral sensitivity characteristic of each pixel whereas the use of the slit method results in the aperture function of each pixel. The position of the middle of pixels and of channel-stop areas can be calculated from these measured values. These positions are regarded as graduation of a measure of length analogous to DIN 2268. The graduation errors are calculated.

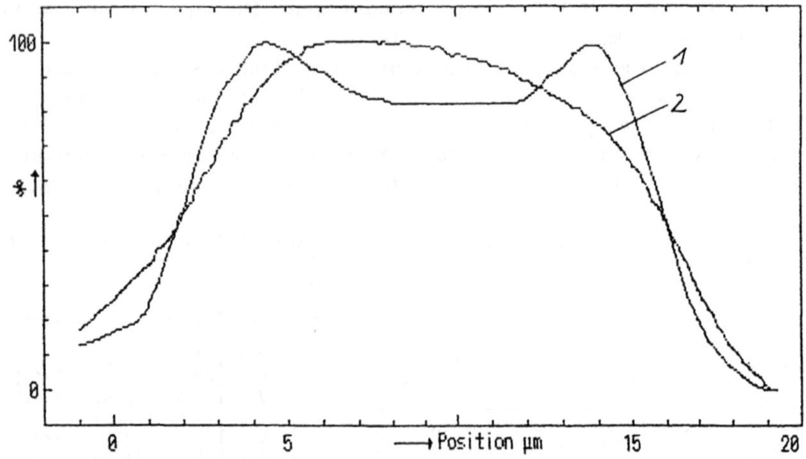

Fig. 1. Scanning of a pixel with a image of a slit of about 2 /um-width (aperture function) at ligthing with a green interference filter (1) and a heat protecting filter (2).

The aperture functions principally let expect less measuring uncertainties at the determination of graduation errors than the measured integral sensivity characteristics. The shape of the

measured aperture function is decisively influenced by the special type of CCD sensor, the wavelength of the used ligthing and the quality of slit image. Two aperture functions measured with ligthing of different wavelength are shown in fig.1. Therefore the use of CCD sensors to precise measurements should be carried out at the same measuring conditions as at theirs calibration.
It was reached a standard deviation of 30 nm to determinade the graduation of CCD sensors and it is considerably less than the measured graduation errors.

References
1.Christoph, R.: Optoelktronische Bildauswertung zur Objektvermessung. Bild und Ton, Leipzig, 43(1990) 133-143
2.Hantke, D.; Philipp, H.; Sparrer, G.; Tschirnich, J.:Meß- und Auswerteverfahren für Teilungsmessungen an CCD-Zeilen. Bild und Ton, Leipzig, 43(1990) 237-243

Optical Linewidth Measurement on Masks and Wafers in the Micrometre and Submicrometre Range

M. CZASKE, W. MIRANDÉ, M. FRAATZ
Physikalisch-Technische Bundesanstalt (PTB), Braunschweig

In order to be able to perform optical linewidth measurements with high accuracy on structures whose size is in the range of between 0.3 μm and 150 μm, a computer-controlled measuring system was developed at the PTB optics division. It consists chiefly of a modified microscope working with both transmitted and reflected light. The precision measurements are carried out under clean-room conditions (class 100). For the linewidth measurement, the masks or wafers with their edges are translated rectangularly to a stationary slit in the image plane. Twin parallel springs with a very accurate parallel translation and driven by piezo-electric translators are used for this purpose. The displacement of the object is measured with calibrated inductive transducers with an accuracy of a few nanometres. The stationary scanning slit is arranged in the image plane of the microscope parallel to the edges. The luminous flux passing through this slit is registered by a photo multiplier and the resulting photocurrent is represented as a function of the object displacement. For exact focusing a piezo translator is also used. The focusing accuracy is better than 40 nm, which is sufficient even in the most critical case when highly resolving objectives with numerical apertures of 0.95 are used. The maximum slope of the edge image serves as a focusing criterion in most cases. The precision of the measurement amounts to a few nanometres for all imaging techniques employed.

The uncertainty achieved here for the linewidth measurement is limited by unknown systematic errors, amounts up to ±20 nm, depending on the edge quality. It depends very much on the criteria used for the determination of the edge position and on the illumination of the structure. Edge detection criteria can be derived from the theory of partially coherent imaging. For measurements on chromium photomasks a threshold criterion is used. The threshold is influenced by the wavelength, the coherence parameter (ratio of numerical aperture of condenser and objective) and the linewidth. If the distance between the edges of a slit or line is greater than three times the wavelength, the dependence of the threshold on the linewidth is negligible. In the other case the linewidth is determined by fitting the theoretical curves to the measured curves by varying in an iteration process the linewidths taken as the basis for the calculation. This fit, which is made for bright-field illuminations in transmitted and reflected light, results in a threshold for the relative photocurrent, which then serves to determine the linewidth from the measured curve.

With this technique it is possible to measure linewidths down to half of the wavelength if a well-corrected microscope objective with a high numerical aperture is used. In fig.1 the theoretical and measured curves of the image of a 0.29 μm chromium line in transmitted light with bright-field illumination are shown. The focusing criterion mentioned above cannot be applied to such small structures, and focusing was done using a broader structure in the neighbourhood of the small structure. The measurements were repeated at different wavelengths. There was good agreement between model calculations and the respective measurement.

For bright-field illumination in transmitted light, the polarization of light usually has only a minor influence on the measured values of the linewidth measurement on different kinds of photomasks. But in the case of imaging photomasks, especially chromium photomasks with an antireflective layer, in reflected light when linearly polarized radiation with an orientation of the electric field parallel or perpendicular to the edges is used, there may be substantial differences in the measurement for both polarization directions [1, 2, 3]. Micrometre and submicrometre structures may show this effect, as we have demonstrated. In this case the usual scalar theory of microscopical imaging is not applicable.

The width of grooves with vertical edges on silicon wafers was measured. The depth of these grooves does not exceed a quarter of a wavelength, which is why they may be treated as thin, two-dimensional phase objects. In the image the position of the edge corresponds to a local minimum of the relative photocurrent. For a small groove the distance of the minima becomes smaller than the distance of the edges. This deviation can be corrected by fitting the theoretical curve to the measured one as already mentioned. The focus criterion for measurements on such structures is the minimum half-width around the minima of the relative photocurrent.

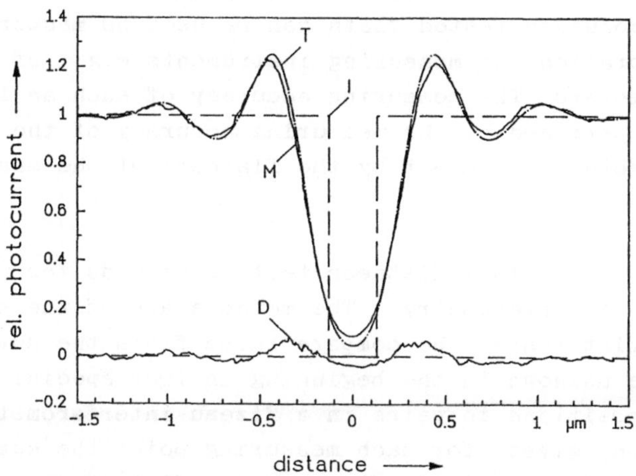

Fig.1 Measured photocurrent in the image of a 0.29 μm chromium line for bright-field illumination in transmitted light with a wavelength of 529 nm, a coherence parameter 0.22 and a numerical aperture of the objective of 0.90. This figure shows measurement (M), calculation (T), difference curve (D=T-M) and ideal curve (I).

References

1. Fraatz, M.; Czaske, M.; Mirandé, W.: PTB Jahresbericht 1989, 164-166

2. Mirandé, W.; Fraatz, M.; Czaske, M.: Effect of polarization on linewidth measurement by reflected light confocal scanning microscopy, 1st International Conference on Confocal Microscopy , Amsterdam, 15.-17.3.1989

3. Mirandé, W.: Jahrestagung der DGaO 1987, Berchtesgaden

Absolute Flatness Testing

K.E.Elssner, J.Grzanna, G.Schulz, A.Vogel

Central Institute for Optics and Spectroscopy,
O-1199 Berlin-Adlershof, Rudower Chaussee 6, Germany

Surface deviations of polished optical flats related to a mathematical plan are determined by absolute flatness testing. Absolutely tested flats can be used as master surfaces for calibration of measuring instruments e.g. of automatic interferometers. The measuring accuracy of such an instrument is then determined by the measuring accuracy of the absolute testing method but n o t by the flatness of the master surface.

We develop absolute flatness testing methods for use with automatic interferometry. The methods are of the kind known as three flat tests. We compare three flats the surfaces of which are unknown in the beginning in four special combinations of positions in pairs in a Fizeau interferometer. Each combination gives for each measuring point the sum of the absolute surface deviations of the two flats combined. These measured values enable the set up of a uniquely solvable system of linear equations if the mathematical reference plane is properly choosen.

The r o t a t i o n m e t h o d [1] which relies on [2] gives absolute surface deviations of the three flats for any number of diameters. Here it is possible to establish subsystems of equations which can be solved one by one in measuring points lying equidistantly on concentric circles that is in a polar grid. This is advantageous for computing especially in case of the big number of measuring points characteristic of the automatic interferometry. On the other side in automatic interferometry the measuring points usually are arranged on a quadratic grid. Therefore interpolation is necessary, introducing additional errors. The r o t a t i o n - s h i f t m e t h o d [3] avoids interpolation. In return experimental and mathematical work is more complicated.

Propagation of measuring errors in the course of computing the surface deviations has been determined by simulating calculations. For a measuring accuracy of the interferometer of /200 the accuracy of the absolute surface deviations is 0.004 um (rotation method) and 0.003 um (rotation-shift method) resp. Our specialized automatic Fizeau interferometer gives absolute surface deviations for surfaces up to 200 mm in diameter. The flats are tested vertically so that inflexion of the plates by graviation is excluded. It is possible to test opaque plates and plates made from glass ceramic material like Zerodure or Sitall. The very low thermal expansion coefficient of the latter garantees a good reproducibility of absolute testing also in routine work. First measurements have indicated a measuring accuracy of = /300.

References

1. Schulz, G.; Grzanna, J.: Appl.Opt. subm. (1990).
2. Schulz, G.: Optica Acta 14 (1967) 375-388.
3. Grzanna, J.; Schulz, G.: Opt.Comm. 77 (1990) 107-112.

Kinetic Factors in Electropolishing

T. Hryniewicz

Technical University of Koszalin
Department of Mechanical Engineering
Racławicka 15-17, 75-620 Koszalin, Poland

A lot of hitherto performed works on electropolishing /ECP/ of metals and alloys, both theoretical and experimental, have been trying the only diffusional conception to take into account when considering the phenomena accompanying the process. The diffusional approach let to solve the diffusion equations [1, 2], taking into account diffusion, natural and/ or forced convections, and the solid film formation. The set of factors affecting the process and the output data are presented in Table 1.

Experimental investigations carried out on various metals and alloys prove that both electrochemical and hydrodynamic conditions are responsible for the positive results to obtain in the process [2]. The overall process of metal dissolution may consist of several reaction steps. If the charge transfer takes place in two or more consecutive reaction steps, the stepwise mechanism of metal dissolution is obtained.

The problem is much more complicated when not a simple dissolution but electropolishing is to take place. For the ECP process to occur, the reactions in specific highly concentrated solutions are to be considered. For instance, in ECP steels a typical solution covers one mole of the total amount of electrolytes /H_2SO_4, H_3PO_4/ dissolved in only 0.6 mole of water which is fixed in the hydrated zone of anions and cations. The fundamental condition for the oxydated metal atom to get into the solution is a possibility of its hydra-

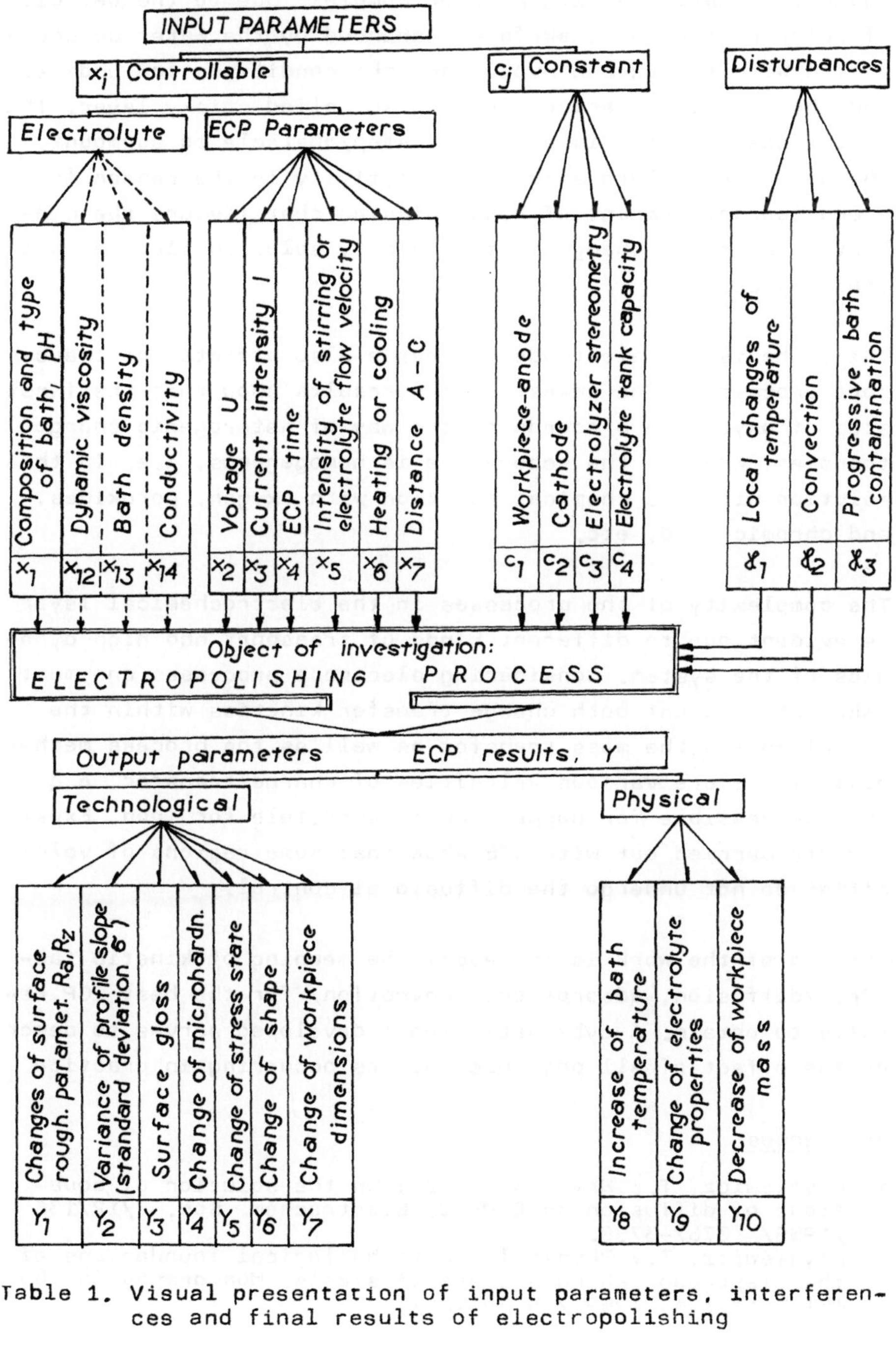

Table 1. Visual presentation of input parameters, interferences and final results of electropolishing

tation, formation an ion pair or complex. Due to the deficit of water on the anode surface, enhanced by the water decomposition with the oxygen evolution, the conditions for undissolved chemical bonds arise, forming so called sticky layer. The of the passive film is a further step/obstacle to surmount. The high electrolyte solution resistivity is the reason for a considerable amount of heat emission that favours the convection process. Change in the density solution also favours the convection.

The mechanism of electrode processes also affects the transport processes. The investigation results show that the first step of oxydation proceeds on the way of heterogenic charge transfer, whereas the second one is homogeneous, i.e. in the solution with the chemical reaction with oxygen, peroxides, and chromic acid, etc.

The complexity of the processes in the electrochemical layer is evident due to different kinds of transport and high dynamics of the system. Considering electrode processes one must take into account both charge transfer kinetics within the interface and the mass transfer as well as the process mechanism. There are various velocities of charge transfer, e.g. quasi-reversible for copper and irreversible for iron. Experiments carried out with RDE show that some regions of velocities do not undergo the diffusional control.

The aim of the work is to reveal the meaning of kinetic factors /diffusion, adsorption, convection/ for the best ECP results to obtain. A cybernetic model developed serves to describe the effect of all possible factors occurring in practice.

References

1. Hryniewicz, T.; Hryniewicz, Z.; On the solution of equations of diffusion in ECP. J. Electrochem. Soc. /12/136 /1989/ 3767-3769.
2. Hryniewicz, T.; Physical and technological foundations of the electropolishing process of steels. Monography, ed. by WSI Koszalin, 1989 /in Polish/.

Precise Performance Tests of Machine Tools with the Use of a Computerised Measuring System

J. Jędrzejewski, W. Kwaśny
Wroclaw Technical University, Poland

Diagnostics of machine tools and especially acceptance tests call for a large number of various signals to be analysed. High accuracy of diagnosis can be then achieved by using multi-sensor computer-controlled measuring systems fitted with appropriate software. Such a system has been worked out and put into service at Wroclaw Technical University.

A large number of sensing probes of various types (see Fig.1) and made a highly developed analog side of the system make it easy to adopt in a programmed fashion various combination of measuring channels that can fulfill best the actual diagnostic needs. A connecting link between the analog and digital parts of the system is formed by three fast A/D converters and one integrating converter. A signal obtained from any of measuring amplifiers "w" may be fed directly to one of A/D converters or may be earlier analog processed by special subsystem such as programmable DC amplifier ("w.DC"), RMS converter, programmable high- ("FG") and low-passed ("FD") filters and differentiating block d/dt.

The program for controlling the measurements is written in Turbo Pascal and permits several various strategies of the measuring process to be employed and the results to visualised [1]. In diagnostic tests of machine tools the measurements which are further processed digitally are of greatest use. These may be: determination of the amplitude-frequency spectrum, long-time measurements and check measurements for comparing some quantities with their stipulated limit various [2].

The present diagnostic system shows great software and hardware flexibility which is likely to keep the system usable for a long time in spite of rapid advances in design and manufacture of machine tools observed at present. Hardware and software structure of the system enables simultaneous analysis of many diagnostic signal, full explanation of their informational content and accurate determination of the malfunction causes and location.

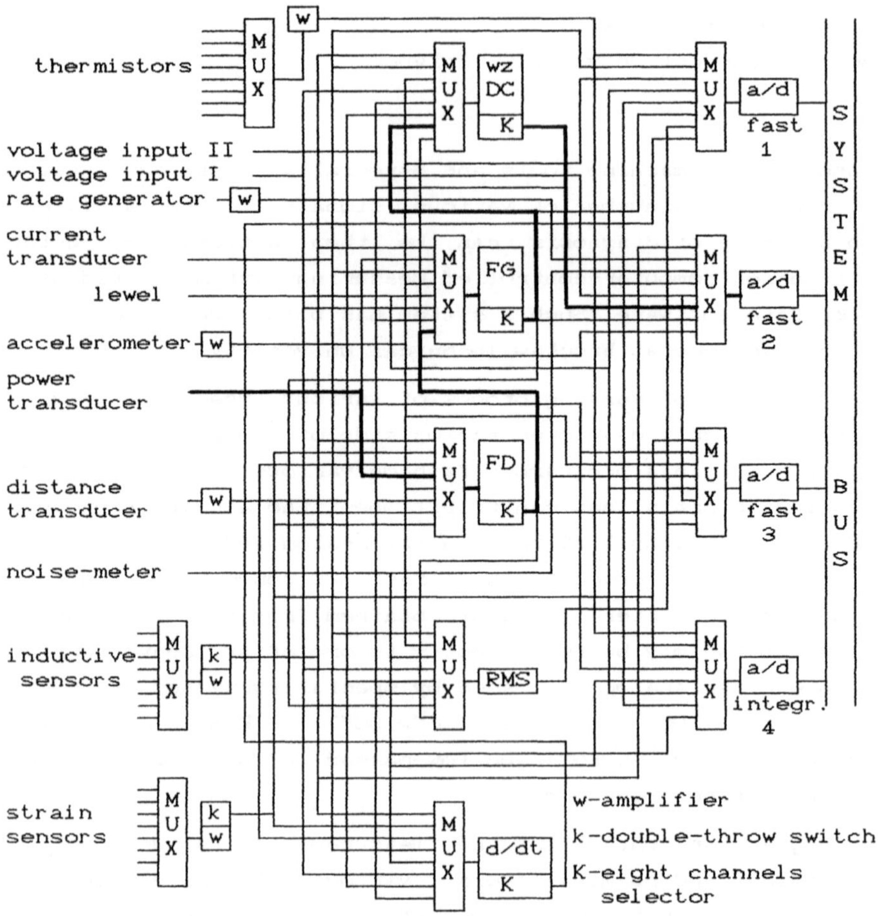

Fig.1. Structure of the analog side of the computer measuring system (MUX-multiplexer, wz.DC-programmable DC amplifier, FG-programmable high-passed filter, FD-programmable low-passed filter, d/dt-differentiating block).

References

[1] Kwaśny, W.;Błażejewski, A.: A multi-sensor measurement system for acceptance diagnostics of machine tools (in Polish), Przegląd Mechaniczny 21-22/1990, 39-41

[2] Jędrzejewski, J.;Kwaśny, W.:Algorithm of measurement strategy in diagnostic tests of machine tool drives. 6th International Symposium IMEKO, Budapest 1987, 219-224

The Real Contact Area Measurement by Means of Scanning Profilometry

J. KACZMAREK, T. KLIMCZAK

Institute of Fundamental Technological Research of Polish Academy of Sciences, Warsaw, Poland

The real contact area (RCA) is essential information for understanding many tribological situations. The methods of RCA evaluating, which have been being applied untill now, were either not accurate or time cosuming.

The presented method is based on three-dimensional copying of a surface topography by means of the stylus scanning profilometry. The applied software allows for detailed analisis of the images of the surface topography. Specially prepared contour map, which reference plane is the plane of a simulated "mating" counter-surface, allows for reckoning of the RCA. The evaluation of the number of contact spots and their individual size - thus the distribution of the size of the local contact spots - is also available.

The method was applied for investigation on the RCA due to a plastic deformation of surface asperities under load. In our experiment turned surface was pressed against flat and rigid counter-surface. After each step of deformation the load was released, the deformed surface was measured by the scanning profilometer and deformed areas were localized (Fig 1). The obtained relation betwen load and RCA due to the asperity plastic deformation is shown in Fig. 2. Figure 3 shows the changes of the number of local contact spots as well as their average size due to the load. We can observe an antithetic trend which results in nearly linear relation betwen RCA and load.

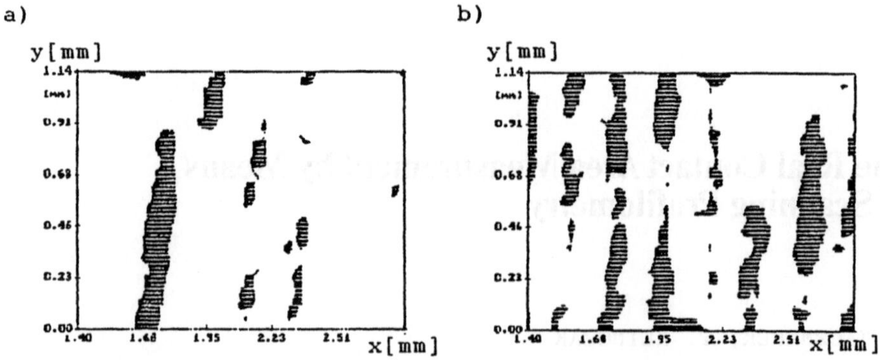

Fig. 1. Turned surface deformed by rigid-flat counter - surface. Black areas represents local contact spots deformed plastically. Nominal load a) 180 MPa b) 400 MPa

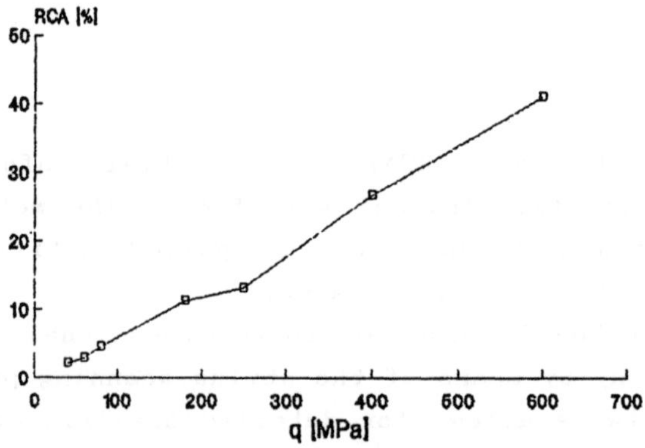

Fig 2. Real contact area RCA versus nominal load q

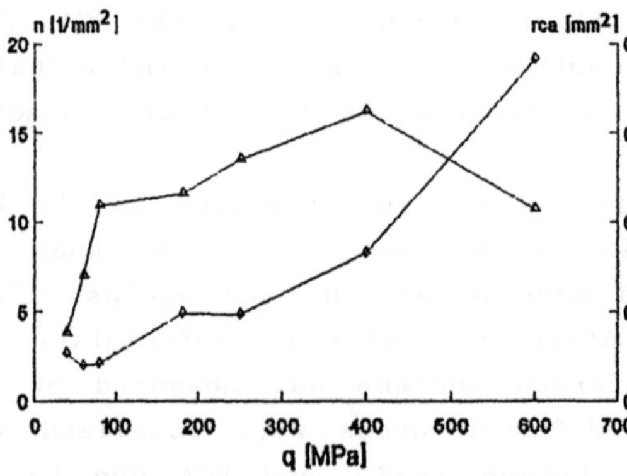

Fig. 3. The number of local contact spots and average size of single spot (rca = RCA/n) versus nominal load q (—△—n —◇—rca)

High-efficiency Polishing of Polycrystalline Diamond

Tsunemoto Kuriyagawa and Katsuo Syoji

Department of Precision Engineering
Faculty of Engineering, Tohoku University, Sendai, Japan

Summary

Polycrystalline diamond (PCD) is extremely hard for mirror finishing and so the large polishing time and the cost become serious problems. Previous wet polishing tests proved that the stock removal was large (0.04-0.05 mm^3/min). However, high polishing pressure was necessary (100-200 MPa) and the surface roughness was too coarse (Ra=40-60 nm). This is because the machining mechanism of PCD is based on a micro fracture (micro cutting) of diamond. In this experiment, a dry polishing of PCD is attempted, for which a polishing machine and new types of wheel (polishing plate) have been manufactured. In the dry polishing process, the polishing temperature between the polishing plate (wheel) and PCD becomes higher, so that thermo-chemical reactions such as graphitization, oxidation and diffusion of diamond will become main mechanisms of polishing instead of the micro cutting action.

Three types of metal bond for diamond wheel were prepared. The first is an iron bond (Fe), which wears diamond thermo-chemically at high temperature. The second is a cobalt bond (Co), which has a wear property similar to the Fe bond and a slight tendency of wetting with diamond. The third is a bronze bond (B). Diamond wheels (grain size; #800, concentrations; 25, 100) were made from these bond materials.

Before polishing tests, precise truing of wheel was carried out on a machine using the GC-cup method. Polishing pressure in operation was controlled by a micro-computer to be constant. Rising of the polishing temperature makes an oxidation of the bond surface progressive. In cases of Fe and Co bonds, especially, overlapping of the oxides on their surfaces induces the reduction of thermo-chemical reaction and the adhesion of the oxides onto the polished surface, so that the stock removal reduces. Therefore, in-process shaving (IS) was carried out to remove the oxide layer on the bond surface. On the conditions of Fe bond, V=6000 rpm and P=2.5 MPa, mirror polished surfaces of Ra=8 nm and flatness =96 nm/mm were obtained.

Acknowledgement

We are grateful to Osaka Diamond Industrial Co., Ltd. for the supply of special-made wheels.

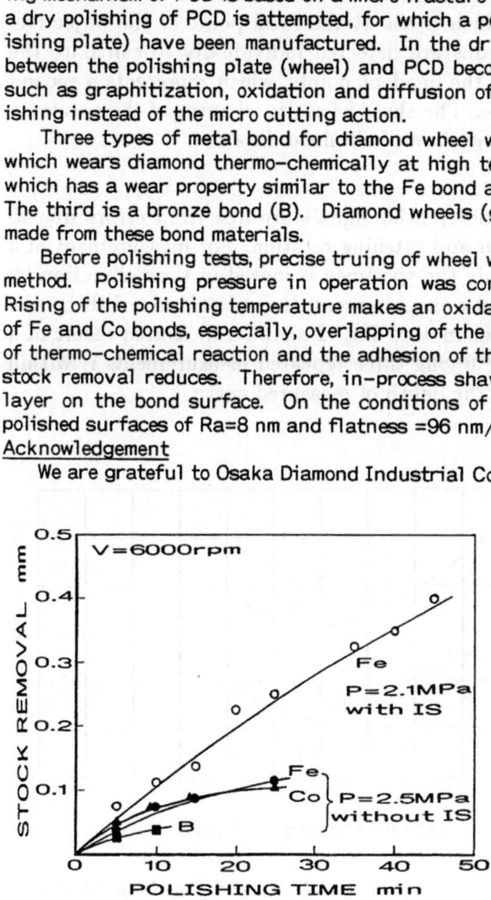

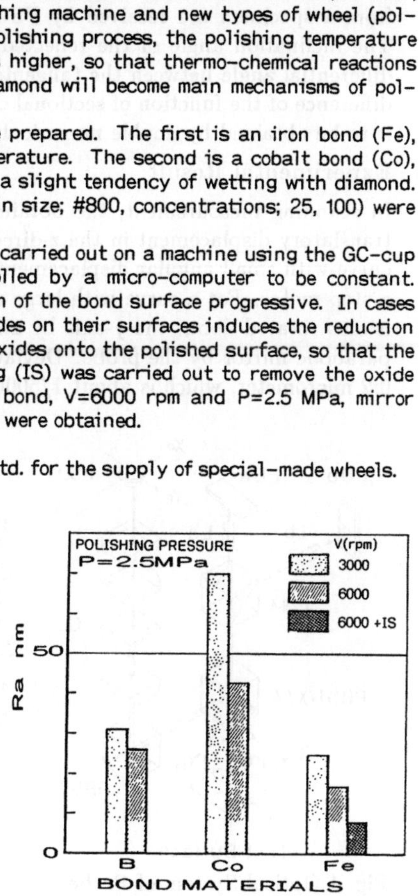

A Differential Laser Autocollimation Probe for On-Machine-Measurement

Satoshi Kiyono, Yoshihiro Asakawa, Kamada Osamu and Tatsuya Kume

Department of Precision Engineering, Faculty of Engineering, Tohoku University, Aoba-ku Sendai, 980 Japan

Principle of Measurement

A differential laser autocollimation probe has been developed, which is so compact as to be able to be mounted on diamond lathes. The probe has been used to measure mirror shapes on an apparatus which can approximately realize the circumferential conditions of an on-machine-measurement system. Fig. 1 shows a schematic illustration of the probe. Light from laser diodes are collimated by selfoc lenses and the incident points of them on a specimen are lined in the scanning direction. The two collimated beams reflected from a specimen are received by PSD which is set at the focal point of the objective. The inclination angle of the reflected beams are detected individually to calculate the differential angle between the tangents at the incident points, which represents a second difference of the function of sectional curve. The shape of sectional curve of the specimen is to be obtained by double numerical integration of obtained data.

Experimental Result

In scanning measurement, the obtained differential signals are free from errors due to translatory displacement in the z-direction and pitching rotation. For measurement of a curve with small angular displacement, only the specimen is moved in the x-direction by a step motor. For a curve with large angular displacement, the optical axis of the probe is also rotated about the y-axis by another motor. Fig. 2 shows a measured result of a parabolic mirror by the probe. Deviation among three repeated measurements is within 0.2 micrometer, which is about 1/5000 of the depth of measured curve.

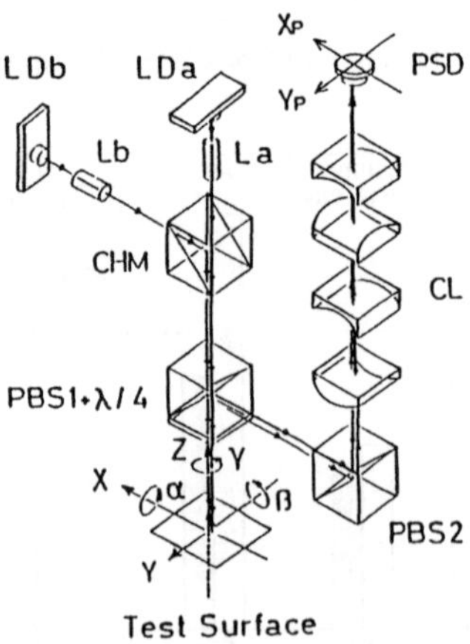

Fig. 1 Optical System of Probe

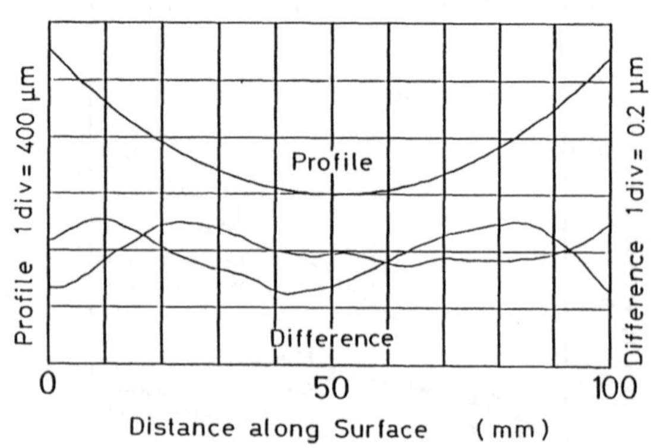

Fig. 2 Measured Parabolic Curve

A New Method of Angle Measurement Based on Internal Reflection Effect

P.S. Huang, Satoshi Kiyono and Osamu Kamada

Department of Precision Engineering, Faculty of Engineering, Tohoku University, Aoba-ku Sendai, Japan

Principle and experimental results

It is well known that a light beam impinging on an air–glass interface internally, undergoes a rapid reflectance change if the beam lies in the vicinity of the critical angle. Using this characteristic, the angle displacement of the incident beam $\delta\theta$ can be readily measured by detecting the corresponding reflectance $R(\delta\theta)$. In this research,however, we relate $\delta\theta$ to the so called linearized reflectance $R(\delta\theta) = \{R(\delta\theta) - R(-\delta\theta)\}/\{R(\delta\theta) + R(-\delta\theta)\}$, in order to reduce the nonlinearity of $R(\delta\theta)$. Figure 1 shows the optical schematic of a prototype sensor, with a laser diode as the light source and a pair of critical angle prisms and photodiodes detecting both $R(\delta\theta)$ and $R(-\delta\theta)$ simultaneously. The dimension of this sensor is 50x50x25 mm, and the weight 70g. Figure 2 displays a calibration curve of $R(\delta\theta)$ vers $\delta\theta$, where $\delta\theta$ is detected by an autocollimator (Nikon-2). The nonlinearity is less than 1% and the resolution is far better than 0.1 second of arc. Theoretical analysis showed that a wide resolution and range of measurement can be designed to optimized by carefully choosing the central incident angle and the number of reflections within the critical angle prism.

Conclusion

The new method for angle measurement has the advantages of compact size, simple structure and high resolution. As one of its applications, an AFM using this method to detect the deflection of its cantilever has been developed. Other applications will be found in the control of moving tables and the profile measurement of optical elements.

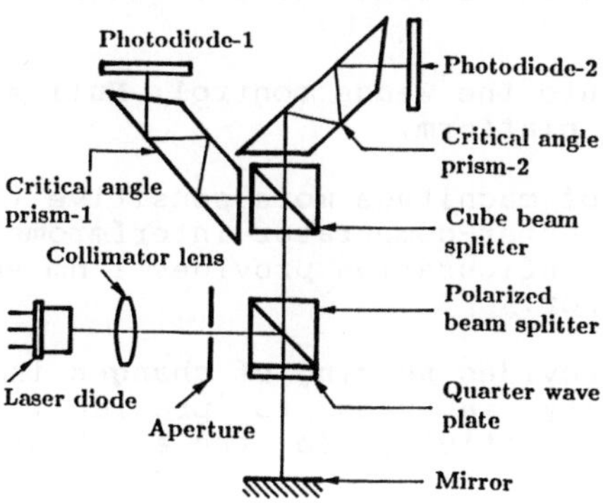

Fig. 1 Optical System of Sensor

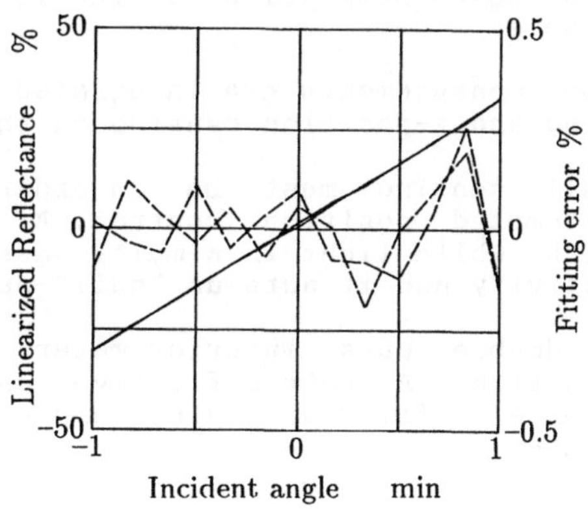

Fig. 2 Calibration Curve of the Sensor

An Ultra-fine Infeed Controller

P. Langenbeck
INTOP, D-7771 Frickingen 2, F R Germany

There is a need to produce ultra-fine motion with a resolution of 0.01 µm within a range of 1 to 2 mm travel at loads between 200 and 2000 N accompanied by stiffness of at least 200 N/µm. This is beyond the reach of piezo-pusher devices and differential lead screws.

The devices should operate independent of elasticity, have no small zone contacts resulting in chatter and exhibit neither stick slip nor backlash. And the position sensing should be of at least 5 nm resolution and be able to be nulled to 3 nm.

These considerations led to reinvestigate wedge control devices, the "Roman-type" fine positioning and lifting control. The need for an ultra-fine positioning device arose from applications where a high load had to be moved in direction of gravity (z) and, therefore, a wedge was designed as a bidirectional air bearing table transmitting load both onto a granite base and onto a platform located above the slideable wedge. The load carrying platform has fixed x, y position and freedom to move in z.

Whilst sensitivity of this motion device by lead screw positioning of the wedge is conceived quickly to be sufficient, there are instant questions as to the stiffness C of the two air bearings riding atop of each other;

$$1/C_{tot} = 1/C_1+1/C_2$$

Air bearing preload by C-clamping was not considered for its volume; preload by permanent magnets for its non-linear characteristic, electro-magnetic for its expense of the controls. Preload with vacuum was found satisfactory. This solution presents the task of finest pressure control. Large lateral size makes the bearings insensitive to the location of load vector, hence tilt-stable.

Two measurements are integrated into the wedge control: Null sensing and z-position sensing of the platform.

Null sensing must be an order of magnitude more sensitive than attempted position control. A. Koester-comparator interferometer with white light in a multi-pass configuration provides 1 nm sensitivity and it acts as "null"-sensor.

A double pass interferometer provides sensing of changes in z-position. A retroreflective cavity is used to provide i. e. lambda/4 fringes giving ample sensitivity to the air bearing table.

Applications will be described like ultra-fine tool positioning and a 3-D positioning table.

Intercomparison Flatness Measurements on a Granite Surface Plate

F. Lüdicke and F.-W. Pietsch
Physikalisch-Technische Bundesanstalt
Bundesallee 100, D-3300 Braunschweig, F R Germany

Flatness comparison measurements were carried out at the PTB in which several European institutes, manufactures of flatness measuring instruments and industrial users of these participated. Within the scope of these measurements the flatness of a granite plate about 1 m^2 in size, with an intentional flatness error of about 10 µm, was to be determined. The plate had been accomodated in a temperature- and humidity-controlled laboratory of the PTB where the measurements had to be carried out by all the participants. 104 points in a 100 mm x 100 mm grid were contacted. In most of the cases electronic levels were used, some participants used autocollimators and laser interferometers. The differences between the measurement results obtained by the various participants were discussed by least squares evaluations.

In the majority of cases the measurement results agreed to about 0.5 µm to 1.0 µm. The differences in the results are probably not so much due to unavoidable deposits on the plate surface but rather to different contacting conditions on the granite plate, since roughness of the micro finish cannot be avoided. Profile measurements of granite plates furnished deviations of some micrometers.

It turned out that, even if comparable equipment was used, the results achieved by industrial firms purely orientated to practical use were affected by greater deviations than those of the manufacturers concerned and of universities and national metrolgy institutes.

Due to the longer measurement time required, flatness measurements by means of autocollimators are more difficult than those for which electronic levels are used. Reliable measurement values can even be obtained with the aid of a laser interferometer providing skilful handling is ensured.

The flatness deviation value according to ISO/DIS 8512 is only conditionally suited to judge flatness measuring methods because of local flatness measurement errors. The flatness deviation values determined by the participants acc. to ISO/DIS 8512 often differed from the values determined from the same data set by PTB. A software comparison would clarify the differences and would be another important step to harmonize the flatness measuring methods.

Is it Possible to use Carbide Tools for Ultra-precision Cutting?

Z. C. LUO, R. Y. CHEN, J. W. XIONG

First Dept. of Mechanical Engineering
Huazhong Univ. of Science and Technology
Wuhan, Hubei 430074, P. R. China

Although diamont is the most suitable tool material for ultra–precision non–ferrous metal cutting, it is rather difficult to use it widely in practice because of the expensive cost, difficulty with lapping and brittleness. To find its substitute, therefor, is an urgent task for those who work in the field of ultra–precision metal cutting.

With the development of powder metallurgical technology a new kind of carbide, i. e. ultra–fine grain carbide has come into use. Owing to the fine grain size (less than 1 μm) it can be ground and lapped to form very sharp edges and very smooth faces, and has potential for precision and ultra–precision cutting.

In order to make clear the possibility of ultra–precision cutting with ultra–fine grain carbide tools, a lot of experimental research work have been done with the help of advanced instruments, such as Scanning Electronic Microscope, X–ray Stress Measuring Instrument, Talysurf 5–120 Profiler and HVA–10A Vicker's Diamond Hardometer. The research work dealt with the lapping of carbide tools, optimisation of tool geometry and cutting parameters and their influences on the quality of machined surface, such as roughness, work–hardening and residual stress when cutting aluminium alloy workpiece with ultra–fine grain carbide tools.

The results have shown that with good lapped ultra–fine grain carbide tools (made in Zhuzhou Carbide Factory, Hunan Province, P. R. China, marked by YM051, YM052 and YD05) and the optimized tool geometry and cutting parameters, a machined surface with smooth roughness (Ra = 0.12–0.22 μm), slight work–hardening (the depth of work–hardened layer is about 0.05 mm, the degree of work–hardening is about 1.5%) and residual stress ($<$+0.5kg / mm^2) can be achieved.

* The project is supported by National Natural Science Foundation of China.

The Straightness Error Compensation Based on the P-integrator Learning Control in High Precision Lathe

Li Shengyi   Liang Jiancheng
Changsha Institute of Technology , P.R.C
James. Li
Columbia University., U.S.A

ABSTRACT

The work described is the result of a research project for developing a straightness error compensation scheme in high-pecision diamond turning processes based on the p-integrator learning controller and an on-line metrology system.

On-line measurement is by means of the 3-sensor technique. So that the errors of slide motion and workpiece straightness can be separated accurately.

Because of the non-linear characteristics of pizoelectric actuator, exact compensation implememted becomes more difficulty. The p-integrator learning control strategy can deal with it much better.

A simple and robust P-Integrator learing controller requires little knowledge of dynamics of the machining process for practical applications, thus, its transfer function is used:

$$Hc(Z) = R(Z)/S(Z) = KrZ^{-p+d} \qquad (1)$$

The characteristic equation $P_{ch}(Z)$ now becomes $A(Z)(1-Z^{-p})+KrB(Z)Z^{-d}=0$. Kr is the repetitive control gain and should be chosen so that $P_{ch}(Z)$ is a symptotically stable. This is the simplest possible P-Integrator. With the Hc specified in eq.1 the control action is

$$U(Z) = e(Z)KrZ^{-p+d} / (1-Z^{-p}) \qquad (2)$$

After in inverse Z-transform, then can be rewritten as the form for computer control, equation now becomes

$$U_j(k) = U_i(k)+K_r\Sigma\, \alpha^{j-1-i} e_i(k+d) \qquad (3)$$

here k is sampling interval

j is number of repetition

α is forgetting factor and $0<\alpha<1$

Therefore, the P-Integrator control algorithm can be implemented by looking at the sum of past errors in each repetition at a point d sampling intevals downstream.

Therefore, the P-Integrator control algorithm can be implemented by looking at the sum of past errors in each repetition at a point d sampling intevals downstream. Equation (3) indicates that the summation is performed over the history of error $e_j(k+d)$ in the repetition domain.
The experimental sysytem which consists of a piezoelectric device and its mechanism, four probe-type inductance sensors, and microcomputer IBM/386.
Piezoelectric device is PZT-La stack which is made of 50 ships , its size is 14×16×0.5mm, made from electrostrictive material.
The straightness error of the experimental workpiece is about 1.89 μm/128mm. After 7 times learning processing, the straightness error is reduced to 0.87 μm/128mm. For pizotool tracking curve, the sum of absolute error of 64 points is reduced from 4.387 μm used PD controller to 0.6801 μm after 7 times learning processing. So that the improvement of straightness only is limited by measurement noise.
Compare with other modeling method, learning controller requires little knowledge about the controlled system, and its algorithm is the simplest one. Especially in the application on ultra-precision engineering.

The Research of Super-precision Grinding Control System for Axital Dimension of Hydro Servo Valve Rod

Li Shengyi, Wang Shimin, Tang Bingyang
Changsha Institute of Technology, Changsha, Hunan, P.R.C

ABSTRACT
Hydro-servo valve is one of very improtant part in hydro-cintrol system. Becauce the limiting of its manufacturing accuracy, some serious influence can be brought into all of the cintrol system, therefore make its control accuracy bad.
The new method and instrument have been introduced in this paper. The construction of all of this system can be showed in Fig.1 . There are three steps in the worhpiece. The thikness of steps needs control accurately. Its allowance must be less than 2 or 3 μm.

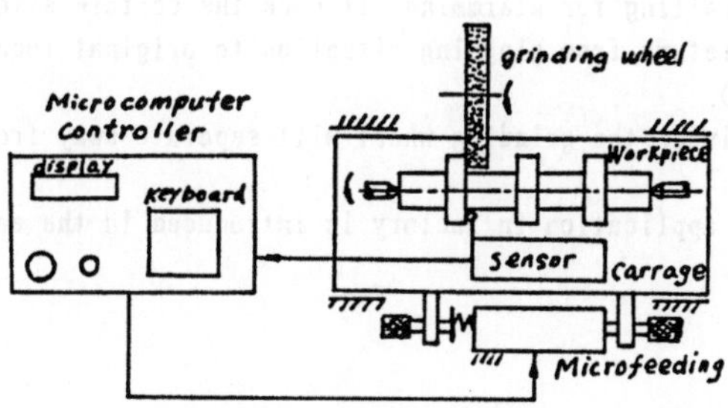

Fig.1 Construction and principle

The workpiece is set on a carrage of VS2305 Universal Tool Grinder and used to be controlled by manual of worker. Now, our system can achieve high accuracy and automation, and the grinder needs not any amentment. It consists of three parts, a) Piezodriver. b) Dimension measurement device. c) Microcomputer-based controller. All of these as appendages can be added to the machine.

There are two different ranges in automatical grinding control, 1) in range within 0 to 20μm, its resolution can be up to 0.01μm and accuracy is 0.05μm. 2) grinding measurement range,0 to 200μm, its resolution can be up to 0. 1μm, accuracy is 0.3μm. Fig.2 shows the process about grinding. One hundred types of grinding speeds can be chosen by input with thekeyboard. The grinding allowance can be input with the key-board too.

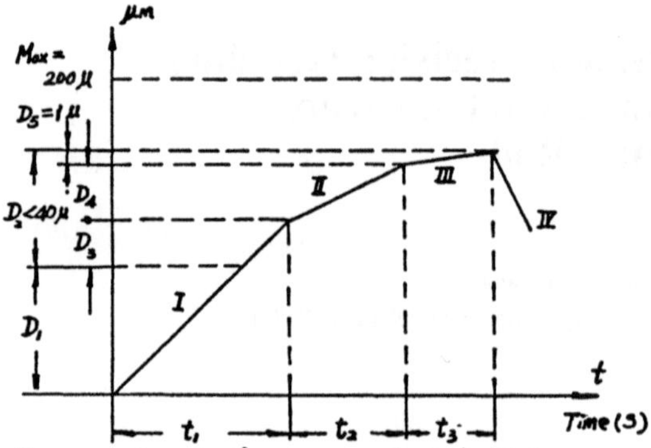

Fig 2 Process of grinding

Within 0 to 200μm the starting point for grinding can be arbitrary.The rough grining is used by manusl feeding(D1 + D3). D1 is empty travel in which the grinding wheel does not touch with the worhpiece. At moment of t1 the automatic grinding is behining after pessing gring key, so that the process is entered into automatic stage II. In order to complete gine polishing grinding and inprove the roughness of worked surface, in last stage III where is away from the allowance only 1μm , the microdriver stops feeding a loudapeakor is calling for alarmming. If once the certain allowance is reached, the microdriver retums from gingding situation to original location an all process finishes(stage IV).
When the allowance is achived, the grinding wheel will seperate away from workpiece automatally.
The information about its application in factory is introduced in the end of this paper.

Bluntness of the Tool and Process Forces in High-precision Cutting

D.A. Taminiau, F.A.C.M. Habraken, J.H. Dautzenberg
Eindhoven University of Technology
Eindhoven, The Netherlands

Summary

It is generally recognized that tool wear has a great influence on the cutting forces and the quality of the machined surface. Especially when making smaller cuts, for instance in high-precision cutting, this influence is considerable. Besides its effect on the surface roughness it is obvious that a less sharp tool will lead to higher cutting forces. This will lead to more power dissipation in the cutting process, more heat development and to thermal expansion. Cutting with a blunt tool will therefore decrease machining accuracy.

In this study the influence of tool edge bluntness on the cutting forces is investigated. The tool edge bluntness has been approximated by an edge roundness. Flank wear is assumed to be negligible.

Rough cutting experiments have been carried out using tools having varying degrees of edge roundness. Preliminary rough cutting experiments were employed since in high precision cutting a defined edge roundness is very difficult to obtain and to measure. In rough cutting this edge roundness is larger. The experiments were carried out using tool inserts ground with edge radii varying between 0.10 mm and 0.20 mm. Cutting was performed with feedrates varying between 0.04 mm/rev and 0.20 mm/rev, depending on the radius of the tool edge. The workpiece material was steel C45.

High-precision cutting experiments have been carried out using diamond tools ground with an edge radius of 20 μm. Cutting was performed with feedrates varying between 3 μm/rev and 20 μm/rev. The workpiece material was free turning brass (CuZn40Pb3), it is shown that diamond cutting of steel is not possible.

Force measurement results are presented and discussed. The specific cutting forces are plotted against the ratio depth of cut/cutting edge radius. Approximately logarithmic relationships are evidenced.

Design of High Performance Externally Pressurized Gas Bearings

J. M. Wang and P. H. J. Schellekens
Eindhoven University of Technology, The Netherlands

Summary
A new gap shaped externally pressurized gas (EPG) bearing was developed, which has higher load carrying capacity, higher stiffness than other gap shaped EPG pads. Furthermore, since no-flexible part is used, design procedures can be simplified and the additional instability is eliminated.

Introduction

The EPG bearings (pads) with convergent conical shapes have been noticed for years, as summarized in [1], because this kind of bearings can provide high load capacity and high stiffness. This paper presents the numerical and experimental investigation on a new type of conical gap shaped EPG pad - the multi-conical EPG pad. Under the same conditions, multi-conical shaped pads have high stiffness, high load capacity and a wide stable region.

Multi-conical gap shaped, circular EPG pads

Fig. 1 presents the illustration of the gap shape areaof simple conical and multi-conical EPG pads. Our research shows that when the diameter ratio (DC/DD) of a simple conical pad equals unity, both load capacity and stiffness increase with gap difference (hv). In other diameter ratios, there is an optimum gap difference to achieve high stiffness. Fig. 2 shows the changes of the stiffness and the load capacity of simple conical EPG pads for different diameter ratios. It can be seen that optimized diameter ratio is located about 0.8 to 0.9.

However, the choice of hv and DC/DD are limited by stability of EPG pads. The stability criteria of simple conical EPG pads is plotted in Fig. 3, which gives important information on the combination of gap differences and the characteristic diameter ratios, for instance, the optimized diameter ratio for the simple conical pad requires a gap difference of about 5 μm to ensure the stability. If we design a pad gap shape according to the relationship of the diameter ratio and the gap difference, a stable EPG pad with high load capacity and high stiffness can be achieved. Tri-conical gap shape is a simple approach for this.

The results of FEM calculations and experiments for a optimized simple conical pad and a tri-conical circular pad are further illustrated in Fig. 4, from which it can be seen that multi-conical pads are indeed optimized pads.

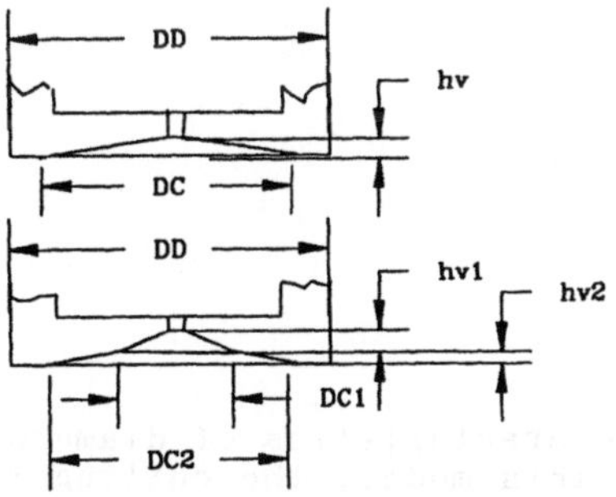

Fig. 1. Illustration of conical and multi-conical EPG pads

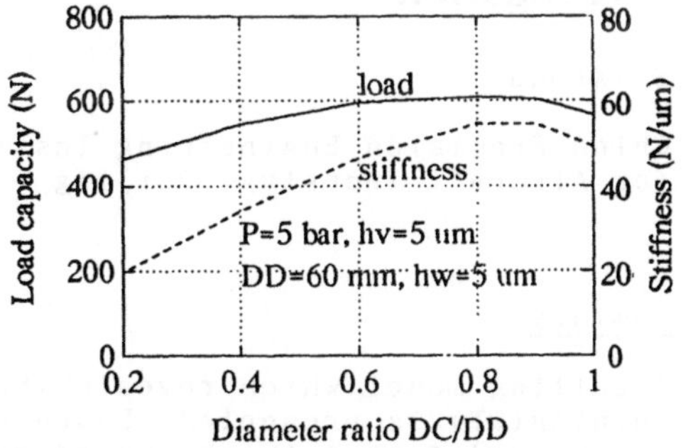

Fig. 2. Static characteristics in terms of diameter ratio

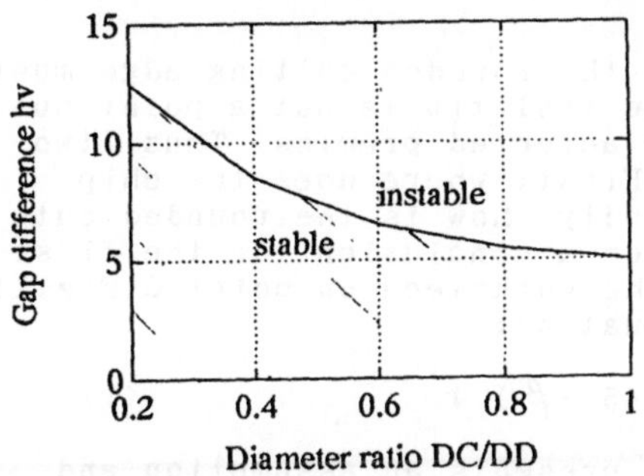

Fig. 3. Stability criteria

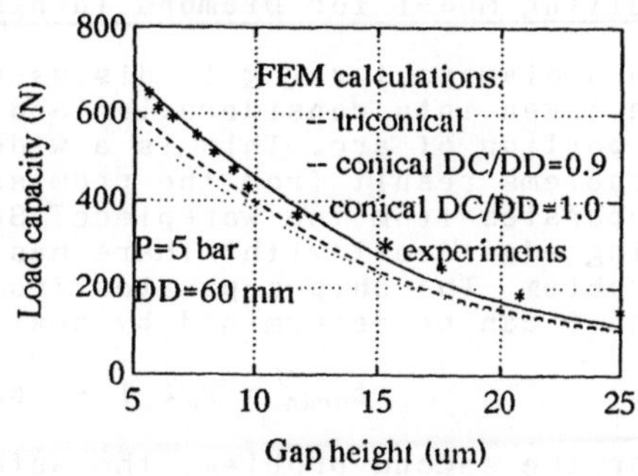

Fig.4. Numerical and experimental results of conical EPG pads

Conclusions

The multi-conical shaped pads can be considered as an approach to design the high stiffness, high load capacity and stable EPG bearing pads. Both FEM calculations and experiments shows that multi-conical shaped EPG pads can indeed reach this goal.

References

1. J. M. Wang : Principles and methods to improve the stiffness of aerostatic bearings. 1988, Master Thesis, Eindhoven University of Technology

Mechanisms and Models of Diamond Turning: Theoretical Analysis and Experimental Investigation

S. XIONG

China Precision Engineering Institute
for Aircraft Industry, Beijing

Summary

A cutting model which reveals the characteristics of diamond turning(DT) is presented. Based on this model, the cutting force versus undeformed thickness of cut relationship is derived. Experimental results is given to support the theoretical analysis. Only two dimentional orthogonal cutting is discussed.

Cutting Model for Diamond Turning

When diamond turning is discussed, the rounded cutting edge must be taken into consideration and the tool tip is not a point but a portion of arc. This is a widely accepted premiss. Thus, two problems result from the premiss. First, where does the chip separates from the workpiece? Secondly, how is the rounded cutting edge dealt with? There has been a conclusion for the first problem. The chip separates from the workpiece on point O(Fig.1) and O can be determined by next equation:

$$a_{cmin} = r_n [1 - \cos (45 - \beta)] \quad (1)$$

For the second problem, the author presents an assumption and get a new cutting model for DT. In Fig.1, $\widehat{AO}$ is a part of the rake ($\widehat{AO}$ take part in material removing) and $\widehat{OB}$ is a part of flank ($\widehat{OB}$ has an effect on the machined surface). Therefore, the rake, flank, and shear zone are not planes but curved surface. And the rake angle and shear angle are not constants but variabls depending on the undeformed thickness of cut. This model can be called curved shear zone model.

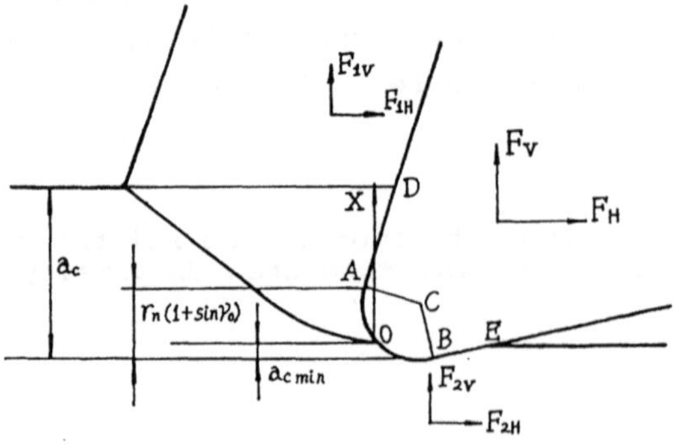

Fig.1
Cutting model for DT:
curved shear zone model

Mechanics of Diamond Turning

The force on tool is analyzed into F_H and F_V, the force on rake is analyzed into F_{1H} and F_{1V}, the force on flank is analyzed into F_{2H} and F_{2V}. Therefore,

$$F_H = F_{1H} + F_{2H} \quad , \quad F_V = F_{1V} + F_{2V} \tag{2}$$

$$F_{1H} = \tau_s b \int_{\widehat{OA}} [(a_c - a_{cmin})\cos(\beta-\gamma_o)/\sin\phi\cos(\phi+\beta-\gamma_o)]dx$$
$$F_{1V} = \tau_s b \int_{\widehat{OA}} [(a_c - a_{cmin})\sin(\beta-\gamma_o)/\sin\phi\cos(\phi+\beta-\gamma_o)]dx \tag{3}$$
$$\gamma_o(x) = \begin{cases} \arcsin[x-(r_n - a_{cmin})/r_n], & 0 < x < r_n(1+\sin\gamma_o) - a_{cmin} \\ \gamma_o, & r_n(1+\sin\gamma_o) - a_{cmin} < x < a_c - a_{cmin} \end{cases} \tag{4}$$
$$\phi = \pi/4 - \beta + \gamma_o \quad \text{(Lee-Shaffer equation)} \tag{5}$$

$$F_{2H} = \int_{\widehat{OBE}} b p_o \cos\alpha \, r_n \, d\alpha + \int_{\widehat{OBE}} b\mu p_o \cos\alpha \, r_n \, d\alpha$$
$$F_{2V} = \int_{\widehat{OBE}} b p_o \sin\alpha \, r_n \, d\alpha + \int_{\widehat{OBE}} b\mu p_o \sin\alpha \, r_n \, d\alpha \tag{6}$$
$$p_o = f(\beta_o)[1+\mu\Phi(\beta_o)](HB/100)^2 + \sin(\beta_o/2+\beta')(\sqrt{1+\mu^2}/2\sin\beta_o)(HB/100) \tag{7}$$

p_o ---- the average normal stress on flank,
μ ---- the average coefficient of friction on flank.

As shown in Fig.2, when $0 < a_c < a_{cmin}$, no material is removed, there are strong pressing and friction effects between tool and workpiece, the force increases sharply with the undeformed thickness of cut. When $a_{cmin} < a_c < r_n(1+\sin\gamma_o)$, rake angle on $\widehat{OA}$ is very small, so the force increases rapidly. When $a_c > r_n(1+\sin\gamma_o)$, the force increases slowly and linearly.

Experimental Results

The cutting forces are measured with a micro-computer based testing system in which piezo-dynamometer is used. The results are shown in Fig.3.

When a_c is large enough, forces vary linearly but the lines do not pass the origin. The force curves intersect the F-axis when extended. That means there does exist a non-linear zone when a_c is very small. F_R is the force when no material is removed and there are only pressing and friction on tool. It is a very important parameter that is associated with tool edge condition and the integrity of machined surface.

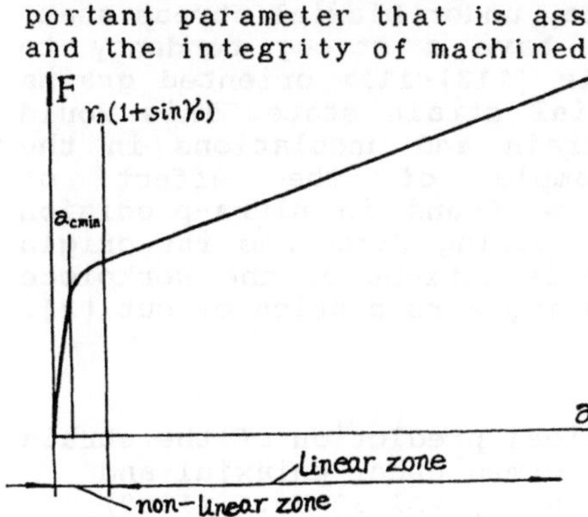

Fig.2 Theoretical F-a_c curve

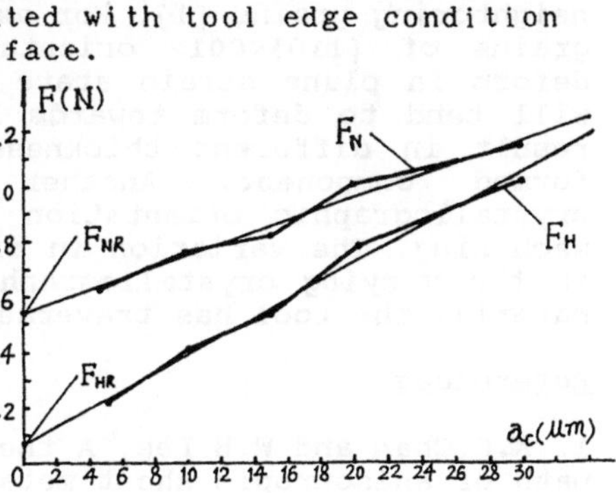

Fig.3 Experimental results

Effect of Crystallographic Orientation on Micro-deformation Processing

W.B.LEE

Department of Manufacturing Engineering,
Hong Kong Polytechnic,
Hong Kong.

Abstract

Deformation processing is widely used in the micro-fabrication of small and precision parts in electronic, optical and computer products. Examples of successful applications include ultra-precision machining of optical components, and precision blanking and forming of high density integrated circuit lead frames, high resolution diffraction gratings, and millimeter wave guide. Micro-deformation processing presents problems not encountered in the conventional bulk deformation processes. The deformation zone geometry is such that only a few grains are involved across the workpiece thickness and the crystallographic orientations of the grains will exert a large influence on the plastic responses of the workpiece materials. Common problems encountered in the micro-deformation processes are accuracy of the formed components, surface undulation in an otherwise smooth surface, and the erratic load imposed on the tool. All these problems have their origin in the plastic strain inhomogeneity caused by the grain anisotropy of the materials. During deformation, the grains will tend to deform independently and take up strain path different from their neighboring grains [1]. For example, under biaxial stress state grains of {110}<001> orientation have a strong tendency to deform in plane strain state while {112}<111> oriented grains will tend to deform towards biaxial strain state. This would result in different thickness strain and undulations in the formed component. Another example of the effect of crystallographic orientation can be found in ultra-precision machining. The variation in micro-cutting force has its origin in the varying crystallographic orientations of the workpiece material the tool has traversed during a revolution of cut [2].

References

1. K.C.Chan and W.B.Lee, A theoretical prediction of the strain path of anisotropic sheet metal deformed under uniaxial and biaxial stress state, Int. J. Mech Sci., Vol.32 No.6 (1989) 497-511

2. W.B.Lee, Prediction of micro-cutting force variation in ultra-precision machining, Precision Engineering, Vol.12 No.1 (1990) 25-28.

Long Travel Optical Servo Probe

Y. OKAZAKI*, K. MITSUI** and N. ODAGIRI***

* Mechanical Engineering Laboratory
Agency of Industrial Science and Technology
** Faculty of Science & Technology
Keio University
***Tokyo Seimitsu Co.,Ltd.

Summary

A non-contact optical displacement sensor was developed for the use of in-process or in-situ measurement of machined surface position referred to tool post/slide, for example. The probe incorporates; 1) a optical microprobe base on astigmatism having high resolution and a limited travel, 2) a linear slide mechanism and drive, and 3) a laser interferometry transducer using a fiber light guide system. The optical microprobe generates a signal related to target surface position referred to the probe. The effects of reflectivity and inclination of the target surface are compensated. The signal is fed to the linear slide controller, and the slide carrying the probe adjusts its carriage position so that the mean of the signal i.e. servo error is converged to zero. The laser interferometer measures the carriage displacement using a retroreflector attached at the rear end of the microprobe. The actual displacement is summation of the output of the microprobe and of the interferometer. The system covers the maximum travel of the slide of 80 mm beyond the travel limitation of the microprobe of some ten microns, while keeping the resolution of 10 nm. The system has automatic approach mode and emergency pull-back mode which operates when excess servo error occurs for preventing the collision.

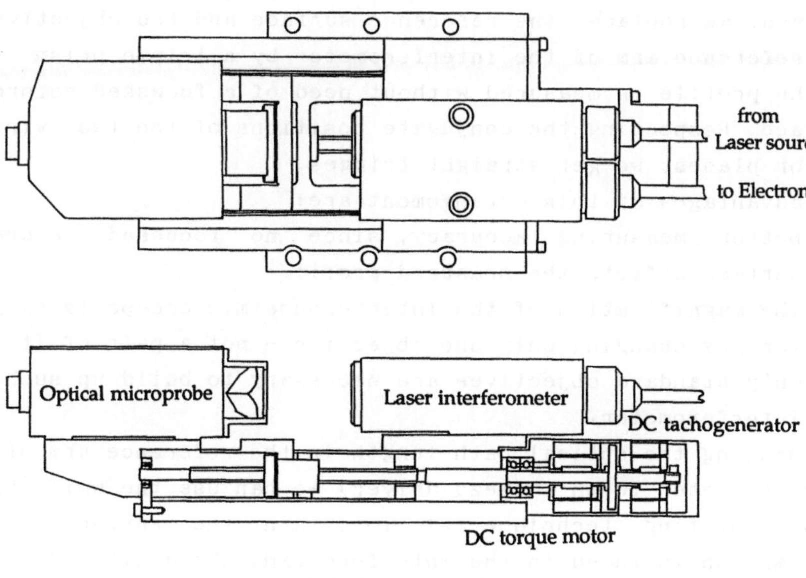

Fig.1. System overview

A New Optical Profilometer for Optical Smooth Surfaces

H. Fritz and K. Körner

Zentrum für wissenschaftlichen Gerätebau
Rudower Chaussee 6
O-1199 Berlin

Summary
In general the accuracy of measuring the surface profile of optical smooth surfaces with an interference microscope is not only limited by vibrations, gray level detection etc., but also by the quality of the reference surface. Therefore we have developed an interference microscope without any sharp focussed reference surface. This allows an improvement of measuring accuracy, even in the case of ultra smooth surfaces with an rms roughness of less then one nanometer.

During recent years several methods of determining surface characteristics of very smooth surfaces have been presented. This kind of surfaces may be very soft and easily damaged, and therefore noncontacting methods are advantageous.

Our development based on a *Linnik*-interferometer but with some changes. We replaced the reference surface and the objective in the reference arm of the interferometer by a triple prism [1]. So the profile is measured without need of a focussed reference surface. Respecting the conjugate positions of the two virtual mirror planes, we get straight fringes.

The advantages of this arrangement are:

- Better measuring accuracy, since no focussed reference surface affects the measured profil.
- The magnification of the interference microscope is easy to vary by changing only one objective - not a pair of it.
- Only standard objectives are necessary to build up such an interferometer.

By changing the optical path length in the reference arm of the interferometer (with a piezo driver) we can use the well known phase shifting technique [2]. To obtain the surface height information included in the interferogram, the digitized intensity data from a CCD-line-camera are fed into a 16 bit IBM-PC.

The following table shows the parameters reached by an experimental realisation of the interferometer described above:

number of CCD-pixels:	1024			
intensity resolution:	8 bit			
height resolution:	0.6 nm			
reproducability (without drift and justification):				
PV:	± 1.5 nm			
RMS:	0.6 nm			
magnification:	3.2x	25x	50x	100x
num. aperture:	0.06	0.50	0.80	1.30
mesuring length / µm:	4160.0	532.48	266.24	133.12
pixel interval / µm:	4.07	0.52	0.26	0.13
lateral resolution / µm:	6.43	0.77	0.48	0.30
max. gradient / nm/µm:	39	304	608	1216

Table: Parameters of the interferometer

The fig. shows the micro profile of a grid, structured by means of holography. The amplitude is about 110 nm, and the period is ≈ 30 µm. Note the absence of overshooting on the flank sides.

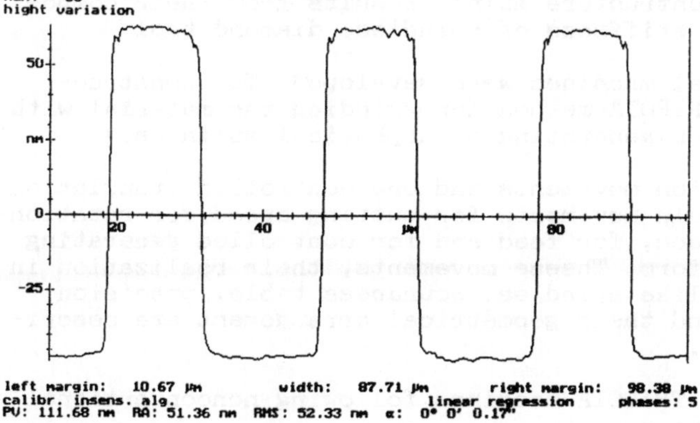

Fig.: Cut-out of the micro profile of a grid

By increasing intensity resolution and stability of the interferometer the height resolution and reproducability can be improved.

References
1. Körner, K.; Fritz, H.: patent DE 3902591 (1989)
2. Greivenkamp, J.E.:Generalized data reduction for heterodyne interferometry. Opt. Eng. 23 (1984) 350 ff.

Precision Engineering of Aspherical Optical Surfaces for Silicate Glass Components

Heiner Lammert,
Central Institute for Optics and Spectroscopy (ZOS)
D-1199 Berlin-Adlershof, Rudower Chaussee 5

Summary
In this paper the principal function of the TRIROTA-method for grinding of aspherical brittle optic-components is described. The main and profitable characteristics of TRIROTA-machines are defined.

Uncomplicated movements, high precision and quasi-point-touch in subpositions for tool/material interaction are the important precision engineering of aspherical surfaces on silicate glass components.The brittle material glass is to work with the grinding process at best. The macro-accuracy of the asphere mainly results from the principle construction of the machine, the workpiece microstructure mainly results from their components for instance stiffness of spindles, diamond tool characteristics.
In Berlin aspherical machines were developed. The construction employs the TRIROTA-method for grinding the material with diamond cup tool for generating of aspherical surfaces.*

Three normal rotation movements and one controlled translation movement are the main movements for cutting speed,for rotation-symmetry of workpiece, for feed and for controlled generating of the aspherical form. Theese movements, their realization in function elements like spindles, roundness table, precision position element and their geometrical arrangement are described with picture 1.

In construction of TRIROTA-machines following nonconventional principles dominate:
- noncomplicated movements for the working process with minimal acceleration forces (precision spindles and roundness table, no slides)
- employment of airostatic bearings
- only one translation movement in the roundness bearings of spindle in an open control system in 100-nm-accuracy
- adjustment of all relationships of the function elements in accurary of 1 micrometer and 1 arcsec with pneumatical measurement methods
- control system engineering with only one open control circuit and with internal timing unit

* TRIROTA-method mainly was developed in the ZOS-Berlin-Adlershof by Rolf Riekher, Günther Minkwitz, Manfred Schubert, Heiner Lammert and Arthur Wiemer ZWG (not published up to date)

- arrangement of main machine components in polar-carthesic coordinates gives the profit of independence of generated aspherical form from material-level-value respectively from workpiece thickness.

In the Science-centre Berlin-Adlershof in the ex-Academy of Science an experimental machine and a prototyp of a production machine for aspherical optical components with diameter up to 180 mm and up to 80 mm were built. The realized accurary is smaller than ±50 nm difference of the ideal aspherical meridian curve.
Projets of different TRIROTA-machines up to diameter to 350 mm are prepared.

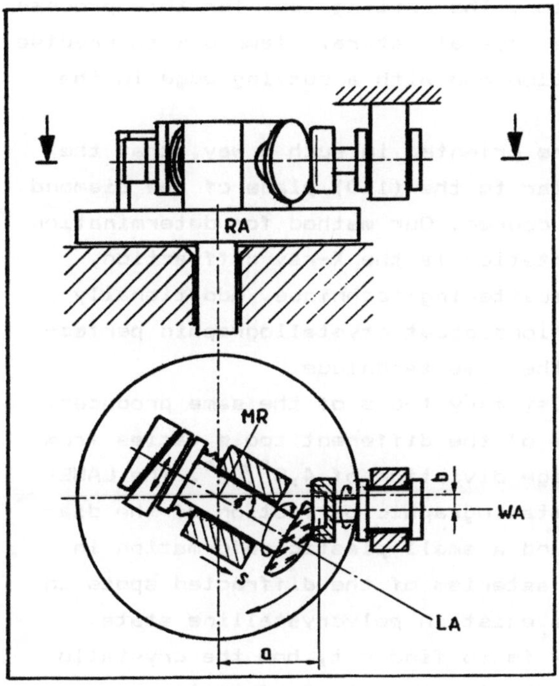

Picture 1:

Principle construction of TRIROTA-machine (front view, top view)

RA-roundness table axis
MR-middle of roundness table
RA crosses rectangulary the workpiece axis LA.
The parameter a and b define the position of the tool axis WA.
s is the translation of rotating workpiece-spindle in steps of 100 nm or smaller, computercontrolled in dependence of the roundness table - position σ (solution up to 1 arcsec)

Reference:
1. Rogal, R.; Lammert, H.; Riekher, R.
Spindelstock mit Luftlagerung und axialer Feinverstellung
Feingerätetechnik 31 (1982) H.12, p.555-558.

Investigations on Single Crystal Diamond Tools for the Ultra-precision Manufacturing Process

G. Hunsalz; U. Flechsig; D. Kürsten
Zentralinstitut für Optik und Spektroskopie
O-1199 Berlin, Rudower Chaussee 5-6, Germany

In the process of diamond turning and fly cutting the diamond tool is an important factor. The cutting tool for this process is in most cases a single crystal natural diamond with precise crystallographic orientation and with a cutting edge in the dimensions $\lesssim 0{,}1\ \mu m$.
All our diamond tools were oriented in such a way, that the shank axis is perpendicular to the (110) plane of the diamond with a deviation of some degree. Our method for determination of crystallographic orientation is the X-ray diffraction, especially the LAUE-backscattering-technique. Additionally we have got some informations about crystallographic perfection of the diamonds by the same technique.
We investigated 32 new customary tools of the same producer. The deviation of the axes of the different tools varies from 0^c to 14^c , with an everage diviation of $4{,}8^c \pm 1^c$. The LAUE-photographs showed a crystallographic perfection of the diamonds in 22 tools. We found a small plastic deformation in 7 tools and twinning and asterism of the diffracted spots in 3 tools. These 3 diamonds exist in polycrystalline state.
The intention of our work is to find out, how the crystallographic properties influence the live time of tools.
The figures 1 and 2 show two LAUE-photographs.
After using the tools for a certain time (for a cutting distance more than 25 km) we investigated once more 5 of the tools. There were no changes in relation to the plastic deformations.
The edge quality of the used tool is very important for the surface quality of the workpiece in the ultraprecision procedure. It is possible to examine the edge subjectively by light-microscopy or by SEM after gilding the used knife.

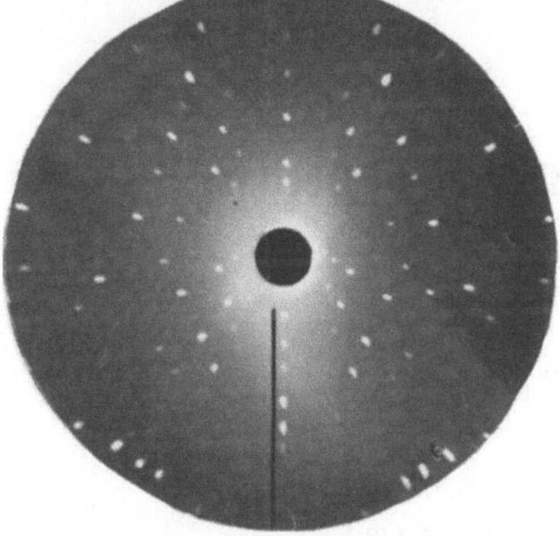

Fig.1. LAUE-photograph with point-like LAUE-pattern (single crystal)

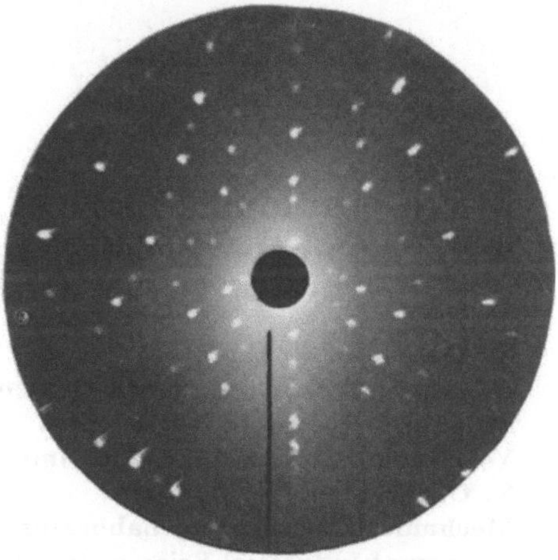

Fig.2. LAUE-photgraph with twinning and asterism of the diffracted spots (crystal with plastic deforformation)

However, these methods are limited in their results.

Plunge cuts are an other way we used to get some imformations about the state of the cutting edge and to examine the tools qualitatively. These cuts are relatively easy to produce.

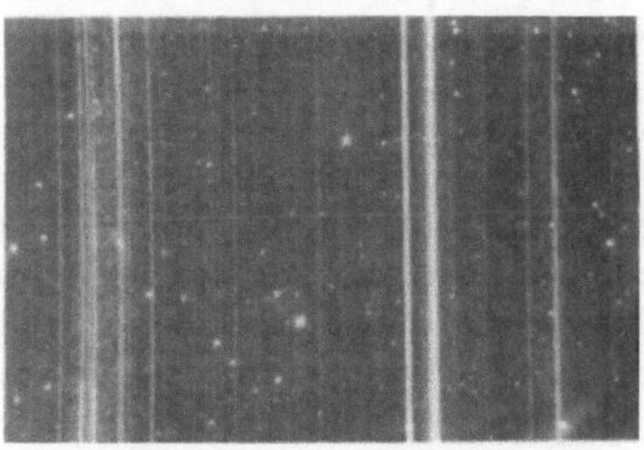

Fig.3. Plunge cut by a used tool (cutting distance more than 45 km)

The result is a groove on the worpiece surface which is an image of the cutting edge. Demages and wear can be verified by light microscopic-photography. The bright lines are a measure for wear and demage. We did not receive such lines by plunch cuts with new tools.

First results of comparisons with profilometry are encouraging.

References
1. Decker, D. L.; Hurt, H. H.; Dancy, J. H.; Fountain, C. W.
   Preselection of diamond single-point tools.
   SPIE Vol. 508 (1984) p.132-139
2. Langenbeck, P.
   Roof Edge Test for Micro-Roughness
   SPIE Vol. 802 (1987) p.89-93

Micro-positioning Stage of 6 Degree-of-freedom with Nanometer Resolution

S. Asano and T. Goto
Advanced Technology Research Center,
Mitsubishi Heavy Industries,Ltd.
Yokohama KANAGAWA 236, Japan
Y. Okazaki
Mechanical Engineering Laboratory,
Agency of Industrial Science and Technology, M.I.T.I.
Tsukuba IBARAKI 305, Japan

A micro-positioning stage comprising PZT actuators and consisted of monolithic flexure hinges as guides, formed by means of wire cut EDM, was developed (Fig. 1). The stage position is determined by capacitance gages with a resolution of 5nm, using the closed-loop control system. This stage can move with respect to 6 degree-of-freedom $(X, Y, Z, \theta_x, \theta_y, \theta_z)$, so it can compensate errors of generated by pitching, rolling and yawing.
Furthermore, a dual servo control system was developed in order to be capable of traveling longer distance with a resolution of 5nm, and applied to a positioning system consisted of a coarse-positioning stage and the abovementioned micro-positioning stage. The coarse-positioning stage consists of ball-screw driving mechanism and DC servo motors, and can move a range of 100mm with a resolution of 0.1μm. The micro-positioning stage is mounted on coarse-positioning stage parallel to it, and compensates the positioning error of the coarse-positioning stage. In result, this coarse/micro-mechanism with the dual servo control system can travel a range of 100mm with a resolution of 5nm.

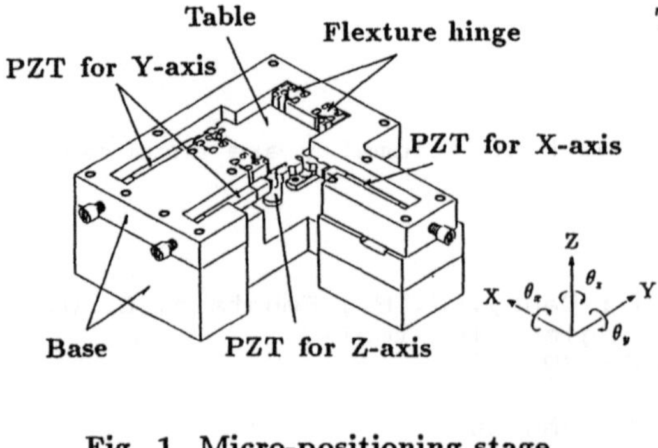

Fig. 1. Micro-positioning stage

Table 1. Specifications of micro-positioning stage

Item	Performance	
Travel	X	30μm
	Y	36μm
	Z	5μm
	θ_x	3.1×10^{-5}rad
	θ_y	5.6×10^{-5}rad
	θ_z	3.6×10^{-4}rad
Resolution	5nm	
Band width	1kHz	

Measurement of Flatness of Technical Surfaces Using Anamorphotic Planinterferometer

Ľudovít N Y Á R S I K

Institute of Measurement Science, Slovak Academy of Sciences
Dúbravská cesta 9, 842 19 BRATISLAVA, CSFR

With steady increasing requirements concerning product quality and accuracy in individual stages of technological processes the requirements concerning measuring technique increase as well. Non-destructive testing of parameters with computer-processed measurement results is highly required. The measuring methods applying interference of laser beam are of special suitability for this purpose. The advantage of interferometry by the measurement of macro-profiles of objects in micro- and nanometer range of shape deviations is that all-surface measurement can be performed with high accuracy and at high speed.

There are various industrial products with planar surfaces produced from various materials. So-called technical flat-surfaces are concerned, with either glossy or rough surfaces and with shape deviations in the micrometer range (1 to 100 μm). Flatness of such surfaces is mostly measured by help of prism planinterferometer designed optically by Abramson (see Optik **30**, 56, 1969). The main disadvantage of such an arrangement is in small air gap between measured and reference surfaces (approx. 100 μm).

In this paper a new optical scheme for an oblique incidence planinterferometer is shown, using anamorphotic optical elements (above all prisms) to expand the collimated beam to the aperture required. The interferometer is of the Mach-Zehnder type with separate reference and measuring beam paths thus allowing positioning and levelling of the object measured without the risk of reference surface damage. Moreover, phase sampling of the interference pattern on a small aperture in separate reference arm can be obtained in easier way.

Expanding of laser beam to the aperture required in the measuring arm of the interferometer is attained by double anamorphotic expanding. First, the parallel beam is expanded in one direction using an anamorphotic optical element (prism), and thus parallel beam with elliptic cross section is formed. To attain the expanding in the other direction, oblique beam incidence on the flat-surface measured is applied directly. With proper choice of both enlargement factors circular laser spot is obtained on the object measured. Reverse transformation of the aperture is performed in similar way. The anamorphotic surface (prism hypotenuse face) is also used to divide the beams into reference and measuring arms.

The function of the anamorphotic planinterferometer was proved in an experimental setup with 100 mm aperture. The interference pattern is scanned by matrix-type CCD camera (256x256), the output signal is digitalized to 256 level. For computer-aided processing of the interference pattern either phase-sampling or fringe pattern methods are used.

Anamorphotic planinterferometer can be used for measurement the glossy and part-glossy flat-surfaces in precision and ultra precision engineering, in microelectronics (e.g. to measure flatness of substrates), in optics and everywhere where the shape deviation of the flat-surfaces measured are approx. within 1 and 100 μm.

Measurement of Gap and Parameters in the Polishing Process of Optical Glass

Haberland R.; Blume-Werry, U. - Feinwerktechnik, Universität Kaiserslautern

High quality optical glass surfaces like lenses, prisms, substrates etc. are manufactured nearly without exception on lever polishing machines. The lapped workpiece is pressed on a rotating polisher under a certain force P and the polishing agent is applied by a free outlet (Fig. 1).

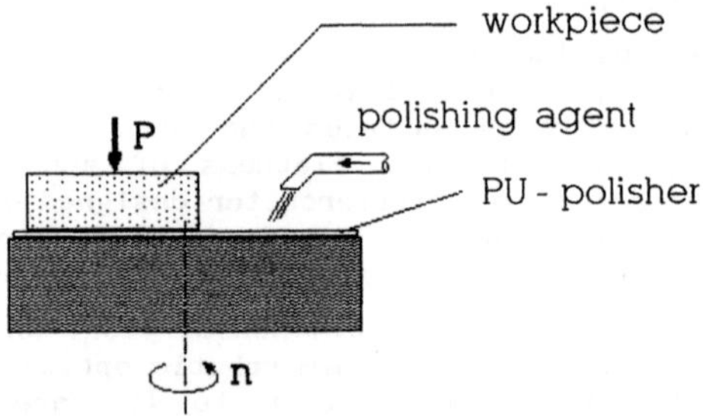

Fig. 1 Principle of polishing process

This polishing process has, so far, seen no significant or essential changes and quality of polish still depends on empiric elicted process parameters, like spindle speed, force P, kinematics of levers etc. This is understandable when considering that polishing is a complex manufacturing process in which all compounds involved - glass, polisher, polishing agent and machine - interact with their own special mechanical, physical and chemical characteristics.

Thus a main reason for poor reproducability and predictability of the resulting surface quality - or even a non-empirical optimization of the process - is the lack of insight into the action that takes place in the gap between glass and polisher.

As a contribution to clear these mechanisms the following measurements were accomplished:

1. Measurement of gapheight

A sensor was developed which by utilizing the conductivity of the polishing suspension measures the resistance of the liquid-filled volume between its own and the workpiece surface. For little variation of the gapheight h (few microns) a good linearity to the output signal can be shown. Fig. 2 gives a scetch of the sensor mounted in the polisher, Fig.3 the electrical scheme. The signal resolution achieved was 50 nm.

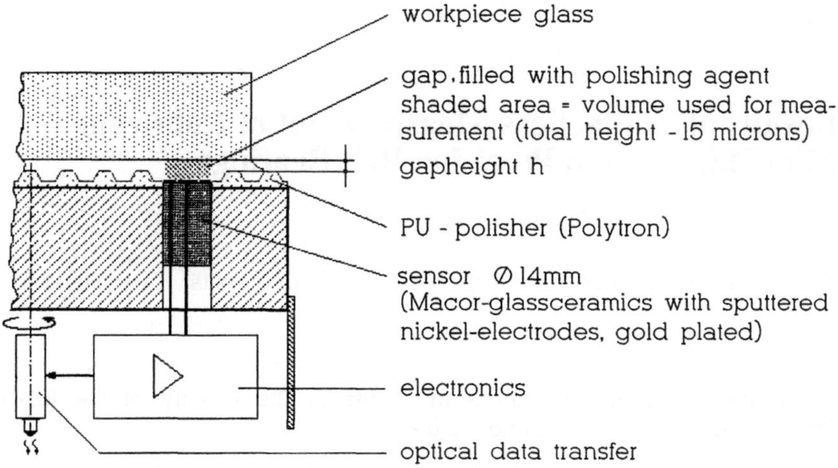

Fig. 2 Scetch of sensor, mounted in the polisher

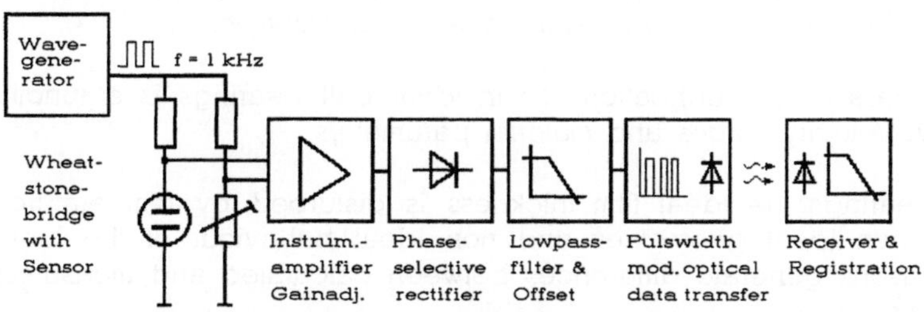

Fig.3 Electrical scheme of gap measurement

2. Measurement of pressure rates

To obtain measurements of the local and time variance of the pressure rates in the gap some miniature pressure sensors with a frontmembrane diameter of 2 mm were coupled to the gap via capillary tubes of 60 - 200 microns.
The resolutions achieved were: local: 100 microns ; time: 1 ms and signal: 1 mbar.

3. Measurement of temperature fluctuation

For the measurements of temperature fluctuation thermocouples of 20 microns in diameter were mounted in both the surface of the workpiece and the polisher. The resolutions achieved were: local: 20 microns , time: 2ms and signal: 0.1 K. The data transfer from the rotating polisher to the registration devices took place optically by means of a pulswidth modulated LED (Fig. 3).

Results of Contactless Measurement of Lubricant Film Thickness in Precision Ball Bearings

Haberland, R. ; Bethke, J. Feinwerktechnik, Universität Kaiserslautern

Especially in precision ball bearing design the quality of the lubrication film thickness is an interesting value.

The absolute magnitude, its stability and behaviour at different running conditions is very important. Differences in film thickness result in different torques and temperatures which influence its dynamic accuracy. Lubricating film breakdown results in low reliability and poor lifetime.

The thickness of the lubrication film in ideal ball bearings is a function of geometry, velocity, forces and material parameters.

In real bearings the ideal film thickness is disturbed by temperature, dirt, damage, non ideal geometries and non ideal behaviour of the lubricant. These reasons generate differences between calculated and measured film thicknesses.

The lubrication film thickness in several precision ball bearings and bearing units has been examinated by a contactless measurement technique /1/, /2/, where signal generation and transmission have been realised using capacitive principles (fig. 1).

Some observations are:

* Even in a testing apparatus, where one single ball bearing (R3) can be investigated under nearly ideal conditions, the lubricant film is a factor of 2 ... 5 times thinner than theoretically expected.

* Contaminated oil is disturbing lubricating films more at low speed than at high speed (fig. 2) .

* High-speed breakdown is observed, sometimes only at one, sometimes at both raceways.

* Torque bimoding can be related to film thickness bimoding (fig. 3)

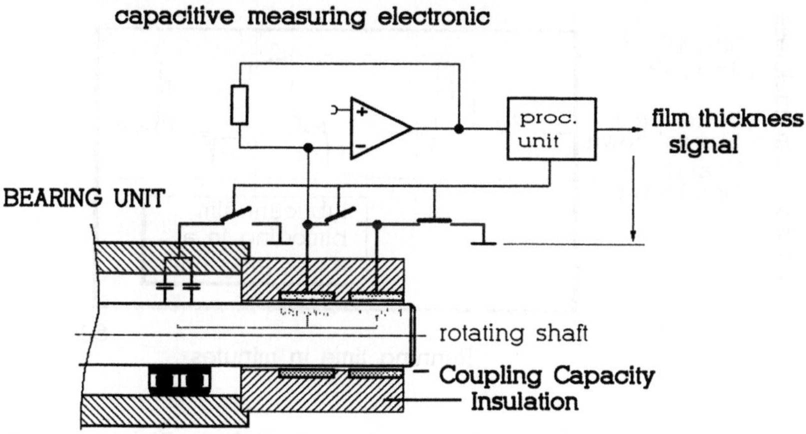

Fig. 1: Capacitive contactless measurement of lubrication film thickness in ball bearings

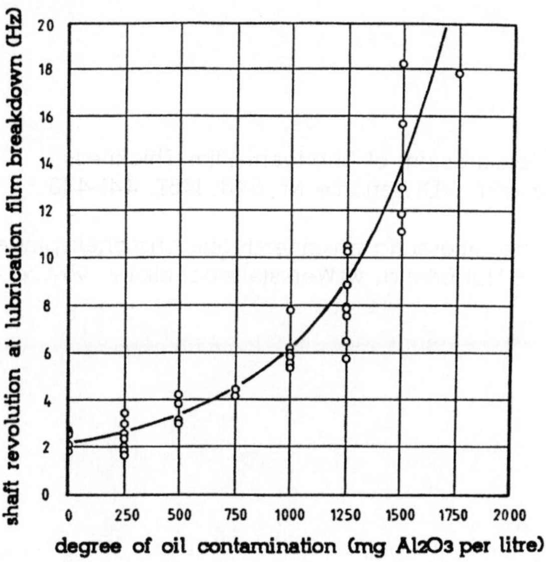

Fig. 2: Low-speed breakdown at different revolutions as a function of lubricant contamination. The lubricant (mineral oil) is contaminated with Al_2O_3 particles with an average diameter of 0.3 µm.

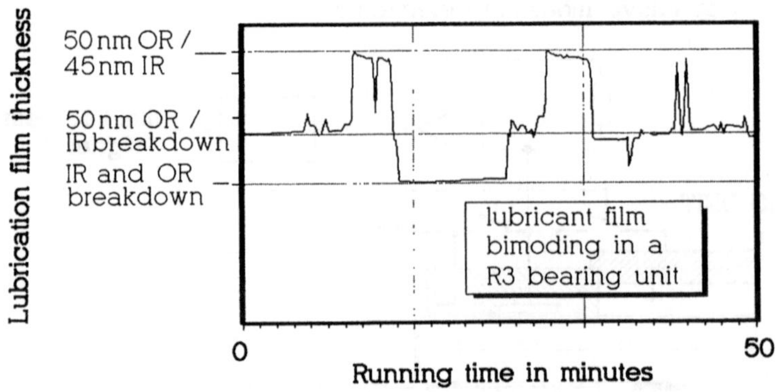

Fig. 3: Partial lubrication film breakdown (bimoding) in a R3-bearing unit. Shaft revolution: 15000 min^{-1}, synthetic oil, OR=outer raceway IR =inner raceway

References:

/1/ Contactless Measurement of Lubricant Film Thickness
Prof. R. Haberland, VDI Berichte Nr. 644, 1987, 441-448

/2/ Schmierfilm-Dickenmessung an einer Präzisionsscheifspindel
J. Bethke und R. Haberland, wt Werkstattstechnik 79, 1989, 361-364

Acceleration and Distance Sensors for Machine Vibration Measurement

o.Prof.Dr.Ing R. Haberland / G. Lauer

FWT Universität Kaiserslautern

Typical application : grinding spindle

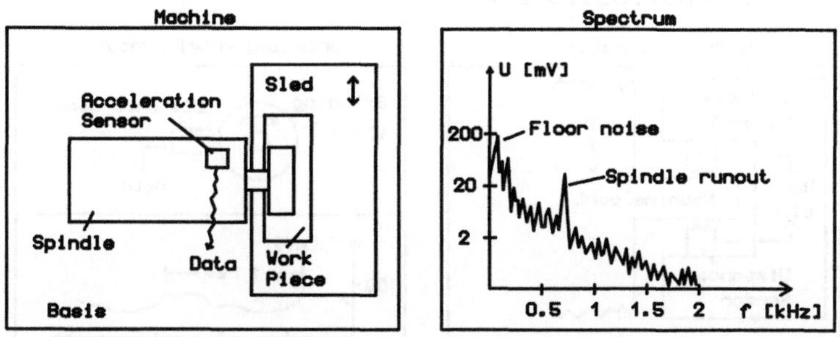

Capacitive Acceleration Sensor

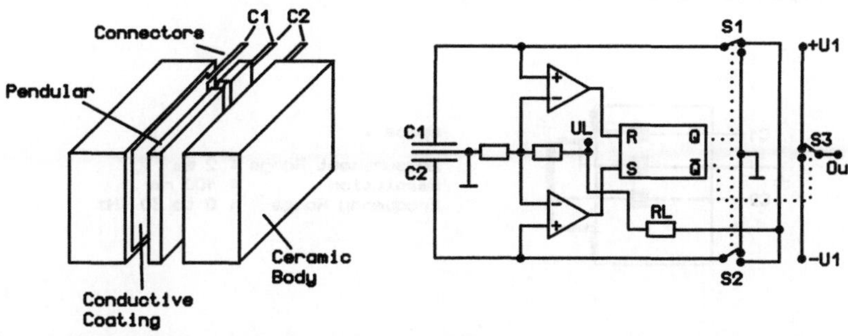

Specs :

Measurement Range	:	+/- 1g to +/- 100g
Resolution	:	10 E-5 of M.Range
Frequency Range	:	0 to 1000 Hz
Max. Overload	:	+/- 1000g

Acceleration and distance sensors for machine vibration measurement

o.Prof.Dr.Ing R. Haberland / G. Lauer
FWT Universität Kaiserslautern

Typical applications :

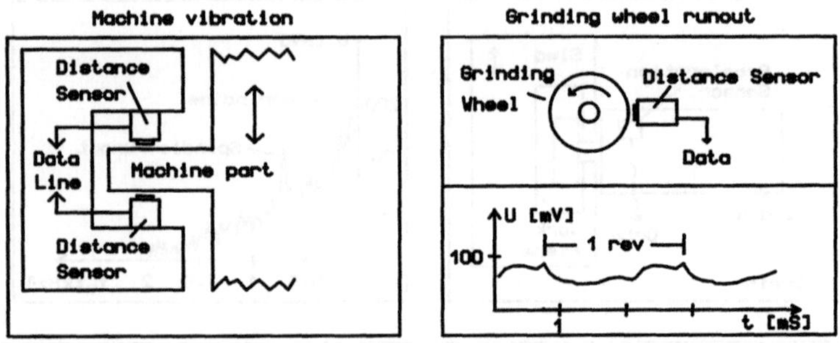

Capacitive Distance Sensor

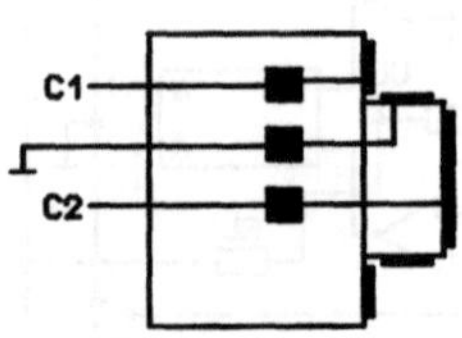

Specs :

Measurement Range	: 2 mm
Resolution	: 100 nm
Frequency Range	: 0 to 30 KHz

Benefits from machine vibration analysis :

- Improvement of existing machines
- Very early recognition of machine failures
- Diagnosis of tool wear

Design and Manufacture of Ultra-precision Three-Axis Test System with Air Bearing Supports

Dong Shen, Wang Zhiyong, Yuan Zhejun

Dept. of Mechanical Engineering
Harbin Institute of Technology
Harbin 150006, China

Summary

This paper deals with the principle and characteristics of aerostatic bearing series designed by air lubrication theory and utmost stiffnessoptimization method. Developed Ultra-Precision angle testing and control system and fast data auto-collecting hardwave system.

Desige of Composed Air-Static bear Series

The table is a bear series with five groupes of radial and axial bears (fig. 1), pivided air lubrication area Ω of bears into finite elements, and according to Reynold's equation. FEM analysis and optimization theory, we designed the table with revolution error between (0.03~0.25)μ.

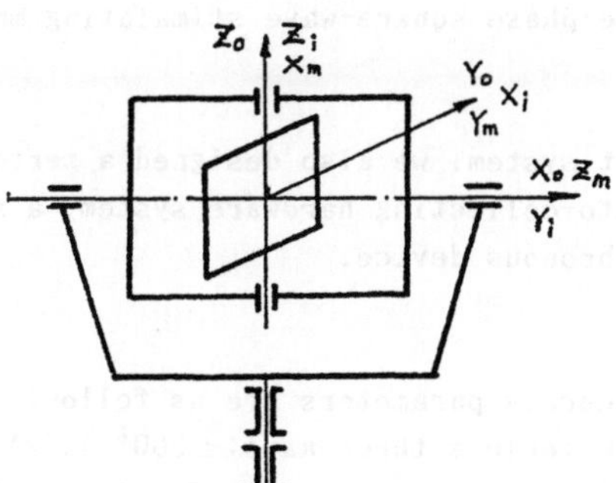

Fig.1 Diagrammatic sketch of the table's bear series

Development of Test and Control System

The accuracy of the system is about ±1 arc sec. working rate is 200°/s. When angle θ is small, we can use model in Fig.2 to show it, the principle is show in Fig.2.

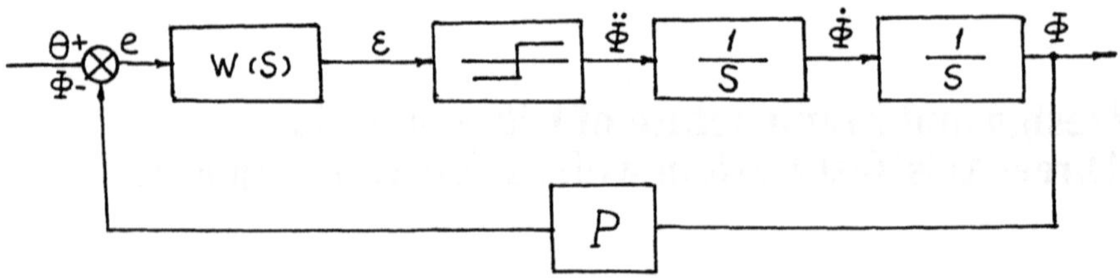

Fig.2 Mathematical model close-loop angular test system of double-phase square-wave stimulating magnet

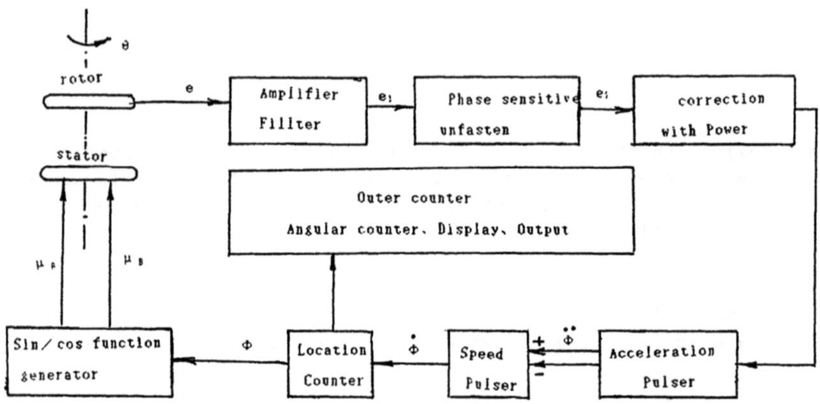

Fig.3 Principle frame of closed-loop anqular test system of double-phase square-wave stimalating magnet

Design of ther Parts

Besides this angular test system. we also designed a servo system without over feedback. a data auto-collecting hardware system, a speed-test generator and an inducing synchronous device.

Conclusions

The main function and accuracy parameters are as follow:
(1) Rotating range of the table's three axes: $\pm 360°$; (2) Rotating accuracy of the table's axes: 1σ 0.3″ ; servo accuracy for inner ring axis: 1″ (3σ), for middle ring axis: 1″ (3σ), for outer ring axis: 1″ (3σ); Position accuracy of three axes: 1″ ~ 2″ ; Resolving power of angle rate: 0.0001° ; Othogonalness of axis: $\pm 1''$.

Computersimulation for Determining the Machined Surface Using Process Parameters in Single Point Diamond Turning

F. SCHILLKE, I. STOBBE
Institute of Automation Berlin

Abstract

Preparing suitable models for monitoring and controlling of machine and surface parameters in single point diamond turning processes is one of the most complicated tasks because of the complexity of the process on one hand and the needs of simple methods for computer aided controlling in real time on the other hand.

The quality of the real surface is determined by the displacements between the tool and the workpiece. The machine and the tool-workpiece-system have to be modeled as a complicate multibody system with a large number of degrees of freedom, of stiffnesses and damping elements, several drives causing exiting forces. Self- and parameter-exited vibrations are caused by the varying stiffness of the cutting process. The analysis of such systems demands numerical methods like FEM /1/. The aim of evaluating these dynamical models and optical surface measurements consists in defining parameters of more simplified models.

Reduced systems or simple kinetical models with only two or three degrees of freedom and with a quasi static behaviour can give sufficient possibilities for monitoring and controlling the process. The measured cutting forces could be enclosed in a direct way into the model.

A quasi static model of forces and distances between tool and surface and some experiments to define the system parameters are presented. The model is defined by

- technological defined motion of the tool
- vibrations, caused by machine and environment
- effects of varying conditions of friction forces during cutting
- the local form of the surface finished before
- geometry of the cutting tool shape.

Using the simplifying assumtions, that in single point diamond turning the form of the machined surface can be represented by the geometry of the tool, the surface is a function of the tool shape and the machine cutting forces /2/. The vibrations of machine influence the deflection between tool and workpiece and the forces too.

Computer simulation was done to determine the machined surface and the components of cutting force using the process parameters, the figure of the cutting edge and the mentioned displacements. In fig. 1 a simulated surface profile is given for following machine parameters:

Cylindrical turning with spindle speed: 550 rev/min,

feed velocity	: 19.25 mm/min,	feed	: 35.0 µm,
cutting depth	: 12.0 µm,	cutting speed	: 345.6 m/min,

Vibrations between tool and work piece from the machine :

direction	frequency	amplitude	phase
normal	503.07 Hz	20 nm	0°
normal	2053.01 Hz	60 nm	30°
feed	42.09 Hz	4100 nm	90°

Parameters of the workpiece:
radius : 100.0 mm,
in feed direction :
    sim. length : 300 µm, sampling rate : 1 µm,
    number of feed marks : 8,
in cutting direction :
    sim. length : 500 µm, sampling rate : 8.5 µm,
    sampling steps: 58,
P_V_min : 30.6 nm.

Parameters of the tool:
tool edge radius : 5.0 mm
lamelling frequency : 50 kHz

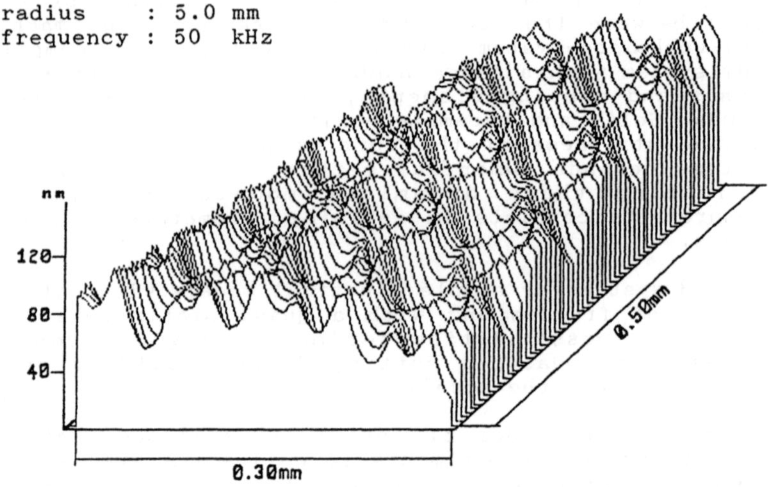

Figure 1 : Surface profile, simulated by the quasi static model

References

/1/ Hill, A,; Tischer, W.: Dynamical Behaviour of an Ultra Precision Turning Machine. 2. International Conference on Ultraprecision in Manufacturing Engineering, Braunschweig, Mai 1991 (in this edition)

/2/ Lo-A-Foe, T.C.G.: Single Point Diamond Turning; Dissertation Eindhoven 1988

Dynamical Behaviour of an Ultra-precision Turning Machine

Achim Hill, Dr. Wolfgang Tischer
Institute of Automation, Berlin

Abstract:

The dynamical behaviour of an ultra-precision turning machine is investigated by using the Finite Element Method. In the centre of interest are processing irregularities, induced by machine oscillations.

The lathe consists of several, partly as welding-steel-constructions produced sections and a natural stone serving as machine base. The latter is connected with the floor by 6 air-springs. Low spring-stiffnesses give rise to an extensively decoupling from ground vibrations. All movable parts of the machine are supported by static air bearings.

Corresponding to the complicated structure of that universal machine, an expensive F.E.M.-model is required with more than 3000 degrees of freedom. With respect to the limitations, given due to the available hardware, the dynamical model consists only of one half of the quasi-symmetric machine. By use of appropriate boundery conditions in the symmetry plane, the symmetrical and anti-symmetrical parts of the eigenfrequency spectrum can be calculated separatly.

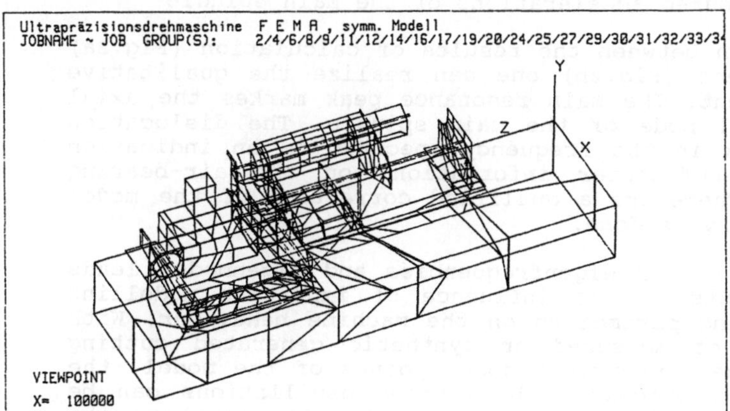

Fig. 1: Simplified Finite-Elemente-Model of the Machine

As a consequence of the complicated structure it was found, that eigenfrequencies, based on elastical deformations of the machine appear already at 30 Hz. This result agrees with measurements, carried out at the machine we are looking upon. However, not all calculated eigenfrequencies are relevant in the working process. This is because the rate of excitation due to the cutting forces is strongly dependent on the shape of the particular vibration modes.

By means of experiments it could be find out, that the F.E.-model is representing the machine in a satisfactory manner. As an example in Fig. 2 is presented the dynamic response of the main spindle due to a pulse excitation in axial direction.

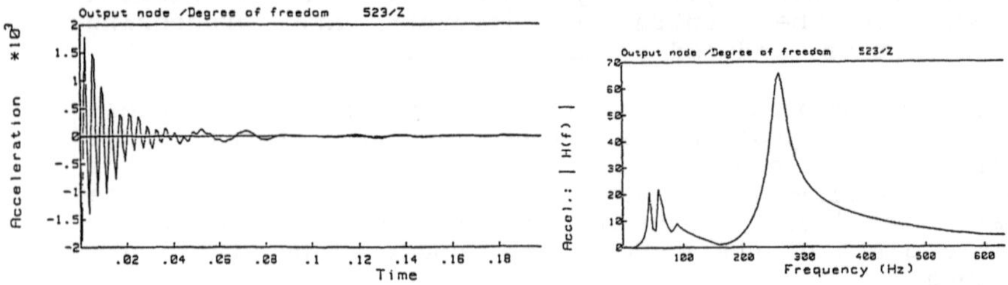

Fig. 2a: Calculated accelerations of the main spindle

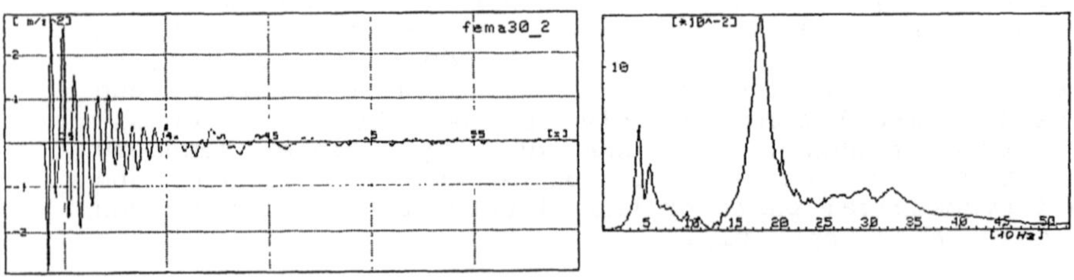

Fig. 2b: Measured accelerations of the main spindle

By comparison between the results of calculation (Fig.2a) and experiment (Fig.2b) one can realize the qualitative good agreement. The main resonance peak markes the axial translational mode of the main spindle. The dislocation of that peek in the frequency spectrum is an indication to wrong manufacturer informations on the air-bearing stiffness. Therefore a qualified correction of the model input data may be done.

The calculation of eigenfrequencies and eigenmodes leads to first hints on the influence of important modelling and processing parameters on the machine behaviour. With introducing of measured or synthetic generated cutting forces at the appropriate nodal points of the model, the modulation of processing by machine oscillations can be simulated. In this way it is possible to attribute the processing irregularities to model properties and finally to machine parameters.

Advanced Wheel Conditioning Technique for Form Wheel Diamond Grinding of Ceramic and Glass Materials

SHORE P, COMLEY P*, JONES D G**,

Cranfield Unit for Precision Engineering
Cranfield Institute of Technology
Bedford, MK43 OAL, England, UK

* Cranfield Precision Engineering Ltd.
** School of Industrial and Manufacturing Science

Summary

Research into the grinding of advanced ceramics and glasses has shown the importance of wheel conditioning. In order to reliably grind these materials in a "low damage" mode, the inherent variability of the grinding process has to be controlled.

Present techniques employed for preparing diamond wheels of a complex shape have both a limited accuracy capability and insufficient control of the grit protrusion level.

This paper briefly discusses the need for a more advanced wheel conditioning system especially when grinding ceramics and glasses and describes work carried out in the development of such a method. This new technique aids the grinding of ceramics and glasses in a "low damage" mode.

Background to Wheel Conditioning

In order that a grinding wheel may perform well, three operations are required:

i) Truing and forming the wheel.
ii) Initial dressing of wheel.
iii) Continuous in-process dressing.

The truing operation ensures that the full circumference of the wheel is all at the same height and thus carries out an equal amount of work. Forming provides the correct wheel profile. Initial dressing is performed to obtain a level of grit protrusion from the wheels bond. This provides the wheel with the required cutting action. Continuous in-process dressing maintains the level of grit protrusion during grinding operations.

Techniques presently used for truing and forming diamond wheels have limited accuracy and repeatability. Dressing techniques have until recently (Nakagawa, ref. 1) not provided a fine control of the grit protrusion.

Advanced Wheel Conditioning Technique

Work on a new technique to give overall control of both wheel form/trueness and topography have been carried out. This new technique utilises the principles of CNC wheel forming complemented by an electro-discharge system (EDS). This EDS reforms the forming wheel as it wears and also ensures the wheel has a highly "sharp" cutting surface. This sharpness permits the diamond grinding wheel to be "cut" to shape as opposed to being "crushed". As a result a high level of form accuracy can be obtained with control over the grinding wheels topography.

This CNC/EDM wheel forming technique (shown schematically in figure 1) can be used on all formable CBN/diamond wheels. The EDS generator used for the reforming of the forming wheel can be employed as a power supply for an electrolytic dressing system. This type of initial and in-process dressing can provide fine control of grit protrusion by preferential erosion of the grinding wheels bond. Bonds are however, required to be electrically conductive. A "low damage" mode of grinding can be offered using these wheel conditioning techniques.

Electro-discharge reforming characteristics:

Wear of electrode - less than 2%
Typical reforming rate-15 μm/min
Power requirement - 10 Amps, 100V

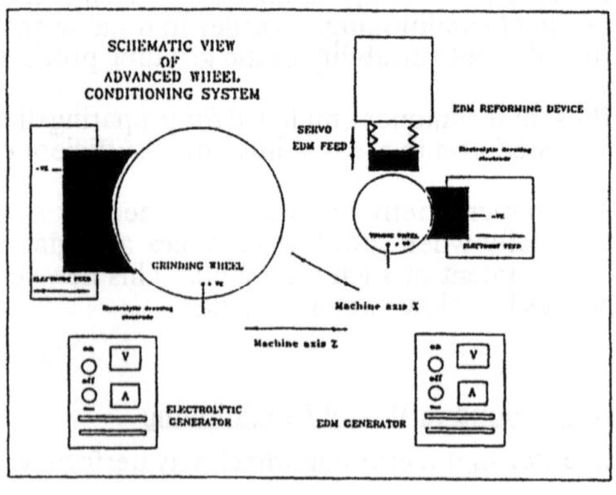

Figure 1

Conclusion

A process for wheel conditioning complexed diamond wheels with a high accuracy capability and control has been developed. This process is based primarily on proven technology which can be readily incorporated into high precision and high production grinding machines.

References

1. Nakagawa T, Ohmori H, Mirror Surface Grinding of Silicon Wafers with electrolytic in-process Dressing Annals of CIRP, Vol 39/1/90, pages 329-332.

Appendix

Some Observations on Tool Wear in Single Point Diamond Turning of Stainless Steel

Chris J. Evans*, Ralph L. Lundin#, Robert S. Polvani*
and Delbert D. Stewart#

* National Institute of Standards and Technology, Precision Engineering Division, Gaithersburg, Md 20899, USA

Los Alamos National Laboratory, Fabrication Development Group, Los Alamos, New Mexico 87545, USA

Single point diamond turning (SPDT) is a cost effective technique for production of certain types of optical, opto-electronic, and mechanical elements. SPDT is often considered a "new" technology, even though single crystal natural diamonds have been used by engravers for millennia and as lathe tools for centuries[1]. The technology which, today, is commonly referred to as single point diamond turning (SPDT) has roots in parallel developments in Europe and the USA starting in the 1940s.

The ultra-precision machining community has generally accepted the premise that only certain materials are "diamond turnable" implying that this is almost as an intrinsic material property. In practice, diamond turnable materials are those where the tool wear rate is low enough that reasonable areas of specular surface can economically be produced. Naturally, final surface finish is also a strong function of machine characteristics, machining parameters and materials related phenomena.

It is well known that ferrous metals are not "diamond turnable"; tool wear is extremely rapid. It is often convenient to think of wear in terms of an individual mechanism. An analysis of the literature[2] suggests that several mechanisms contribute to tool wear and that the dominant mechanism (if there is one) is a sensitive function of operating conditions. Under some conditions, it appears that tribochemical processes are important in some operating regimes although there is little direct experimental evidence.

It is well known that all chemical reaction rates are strong functions of temperature. Thus we would provide strong evidence that tribochemical processes are dominant if we showed that reducing the operating temperature (without altering any other parameters) lead to substantial reductions in tool wear.

This presentation will describe in detail a "proof of principle" cryogenic machining apparatus using closed flows of liquid nitrogen. This system has been designed, built, and installed on a diamond turning machine at Los Alamos National Laboratory. Tool tip temperature and workpiece surface temperatures less than 150K are obtained.

Machining trials have been performed using 440V stainless steel and 99.998% purity iron. Fig 1 is indicates the surface finish achieved on 35 mm diameter 440V flats. Fig 2 shows that figure accuracy is better than 1 wave (HeNe) peak to valley. Experimental results will be discussed in detail.

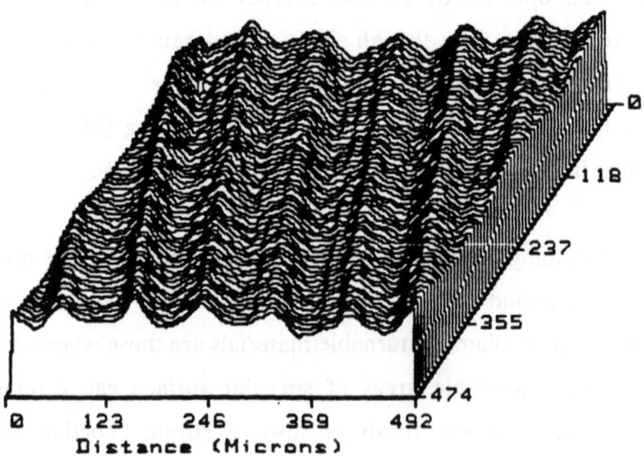

Fig 1. Surface finish of a stainless steel flat (20.9 nm RMS, 16.9 nm Ra)

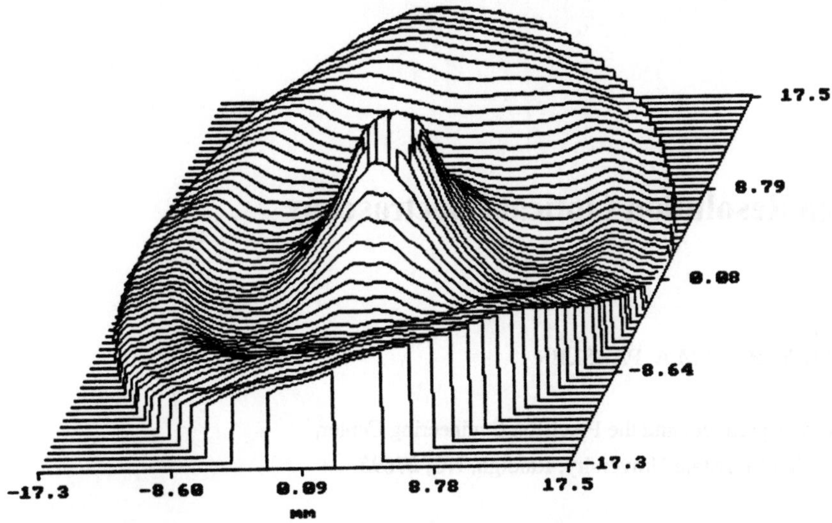

Fig 2. Surface figure of 35 mm flat (0.94 waves P-V, 0.168 waves RMS (He-Ne)

Acknowledgements

Crucible Steel Company provided samples of 440V, a powder metallurgy stainless steel. W. T. Estler (NIST), R. L. Rhorer and P. Hannah (LANL) have provided encouragement, suggestions, help and support.

References

1. Evans, C.; Precision Engineering: an Evolutionary View. Cranfield Press, Cranfield, 1989

2. Evans, C.: Cryogenic Diamond Turning of Stainless Steel, CIRP Annals 1991, in press

High Resolution Optical Spectroscopy

P.J. MOYER and M.A. PAESLER

Physics Department and the Precision Engineering Center,
North Carolina State University, Raleigh, NC, 27695

Summary

In precision engineering applications, optical spectroscopies can be used to determine properties of interest of machined surfaces. These might include composition, crystallographic structure, amorphicity and surface stress. For some applications, an ability to spatially resolve surface features smaller than the wavelength of the probe radiation is required. For these nanostructures, an analytical photon scanning tunneling microscope (APSTM) has been developed. The APSTM can operated in spectroscopic mode to both image the topography and analyze - for example - stress features on a micro-indented, chromium-implanted sapphire surface. Light originating primarily from the evanescent field generated by an internally reflected beam (with some contribution from scattering by roughness features and radiation from fluorescence and luminescence) was coupled into the probe tip. The topographic resolution observed (~50 nm) is appreciably finer than the diffraction limit. Stresses of 3 kbar around micro-indents were measured by monitoring shifts in the photoluminescence peaks. A discussion of the spectroscopic resolution obtainable with this "analytical PSTM" and possible applications are presented.

Discussion

Optical spectroscopic techniques have long been employed to investigate sample properties in precision engineering applications. The wealth of information that can be obtained from spectral analysis of light transmitted through, reflected from, or scattered off of a sample can be of assistance in analyzing sample properties. As precision engineering processes result in samples of smaller dimensions, however, the limit of spatial resolution of these spectroscopies has presented a problem. Conventional optical spectroscopies are limited in their ability to resolve spatial features to distances on the order of the wavelength of the probe radiation. This barrier is referred to as the diffraction limit. With new technologies that have recently been developed, non-conventional optical microscopes have been employed to investigate samples with new lower limits on spatial resolution. In the following, use of one of these newly developed techniques is demonstrated.

The scanning tunneling microscope (STM)[1] has proven to be a powerful analytical tool for electronic microscopy and spectroscopy. It has spawned a family of similar instruments which exploit optical, thermal, and mechanical interactions to image small structures. Among the optical devices included in this family is the photon scanning tunneling microscope (PSTM)[2]. The PSTM images by sensing an exponential dependence of tip-to-sample separation, like its electronic counterpart. The STM inherits its exponential dependence from the quantum mechanical probability of electrons tunneling through a potential barrier. The barrier is created by placing a potential bias between tip and sample. For this reason, only conductors and selected semiconductors are valid imaging samples. The exponential dependence of the PSTM is a result of total internal reflection of a laser at the image surface from the inside of the material. For this reason, it is only possible to image samples that are transparent to the particular energy of the electromagnetic radiation being used. The PSTM has been shown to be one of the emerging optical microscopies that allow optical imaging of surfaces with resolution beyond that predicted by the diffraction limit. By adding a spectroscopic mode to the PSTM, one can perform such analyses as photoluminescence and Raman spectroscopy.[3] Experiments thus far have indicated that lateral spatial resolution of this instrument is finer than that of the best diffraction-limited conventional spectroscopic microscopes available.[4]

The experiments we have performed have all been done with visible laser sources. Current experiments make use of an IR laser so that materials such as Si and Ge are transparent to the radiation. Of interest to us in this respect is the analysis of diamond-turned Si and Ge samples. The Precision Engineering Center at North Carolina State University is concerned with the fabrication and analysis of quality optical surfaces for the use of IR optics. Among the analytical tools in our laboratory is a micro-Raman system which is used to perform depth-profiling of machined samples to determine the sub-surface damage due to machining. Also of importance is the residual stress of the sample across an individual tool cut or tool shoulder. The high lateral resolution of the spectroscopic mode of the PSTM will allow for the probing of these features so that useful information regarding the machining process may be extracted.

To demonstrate the operation of the analytical PSTM, a cylindrical rod of sapphire was cut with a diamond saw in a plane along its axis so as to have two parallel flat faces. The end of the cylinder was cut along a plane with normal tilted 45° with respect to the cylinder's axis. The schematic drawing of the experimental set-up in figure 1 shows a cross-sectional view of the sample.

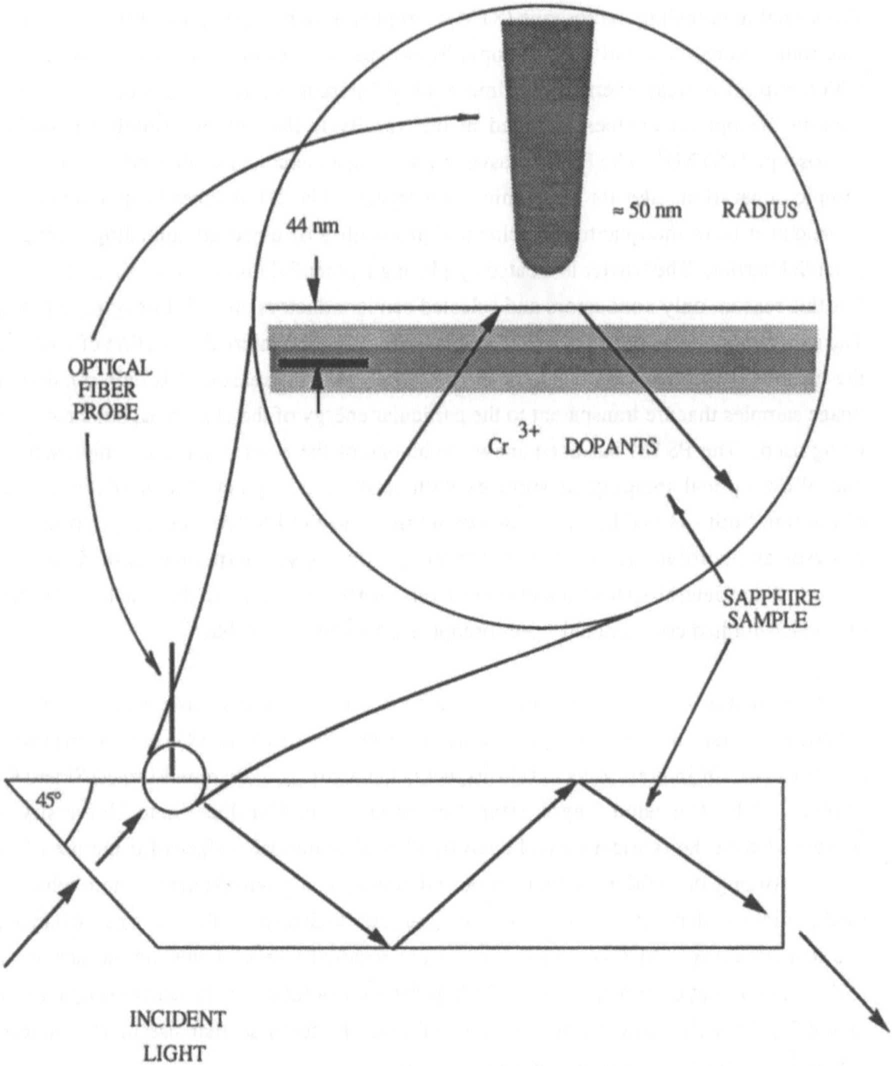

Fig. 1. Schematic diagram of the sapphire sample and microscope probe. The chemically etched optical fiber which serves as the microscope probe tip has a radius of curvature on the order of 100 nm. A layer of Cr^{3+} ions implanted into the sample surface result in a thin layer of ruby beneath the sapphire surface. The peak in concentration of the Cr^{3+} ions occurs at approximately 44 nm beneath the surface.

All flat faces of the sapphire were polished with a series of diamond grits, the final grit size being 250 nm. The longest face of the sample was then implanted with Cr^{3+} ions so as to achieve a profile of ruby (Cr^{3+} doped sapphire) with peak dopant concentration approximately 44 nm beneath the sample surface and a profile width on the order of 25 nm. The sample was annealed to relieve any implantation-induced damage. To create surface features that might be used to determine the lateral resolution of the analytical PSTM, an array of micro-indents was then placed in the implanted surface of the sample. Each indent creates a stress field that decreases with radial distance from the indent. In ruby, a photoluminescence doublet occurs at 692.7 and 694.2 nm and shifts linearly to lower energy with increased hydrostatic pressure - i.e. compressive stress - by 0.0365 nm/kbar. This luminescence doublet provides the spectroscopic signal measured.

The sapphire was put on the sample stage of an existing PSTM[1] which uses as a sensing probe an optical fiber chemically etched to a tip radius on the order of 100 nm. Using the 442 nm line of a He-Cd laser, light was made to suffer total internal reflection at the polished and implanted surface in the region of the indent array. The orientation of the sample surface, the probe tip, and the ruby implantation are shown in the inset of figure 1.

A beam-splitting Y was put in the sensing fiber of the PSTM so as to allow the light sensed by the tip to proceed down two identical channels. One of these channels was returned to the z-axis (vertical) control of the PSTM while light emerging from the fiber comprising the second channel of the Y was dispersed in a monochromator (dispersion of 2.5 nm/mm and resolution 0.5 nm) and detected with a cooled photomultiplier. The output of the photomultiplier was plotted versus the wavelength setting on the monochromator to obtain photoluminescence spectra.

The PSTM image of a region of the sample surface near a micro-indent is shown in figure 2. This grey-scale contour map represents a region 10 µm x 10 µm, and features a micro-indent (in the lower left corner) which extends to a depth of 250 nm below the average plane of the sample surface. Features on the polished sample surface (towards the upper right corner) appear to be several tens of nanometers across, thus indicating topographical imaging resolution to be less than 100 nm.

Three photoluminescence spectra were taken at positions marked by the numbers 1, 2, and 3 in figure 2. One might expect the spectrum taken at position 3 (spectrum 3) to appear as the well known ruby photoluminescence doublet s indicative of a relatively low stress region. The distance from the center of the indent is approximately 8.6 μm. As the probe tip approaches the micro-indent in taking spectra 2 and 1, the rather large compressive stresses known to exist in the region around the micro-indent manifest themselves in a shifted and broadened photoluminescence spectrum. The luminescence peaks taken at positions 1 and 3 are shown in figure 3.

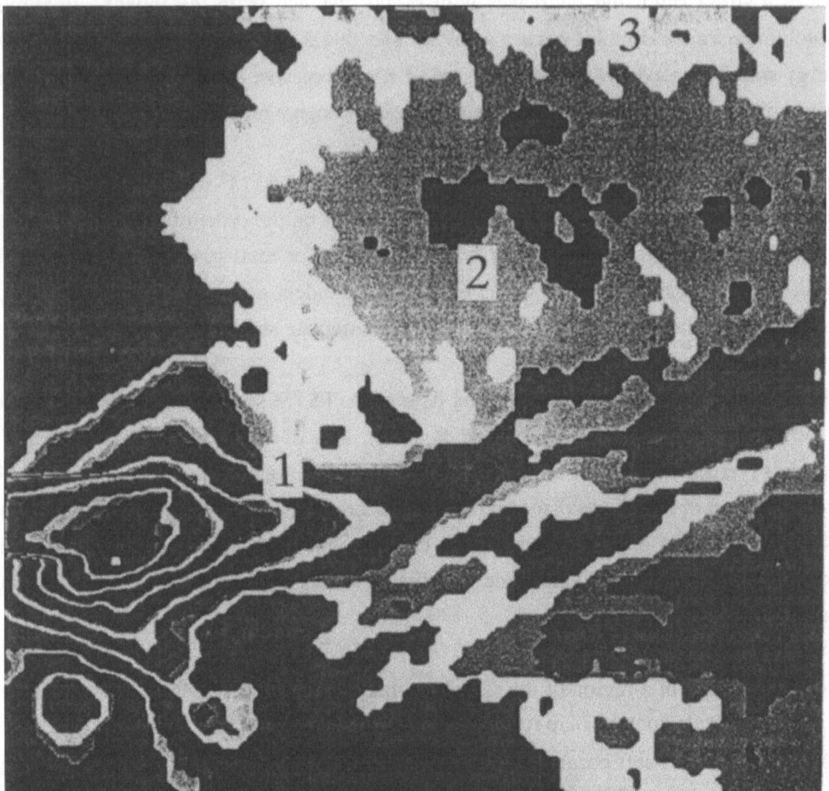

Fig. 2. Grey-scale contour map of the surface of the indented sapphire sample taken with the photon scanning tunneling microscope, or PSTM. This micrograph represents a region on the sample surface 10 μm x 10 μm. Each grey scale cycle represent a 50 nm height change. The indent extends to a depth of 250 nm below the sapphire surface plane. The wavelength of the radiation which was used to make this microgram was 442 nm, and resolution of surface features is ~ 50 nm.

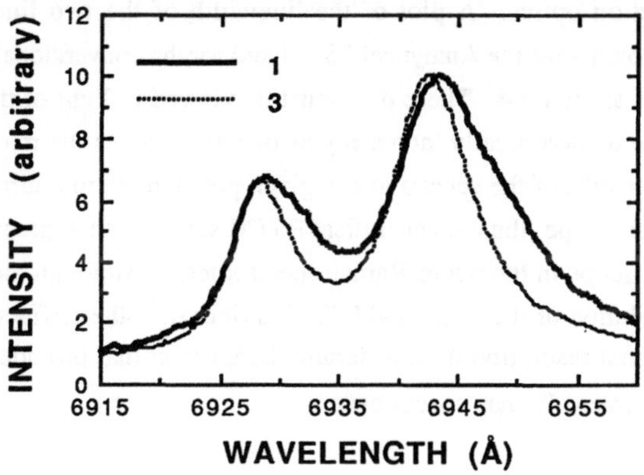

Fig. 3. Ruby photoluminescence spectra taken with the analytical photon scanning tunneling microscope, or analytical PSTM, at positions 1 and 3 as indicated in figure 2 are shown. The spectrum taken on the surface plane (spectrum 3) indicates a region of low stress with relatively unshifted and unbroadened luminescence lines. As the probe tip approaches the indent, the spectra shift slightly and broaden significantly thus indicating increased compressive stress in the region of the indent.

The shift of the peaks to lower energy indicates the higher region of compressive stress, as expected closer to the indent. Perhaps more important than the shift in the peak position is the significant broadening toward compressive stress that occurs as the probe is moved closer to the indent (i.e. for spectrum 1.)

For these micro-indented sapphire samples the lateral resolution of the PSTM is clearly finer than the diffraction limit, but it is difficult to accurately determine the resolution of the analytical PSTM. It is straightforward, however, to show that the lateral resolution of the analytical PSTM is significantly finer than that obtainable in conventional microscopic spectrometers. Using the 40x, 0.6 numerical aperture microscope objective on one such instrument,[4] a spot size on the order of 1 μm is obtainable. Conventional photoluminescence spectra taken with this instrument at positions 1, 2 and 3 on the sample shown in figure 2 result in linewidths considerably larger than those obtained in the analytical PSTM. Care was taken to duplicate the instrument resolution of the spectrometer used in the analytical PSTM experiment so that the measured linewidth was determined solely by the collection optics. A plot of the linewidth of the two lines of the ruby photoluminescence doublet for the Analytical PSTM and for the conventional micro-Raman spectrometer appears as figure 4. These data must be viewed in light of the fact that the stress is known to be monotonically increasing as one approaches the micro-indent. As seen in figure 4, the widths of the spectra at any given position are measurably greater for the conventional microscope, thus the analytical PSTM samples an appreciably narrower region than the microscope in the micro-Raman spectrometer. Although these data do not confirm that the resolution of the analytical PSTM is finer than the diffraction limit, they clearly demonstrate that resolution is considerably better than that provided by the 1 μm spot obtainable in the micro-Raman spectrometer.

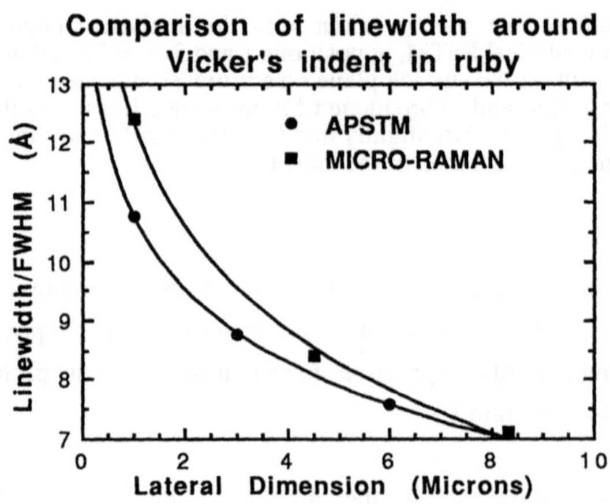

Fig. 4. The linewidth of the photoluminescence spectra taken at positions 1, 2 and 3 shown in figure 2 for the analytical PSTM (solid dots) and for a conventional micro-Raman spectrometer (open dots). The conventional microscope on the micro-Raman spectrometer was focused to a 1 μm probe spot. The decreased linewidth of the analytical PSTM spectra taken at the same positions on the sample indicates considerable improvement in spatial resolution for the analytical PSTM over the conventional microscope.

Conclusions

Successful use of the analytical photon scanning tunneling microscope - or APSTM - has been shown through measurement of surface stresses in the vicinity of microindents on the surface of sapphire. This demonstrates that optical spectroscopic studies of nanostructures are now possible. Thus the APSTM has proven itself to be a powerful tool that may be used to study nanostructures of interest in precision engineering applications

References

1. G. Binnig and H. Rohrer, Physica **127B**, 37 (1984).
2. R.C. Reddick, R.J. Warmack, and T.L. Ferrell, Phys. Rev. B **39**, 767 (1989).
3. M.A. Paesler, P.J. Moyer, C.J. Jahncke, C.E. Johnson, R.C. Reddick, R.J. Warmack, and T.L. Ferrell, Phys. Rev. B **42**, 6750 (1990).
4. The micro-Raman spectrometer used was the Spex Industries Micramate which is discussed in, for example, R.J. Thibeau, C.W. Goldfarb, and A.Z. Heidersbach, J. Electrochem. Soc. **127**, 37 (1980).

Consequence of Orientation on the Single Crystal Diamond Cutting Tool

K.UEGAMI and K.TAMAMURA

Faculty of Engineering,
Osaka City University,
Sugimoto, Sumiyosi-ku,
Osaka, 558 Japan

Summary

The single crystal diamond tools having various orientation are prepared in which the rake face is lapped to identify with the {100} and {110} plane respectively, and the direction of cutting edge is coincided with the direction coresponding to the max. or minimum values of the coefficient of friction. Cutting aluminum alloy by these diamond tools are examined from the view points of the cutting force and tool wear.

1. Introduction

For the ultra precision machining, the single crystal diamond cutting tool is an indispensable material. The results of cutting effectively depend on frictional properties of the tool face and flank. However, the single crystal diamond is known as an anisotropic substance on the lapping and the friction characteristics as to the crystallographic orientaion. Many works on the lapping and the friction of the diamond have been studing.[1)~4)] But there are only few studies on the effect of orientation of single crystal diamond as cutting tool.[5) 6)] From these facts, it might be considered that when using single crystal diamond tools (hereafter referred to as the diamond tool), the cutting behavior would largely depend on the crystallographic orientation of the rake face and the direction in which the chip is flowed out on the diamond tool. Furthermore, for the single crystal, if a rake face and a tool edge are determined, the crystallographic direction of the flank should be determined inevitably according to the tool geometry. Since the crystallographic direction of the flank might have effects on the tool wear and the finished surface quality, an

attention should be paid to setting the tool for the precision machining.
Mercator method which is used to the map is proposed to designate the crystallographic orientation of the diamond tool. Since the reference sphere of a single crystal diamond is divided 48 equivalent units spherical triangle zone that enclosed by the points of (100), (110) and (111) plane on this reference sphere, the crystallographic orientation of a rake face and a flank of each diamond tool can be indicated by the points in this unit zone as the stereographic method. Moreover, the direction of tool edge is expressed by the arrow head putting on the above points. In the same way, this Mercator method is capable to apply for the study of direction on the friction and the lapping of the tool.
The cutting forces and tool wear are measured by the four kind of tool having a combination of {100} or {110} rake face and <100> or <110> tool edge direction respectively, when machining the aluminum hyper-eutectic Si alloy.

2. Using the Mercator method for representing crystallographic direction of diamond tool

When using the stereographic projection to represent the crystallographic direction of the diamond tool a large number of drawings are required, because of the station point (light source) is not the center of reference sphere. Integration into a single drawing would be much more convenient to compare the various characteristics relating with the orientation of the tool rake and flank. For these reasons, the authors propose using the Mercator method as following.

Figure 1 (a) shows a spherical projection whose center is coinsided with a diamond crystal's one. The equator, the circles of longitude of 0°, and 90° are shown as solid lines with setting the north pole on 001. These lines should be the projection of the [001], [010], and [100] crystal zones. The representative plane of the crystal are projected on 1/8 surface area of the sphere. The same projection is applicable to remaining 7/8 of the sphere.

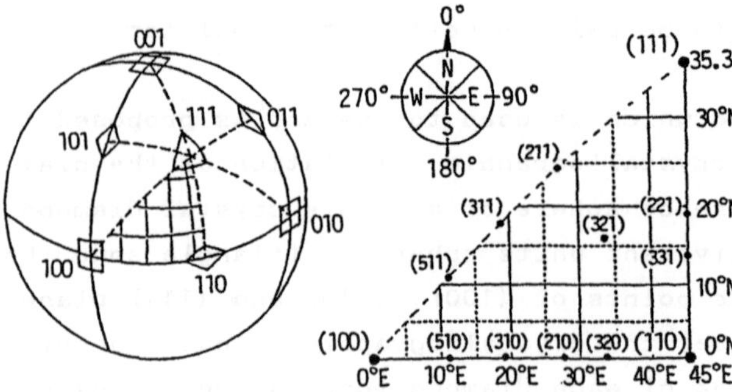

(a) Reference Sphere (b) Mercator map

Fig.1. Reference sphere and Mercator map of diamond crystal

The six spherical triangles which are crystallographically equivarent as shown in Fig.1 (a). Therefore, crystallographical anisotropy can be discussed with the spherical triangle enclosed by (100), (110) and (111) planes only, which corresponds to 1/48 of the whole spherical surface. In the Mercator method, a spherical triangle may be expressed with describing the latitudes and longitudes as shown in Fig.1 (b) in which the position of crystal planes having a main index number are given with each point. So the specific crystal plane to be subjected to lapping, friction and to use in cutting may be expressed as a point specified in Fig.1 (b).
The other hand, it is well known that in friction or lapping, the diamond reveals a different character with the working direction, even if, it is on the same crystallographic plane. Then, these characteristics may be easily compared with on the spherical unit triangle where arrow heads attached to the points representing the diamond planes shown in Fig.1 (b), agree with the frictional direction, the lapping direction, and the inverse direction to the tool edge (equal to flow-out direction of the chip on the rake face, hereafter referred to as "edge inverse direction"). So that, characteristics depending on the orientation of the diamond tool are able to study using only 1/48 of the spherical projection.

fig.2

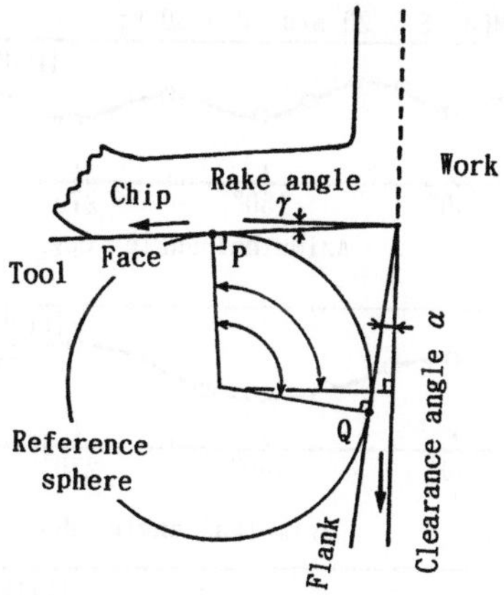

Fig.2. Angular relationship of the tool to the reference sphere

In Fig.2, if a point P representing the rake face is moved along the circle for an angle $(\pi/2 + \gamma + \alpha)$, then it reachs the point Q representing the flank. In this manner, an orientation of the flank is decided easely.
Some of the points obtained in this manner are shown in Fig.4, however it is noted that the direction of a vector do not show the movement of the diamond tool but the chip or the finished surface.
The edge inverse direction on the flank (nearly equal to the frictional direction by finished surface) may be acquired by reversing the direction of the arrow head on the rake face. Fig. 4 (a) and (b) show the tool's rake face and flank determined by the methods stated above. The crystallographic orientation parallel to the finished surface may also be determined in the same way.

3. Experimental results and discussion

3.1. Friction

The friction characteristics of the single crystal diamond are differ according to the plane and its direction. The diamond has

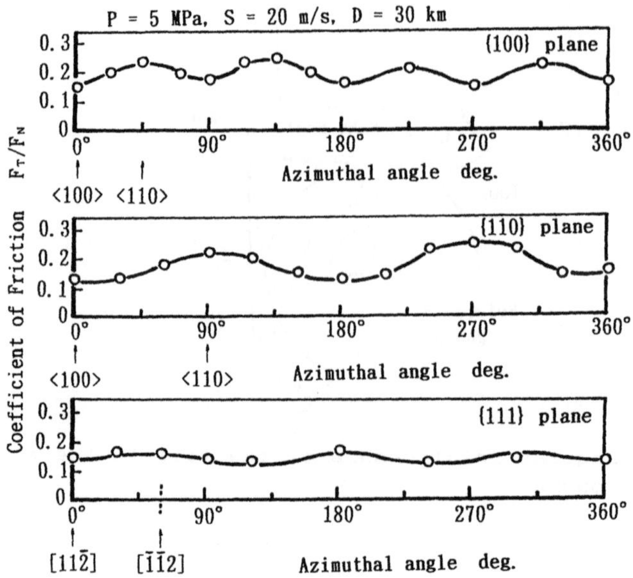

Fig.3. Frictional characteristics of representative {100}, {110}, and {111} planes as a function of azimuthal direction with cast iron wheel
Frictional conditions
Friction speed: 20 m/s, Load: 5 MPa, Cast iron wheel, Dry friction in air,

three typical plane of the crystallographic orientation and its highly anisotropic characteristics. Fig.3 shows the friction properties of the {100}, {110}, and {111} plane of the diamond as a function of azimuthal angle on the frictional direction. These variations of the properties reflect to the fourfold symmetry of the {100} plane, twofold of the {110} plane and threefold of the {111} plane on the arrangement of carbon atom in diamond. It is well known that the {110} plane is a plane lapping easely and the {111} plane is very hard lapping one, therefore, the {110} plane or {100} plane including their neighborhood was applied to the rake face of the diamond tool. The coefficient of friction on the {100} plane becomes a minimum in the <100> direction, and a maximum value in the <110> direction. The same trends are observed in the {110} plane. From the results of friction tests, therfore, the crystallographic orientation of the desired tool for the cutting was identified as the representative plane as {100} and {110} plane, as rake face, of which the edge inverse direction is set in the direction that the coefficient of friction show a max. and a minimum value.

3. 2. Cutting

3. 2. 1. Cutting experimental method

The diamond tools prepared are a round corner tool having the rake angle 0°, the clearance angle 5° and the corner radious of 1 mm. The tools were made by lapping from natural dodecahedron gemstones. The tools were investigated for exact setting on crystallographic orientation by the X-ray back Laue diffraction method.

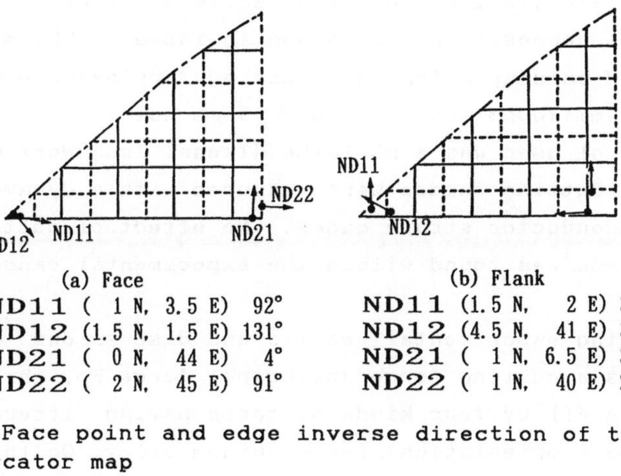

(a) Face (b) Flank

(a) Face			(b) Flank		
ND11	(1 N, 3.5 E)	92°	ND11	(1.5 N, 2 E)	359°
ND12	(1.5 N, 1.5 E)	131°	ND12	(4.5 N, 41 E)	358°
ND21	(0 N, 44 E)	4°	ND21	(1 N, 6.5 E)	302°
ND22	(2 N, 45 E)	91°	ND22	(1 N, 40 E)	268°

Fig.4. Face point and edge inverse direction of tools tested on the Mercator map

Fig.4 shows the crystallographic orientations of the desired tools. The followings are an example of description for the diamond tool used in Fig.4, in which the items marked ahead with ★ mean the description as in [ ].

```
★Representation of diamond gemstone of Natural Dodecahedron
↓ ★Representation of desired tool rake face
↓  ↓  [1: {100} plane, 2: {110} plane]
↓  ↓ ★Representation of edge inverse direction
↓  ↓ ↓  [1: <100> direction for tool with a rake face of
↓  ↓ ↓      {100} or {110} plane,
↓  ↓ ↓   2: <110> direction for tool with a rake face of
↓  ↓ ↓      {100} or {110} plane,]
↓  ↓ ↓
N D 1 1
```

Table 1 Chemical compositions of work

(Al hyper-eutectic Si Alloy, Y413BE-T6) wt.%

Si	Fe	Cu	Mn	Mg	Cr	Zn	Ti	Al
16.6	0.21	4.24	0.01	0.54	0.01	0.01	0.01	Re

Materials cut are aluminum hyper-eutectic Si alloy (Y413BE-T6), their chemical compositions are shown in Table 1. The materials were prepeared as shaped for face turning, apploximately in a shape of ϕ 75 mm×100mm length with ϕ 15 mm hole.
The machine tool used was a NC lathe (Ikegai Iron Work Co. Ltd., ANC25 type). Cutting forces were measured with a dynamometer based on semiconductor strain gages. No effect of cutting speed on cutting force was found within the experimental range.

3. 2. 2. Cutting experimental results and discussions
Figure 5 shows a cutting force (principal force Fp, thrust force Ft, feed force Ff) by four kinds of tools having different crystallographic orientations for Aluminum alloy. On the bigining of cutting, the crystallographic orientation of tools have not affect on the cutting forces, but as increasing of cutting distance an effect of crystallographic orientations appears clearly on the difference of the forces. The values by the tools (ND11 and ND21) with a cutting edge direction <100> are lower than one by the tools (ND12 and ND22) with <110>,

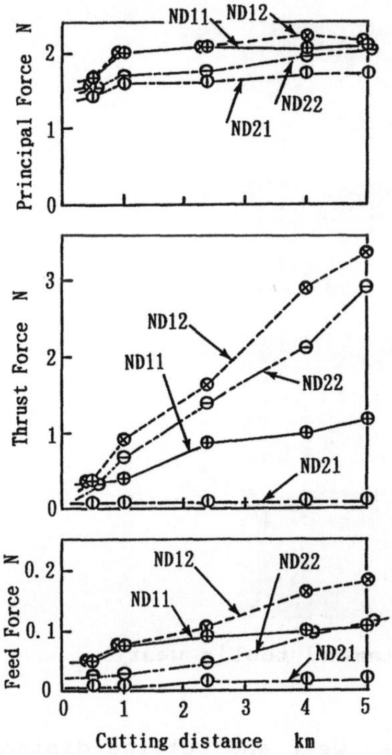

Cutting conditions
Material cut:
Al hyper-eutectic Si alloy,
Cutting speed:
from 7.5 to 3.3 m/s,
Depth of cut:
40 μ m,
Feed:
40 μ m/rev, Dry cutting
Dry cutting.

Fig.5. Effects of cutting distance on cutting force for the diamond tool oriented to various crystallographic direction

ND21 showing the lowest especially. For tools ND12 and ND22, the thrust compornent is larger than the principal compornent of the force. The value of average coefficient of friction between tools and chip agree thoroughly with the ratio of tangent stress to thrust stress which is a kind of the coefficient of friction, when diamond is lapping with a cast iron circular plate[4]. Figure 6 shows photographs of worn tool edge viewed from the flank. Differences of the figure well correspond to the value of cutting force respectively excepting the case of tool ND21. The wear of tool ND21 is relatively large as compared it's value of thrust and feed compornents forces.

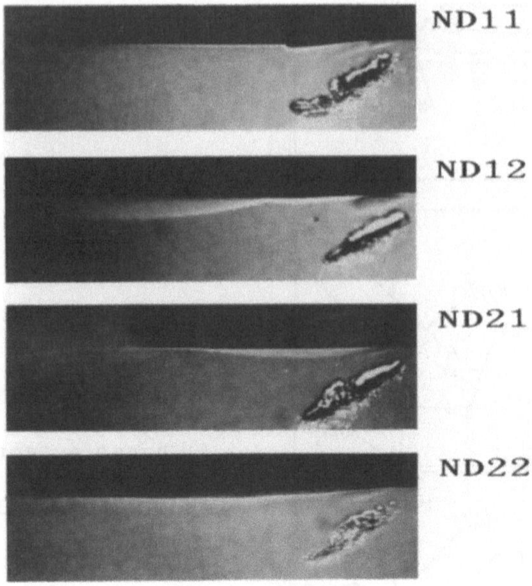

Fig.6. Photographs of diamond tool's wear

Figure 7 shows a relation between wear and cutting distance for various kinds of tools. The wears are measured by the maximum width of flank wear. The tool ND11 followed by ND21, ND22, and ND12 in this order.

The results of Fig.5, 6, and 7 show the facts that cutting forces are mainly effected by crystallographic orientation of the rake face and edge inverse direction on which the friction characteristics are affected by the wear. An effects of the friction of flank become more slight as the wear increasing, however the influence of change of orientation along a course from the rake face to the flank do not known yet. As it is evident from Fig.5, 6, and 7, the tool ND21 is the best in terms of cutting force and the better in wear. The wear of ND21 tool occures relatively large, nevertheless thire thrust and feed forces show the lowest value than that of other tools as increasing the progress of wear. Moreover, the tool is removed with high rate and easy lapping for this orientation.

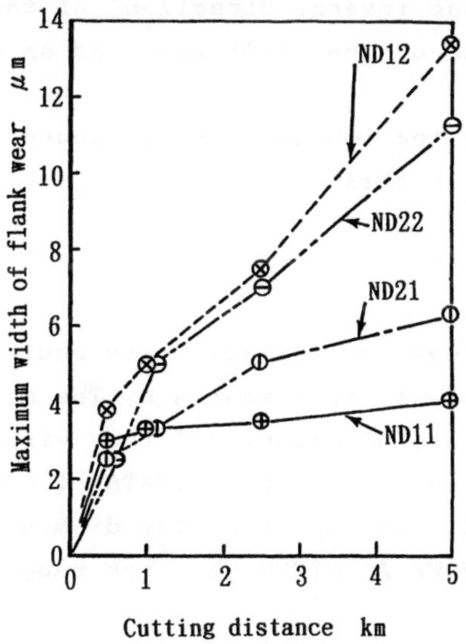

Fig.7. Effects of cutting distance on the tool wear: maximum width of flank wear

4.Conclusions

In cutting Aluminum alloy with the single crystal diamond tools having various crystallographic orientations, the effects on the cutting force and tool wear were observed. The conclusions obtained, are as follow.

(1) The authors proposed that the Mercator method is more convenient for representing crystallographic orientations such as the tool faces of the single crystal diamond tool accompanied by friction and wear rather than the stereo graphic method.

(2) The coefficient of friction on the {100} plane becomes minimum value in the <100> direction, and maximum value in the <110> direction. The same trends are obserbed in the {110} plane.

(3) The cutting forces are affected by the crystallographic orientation. They may be reduced to the lowest by using a

tool having "edge inverse direction" of the <100> direction on the rake face of the {110} plane as an example of tool ND21.

(4) Tool ND11 shows the lowest wear, followed by the tool ND21, ND22, and ND12 in this order.

Acknowlegement

The authors acknowlege the financial aid recieved through The Grant-in-Aid for Scientific Research by The Ministry of Education, and Culture of Japan. And they wish also to express their appreciation to K. OKA and H.OBATA of OSAKA DIAMOND INDUSTRIAL Co., Ltd. for supply of the diamond investigated and, to E. ISOYAMA of SHOWA ALUMINUM Co., for supply of the work material tested.

References

1. F.P.Bowden & C.A.Brookes: Frictional Anisotropy in Nonmetallic Crystals, Proc. Roy. Soc., A 298, (1966) 244.
2. J.E.Field: The Properties of Diamond, Academic Press, London, (1979).
3. Y.Tamamoto & D.Tabor: The Frictional Anisotropy of Diamond, Proc. Roy. Soc., A 373, (1981) 405.
4. K.Uegami, K.K.Jang, K.Tamamura, and H.Tomimori: Study on Lapping of Diamond Tool (1st Report) -Fundamental Lapping Tests by Cast Iron Lap-, J.of JSPE, 54-11, (1988) 2138.
5. Y.Maeda, M.Masuda, T.Nishiguchi, M.Sawa, R.Itoo: A Study on Diamond Turning of Al-Mg Alloy (1st Report) -Influence of Tool Setting Angle on Machined Surface Roughness-, J.of JSPE, 54-10, (1988) 1993.
6. C.J.Wong: Fracture and Wear of Diamond Cutting Tool, J. Engg. Mech. Tech., 103, Oct. (1981) 341.

Ceramics in Finish Machining

S.GOWRI[1] K.NARAYANASAMY[2] R.KRISHNAMURTHY[3]

Summary

Ceramic cutting tools have been used for machining for more than three decades. But ceramics of today are much better than those of past years. This paper includes the role of ceramics in finish machining in a precision, VDF lathe.

The development of alumina ceramic, Silicon nitride based ceramic,Titanium carbide mixed ceramic tools are discussed. The surface finish study reveals the superiority of black ceramic (TiC) mixed tools over other types in machining both Cr-alloyed cast iron and Mo-Cr alloyed cast iron.

The surface texture profile recorded revealed that white ceramic tools had maintained their nose intact over a longer duration of machining as indicated by the near total replication of the cutting nose on the surface texture, while the black ceramic tools did not behave so. But the cutting nose of black ceramic tools had undergone deformation resulting in generation of a land, leaving the surface burnished.

The study on the wear behaviour of different ceramic tools in finish machining revealed the superiority of hot pressed black ceramic tools in resisting flank wear. It was found that black ceramic tools showed more resistance to flank wear than white ceramics. The Silicon nitride based tool suffered an extensive nose deformation.

Introduction

Recent developments in newer class of material have posed considerable machinability problems; this has led to a fast development in cutting tool materials.The developments include improvement micro-structure,reinforcement of micro structure by dispersed particle or whiskered materials. Powder compacted tool materials (HSS), cold compacted black ceramics (Al_2O_3+30% TiC) whiskered ceramics (SiC in Si_3N_4) and transformation toughened Zirconia (TTZ) and Zirconia toughened Alumina (ZIA) are examples

1. Research Scholar, Precision Engineering Section, Department of Mechanical Engineering, Indian Institute of Technology, Madras, INDIA.
2. Assistant Professor, Precision Engg. Section, Dept. of Mechanical Engg., IIT, Madras, INDIA.
3. Professor, Manufacturing Engg. Section, Dept. of Mechanical Engg., IIT, Madras, INDIA.

of such developments. The development of silicon nitride based ceramic tools with their higher level of toughness and resistance to both mechanical and thermal shocks as well as abrasive wear has resulted in easier machining of heat treated Inconel and other harder work materials. Typical performance characteristics of different cutting tools have been presented in Fig.1. It is seen that the newer class of tool materials have been used for higher speeds of machining indicating their superior hot hardness characteristics.

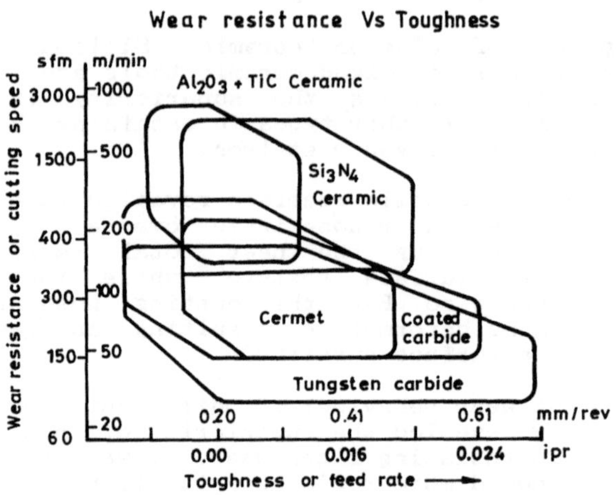

Fig.1 Performance characteristics of cutting tools

Fig.2 is a typical presentation of hot hardness characteristics of tool materials.

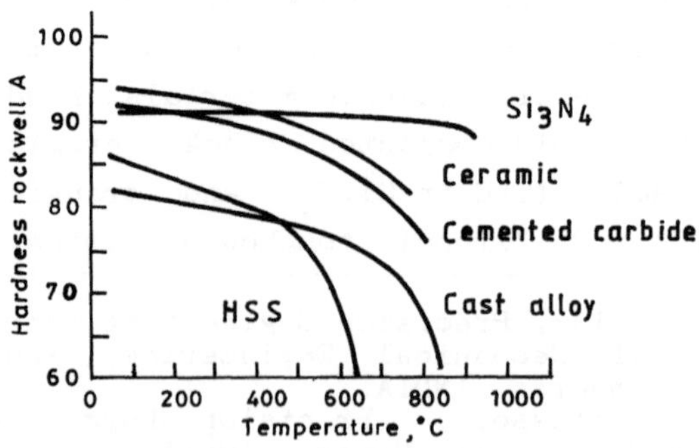

Fig. 2 Hot hardness characteristics of tool materials

This paper presents certain observations in the performance of ceramic tools in finish machining of alloyed cast Iron.

Experimental

Finish machining trials on Mo-Cr alloyed centrifugally cast, cast Iron were carried out in a high precision VDF lathe. The machinability conditions are presented in table I. The performance of different ceramic tools have been evaluated based on the quality of the surface produced, certain observations made on worn out tools using a scanning electron microscope are also presented in this paper.

Results and Discussion

In turning, an ideal surface texture would be one with uniform surface quality with a well defined helical lay. Since the feed motion is a generatrix motion in a lathe, it is but natural that for maintaining a good surface quality in turning an ideal tool is one which would be able to replicate its nose on the machine work surface. Typical illustration of surface production in turning is presented in Fig.3. It is seen that with the occurrence of grooving wear on the secondary cutting edge (resulting in a discontinuity of the cutting edge) the machined surface would be left with "sponzipfil" projection causing surface roughness.

During the machining trials it was found that all the ceramic tools suffered from secondary grooving wear. Occurrence of secondary grooving wear has been attributed to oxidation of the cutting edge behind the nose portion. While machining, the cutting tool experiences higher order temperatures over three specific zones such as depth of cut line zone on the primary flank side, cutting nose zone and the zone of material on the secondary side behind the nose. The cold compact white Al_2O_3 tools experienced a chemical reaction under conditioning. It is known that [Krishnamurthy et al(1,2,3)] Al_2O_3 form a spinel component of FeO - Al_2O_3 during machining of ferrous materials with higher temperatures and in the presence of oxygen. The occurrence of such a spinel has been confirmed by Auger electron spectrum and cathode luminiscence technique [4]. Once spinel is formed, it spells off due to the rubbing action of the finish machined work surface on the secondary flank. This would result in secondary grooving as shown Fig.4

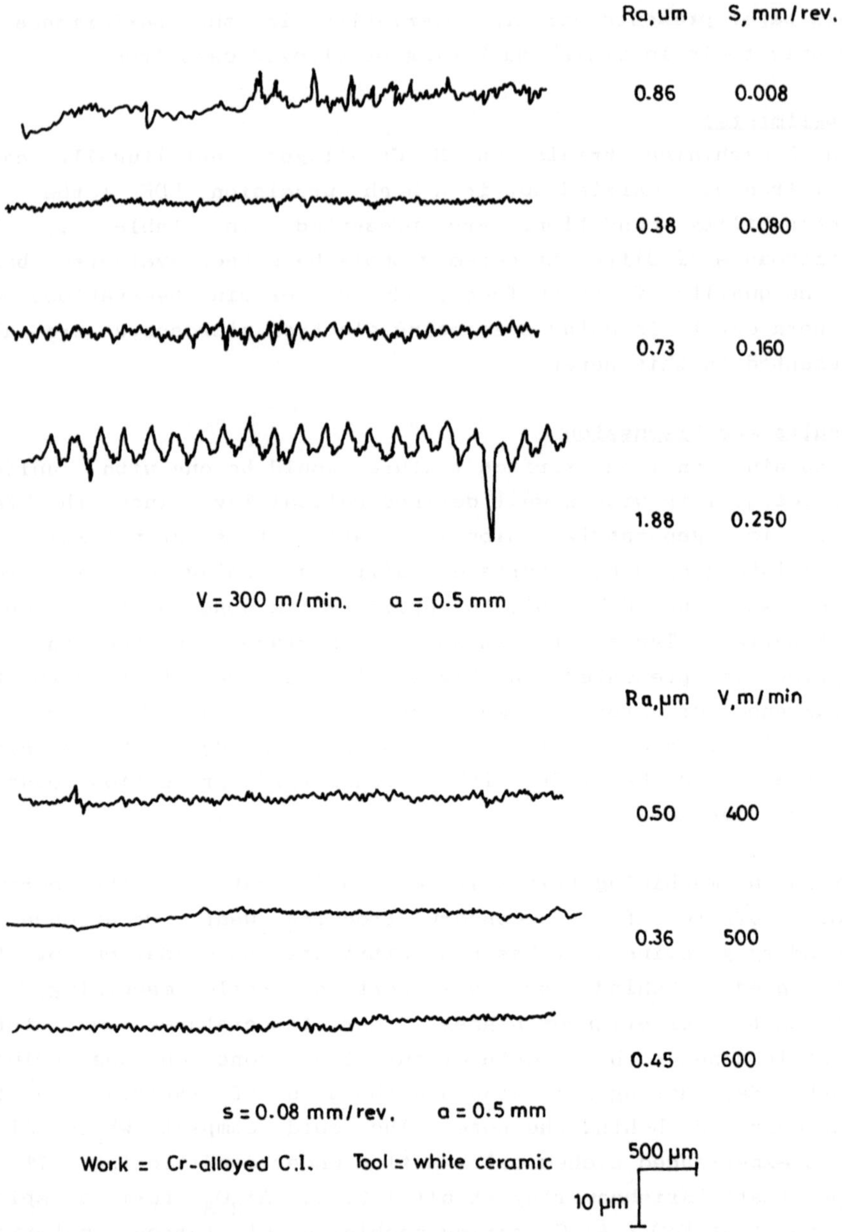

Fig. 3 Influence of feed and speed on surface texture

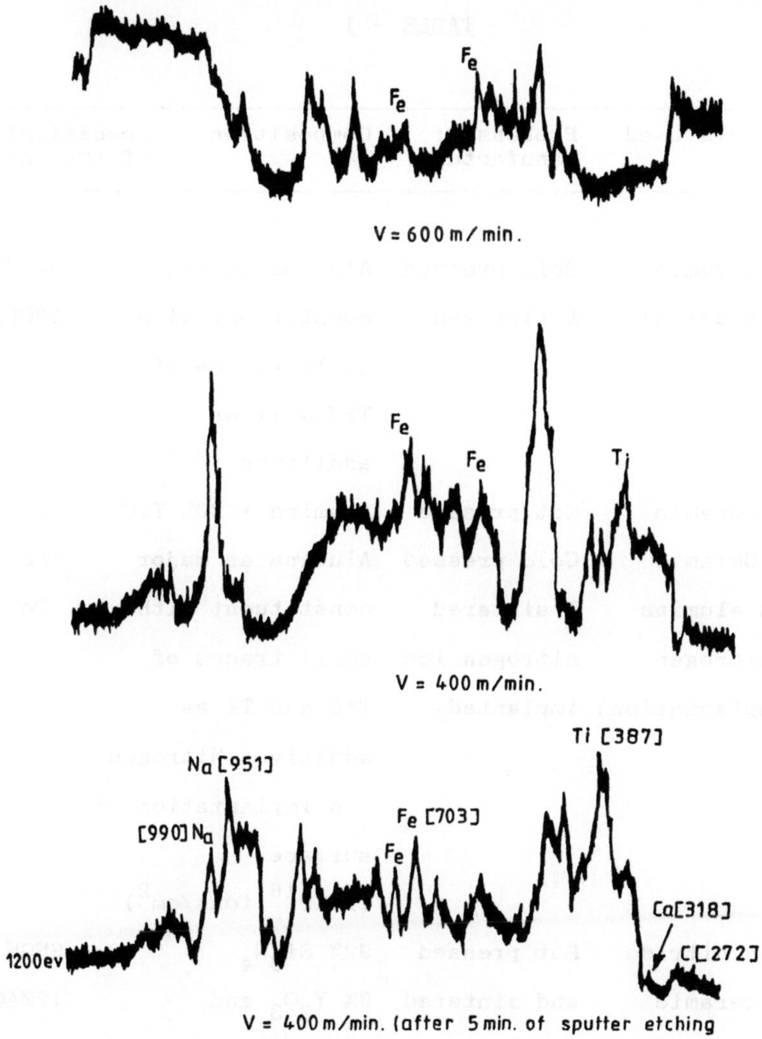

Fig. 4. Typical Auger Electron Spectrum - the pick up of Fe in compound form could be seen — formation of spinel $Fe_2O_3 - Al_2O_3$ confirmed.

TABLE - I

Sl.No.	Tools Used	Process of Manufacture	Composition	Specification of the insert
1.	White ceramic (plain alumina)	Cold pressed & sintered	Alumina as major constituent with small traces of TiC & Ti as additives.	SNGN 120812
2.	Black ceramic	Hot pressed	Alumina + 30% TiC	-Do-
3.	White Ceramic (plain alumina with nitrogen ion implantation)	Cold pressed & sintered nitrogen ion implanted	Alumina as major constituent with small traces of TiC and Ti as additives. Nitrogen ion implantation of surface ($1*10^{16}$ ions/cm^2)	-Do- -Do-
4.	Silicon nitride based ceramic	Hot pressed and sintered	92% Si_3N_4 6% Y_2O_3 and 2% Al_2O_3	SNGN 120408

Tool geometry used :

γ	α	λ	$\varkappa$	ε	
-6°	5°	-5°	75°	90°	r_ε

r_ε = 0.8 mm for Si_3N_4 ceramics

r_ε = 1.2 mm for other ceramics

All the tools have been provided with an edge chamber of about 0.2 mm*20°.

In the case of hot pressed black ceramic tools the secondary grooving wear was minimum. The surface texture obtained was partially burnished. The absence of a clear cut, uniform surface texture with a helical lay indicates that black ceramic is not able to replicate its nose; Owing to this it may be rather difficult to predict the surface quality obtainable with black ceramic during machining of alloyed cast iron. Figs.3 & 5 are the typical illustrations of surface texture obtained with both white and black ceramic tools.

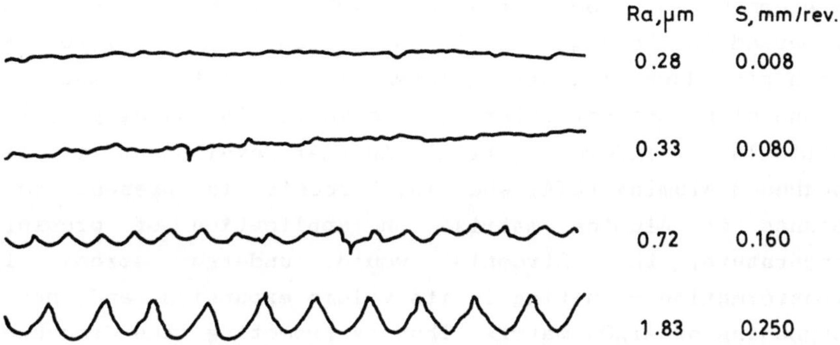

V = 300m/min a = 0.5mm
Work = M -Cr alloyed C.I Tool = Black Ceramic

Typical influence of feed on surface texture profile

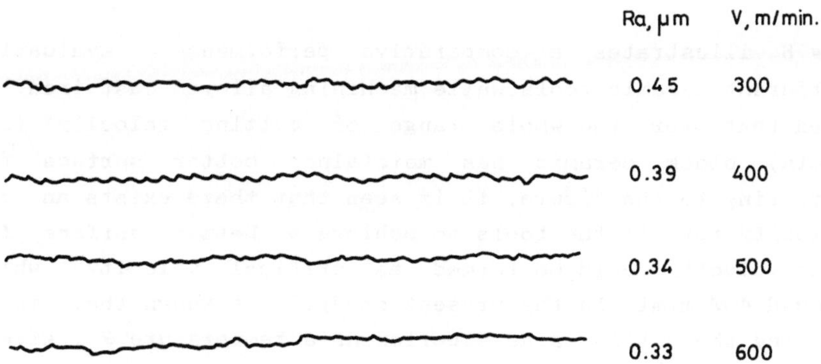

s = 0.08 mm/rev. a = 0.5 mm
Work = Mo-Cr alloyed CI Tool = Block ceramic

Typical influence of cutting speed on surface texture profile

Fig. 5

The occurrence of burnished texture would be attributed to the deformation of the cutting nose of black ceramic tool, resulting in increased nose radius and possible burnishing of the work surface. It is known that the process of chip curling on the rake face could be analogized to an indendation process. When Al_2O_3 ceramic is subjected to indentation process, it experiences macro pile up of dislocation around the indendation which would result in cracking of the tool matrix. For arresting dislocation pile up or diffusing the cracking tendency, suitable modification in the micro structure have to be carried out. Development of black ceramic which contains particles of TiC dispersed in the Al_2O_3 matrix is a case of a micro-structure which pins down dislocation movements and diffuses the cracking by the dispersed particles. Such a matrix is usually termed as particulate reinforced matrix. Another development is Zirconia toughened Alumina (ZTA) wherein, Zirconia is present in small amounts in Alumina matrix; on application of pressure and temperature, the Zirconia would undergo stress induced transformation resulting in its volume expansion and consequent toughening of Al_2O_3 matrix. Thus by promoting the toughness of matrix material or by introducing dispersed particles in the matrix, as in the black ceramics, the cracking and chipping of cutting zone (as in the case of white Al_2O_3) could be minimized. Such ceramic tools exhibited deformation of the nose.

Fig.6 illustrates a comparative performance evaluation of different ceramic tools while machining alloyed cast iron. It is seen that over the whole range of cutting velocity (300-600 m/min) black ceramic has maintained better surface finish. Referring to the figure, it is seen that there exists an optimum velocity for all the tools to achieve a better surface finish; this velocity could be termed as critical velocity, which is around 400 m/min in the present study. It is known that in metal cutting, the cutting zone experiences a temperature θ_c, given as

$$\theta_c = K_1(V)^b \quad \text{where}$$

θ_c - temperature of machining, V - cutting velocity, and b - material dependent exponent.

Depending upon the work- tool material combination, the cutting tool would experience varied thermal influences.

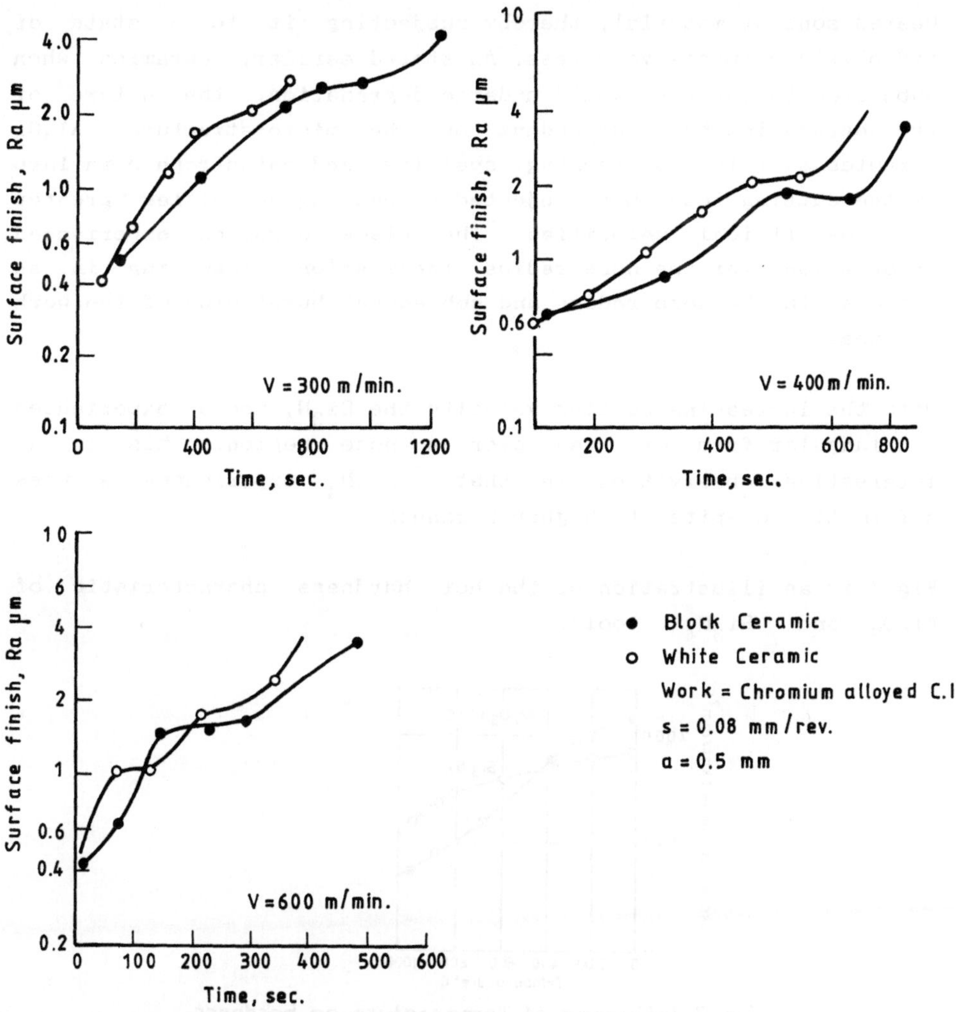

Fig. 6 Typical performance characteristics of ceramic tools in finish-machining Cr-alloyed C.I.

Published temperature on high speed machining of ferrous material with ceramic tools have reported θ_c around 1000-1100°C. It is known that ceramics generally possesses lower thermal conductivity and diffusivity. This would result concentration of heat over a small zone of ceramic material during finish machining. (depth of cut=0.5 mm and feed=0.08 mm/rev). The localized heating could result in volumetric expansion of the heated zone of material, thereby subjecting it to a state of hydrolytic compressive stress. As stated earlier, ceramics when subjected to stresses would undergo degradation, the nature of the degradation being dependent on the micro-structure. Al_2O_3 ceramics experienced cracking, spelling, and catastrophic failure of the cutting nose when subjected to cutting velocities greater than the critical velocities. The black ceramics experienced deformation over the nose radius (depression) resulting in an increase in the nose radius and subsequent burnishing of the work surface.

With the increasing cutting velocity the Si_3N_4 tools experienced a triangular form of wear over a nose region. This is an interesting observation in that Si_3N_4 experiences a nose deformation despite its higher hardness.

Fig.7 is an illustration of the hot hardness characteristic of Al_2O_3 and Si_3N_4 tools.

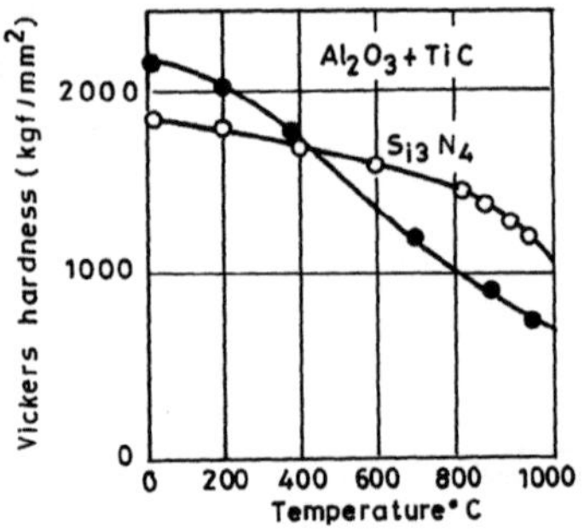

Fig. 7 Influence of Temperature on hardness

Referring to the figure with increasing temperature the hardness of Al_2O_3 drops down considerably, while Si_3N_4 retains its hardness over wide temperature range. The higher hardness of the Si_3N_4 at the elevated temperature, however has not resulted in the chipping of tool as with Al_2O_3 tools.

It is known that Si_3N_4 when heated to a higher order temperature in the presence of Fe, Mn & Oxygen undergoes a chemical reaction, resulting in degradation of the matrix and consequent plastic deformation of the nose material. This has resulted in a deterioration of the surface quality when machined with Si_3N_4. It is seen that all the ceramic tools tried out for finished machining of Mo-Cr alloyed, centrifugally cast, cast-iron exhibited a significant role of cutting velocity for surface quality; in other words production and maintenance of surface quality during finish turning could be a thermal process.

In 1986, the Greenleaf corporation, USA, has introduced ceramic reinforced ceramics- a composite material [5]. These ceramic composites have almost twice the fracture toughness of traditional hot pressed ceramic and a significant increase in resistance to thermal shocks. The reinforcing elements are known as "whiskers" and they are grown under carefully controlled conditions and due to their extremely high purity and lack of grain boundaries, they approach the theoretical maximum strength obtainable for the matrix. The whiskers are very fine and of size Ø.Ø2 microns in diameter and 1 micron in length. These are hexagonal in shape and are dispersed randomly into a matrix of ceramics.

Suitable modification of the micro structure,with the introduction of the SiC whiskers in Si_3N_4, would improve the thermal conductivity of the Si_3N_4 resulting in minimizing the chemical reaction and thereby improving the surface quality of turned work surface.

In case of SiC whiskers in Alumina matrix, the high thermal conductivity of the whiskers conducts heat away from the cutting edge, thereby reducing thermal gradient and resulting stress. It has been found that thermal conductivity is improved up to 4Ø percent and thermal gradient is reduced. Hence the thermal shock resistance is improved.

Hodgson.T., and Trendler.P.H.H., [6] have machined hardened tool steel with CBN inserts and confirmed that tool steels having hardness values exceeding 5Ø R_c, can be machined with CBN polycrystalline cutting tools; the cutting forces involved being well within the capabilities of medium to heavy duty lathes having the required rigidity.

The most economical metal removal rate when machining tool steel

(10 cm^3/min) makes the use of CBN cutting tools an attractive alternative to many cylindrical grinding operations, particularly when large stepped diameters have to be machined. In addition, surface finishes better than 0.6 microns (R_a) can be obtained when turning hardened steel components by means of CBN tools. Cutting speeds between 25 and 100 m/min were generally found to be satisfactory for the machining of hardened tool steel by CBN tools.

Conclusion

From the present investigation, the following major conclusions are drawn.

1. Cold compact, white Al_2O_3 tools can maintain controlled texture during turning, owing to its enhanced replicated capability; however degradation due to thermal and mechanical stresses make them unreliable.
2. Hot compacted, particulate reinforced black-ceramics have maintained better productivity; however, the surface texture was irregular, with localized burnished texture and disordered lay.
3. For medium range of cutting velocities, Si_3N_4 tools can be successfully used for machining alloyed cast iron.

References

1. Krishnamurhy, R.; Sivasankaran, V.: Grey iron machining with ceramics. Int. Jl. of Prod. Res. Vol.25 no.2 (1987) 207-219.

2. Krishnamurthy, R.; Sivasankaran, V.: Performance characteristics of ceramics in machining cast irons. Proc. of 12th AIMTDR conf., IIT, Delhi (1986) 273-276.

3. Krishnamurthy, R.; Sivasankaran, V.: Performance of ceramic tools in finish machining of alloyed cast iron. Proc. of 6th Int. Conf. on Prod. Engg., Osaka, Japan, Nov. (1987).

4. Chadwick, G.A.; Smith, D.A. (eds.) Grain boundary structures and properties. London: Academic Press 1979.

5. Bergstorm, R.P.: Getting a grip on cutting tools. Manufacturing Engg. Jan. (1986) 55-59.

6. Hodgson, T.; Trendler, P.H.H.: Turning hardened steels with

Separation and Correction of the Dynamic Error of the Grating Type Single-flank Mesh Gear Tester

Bai Yongxin and Liu Chao

Shaanxi Institute of Mechanical Engineering, P.R.China

In this paper, by applying the gear integrated error measuring technology and the error separating technique, the gear error signal, exported from a single-flank mesh integrated error measuring instrument, is imported into a microcomputer through an interface, the dynamic errors of the gear grating system, of the master worm grating system and of the master worm have been separated by data processing. The trangential composite error and the profile error of the gear being inspected can be corrected by subtracting the dynamic error which is stored in computer. The mathematical principle of error separation is described. Three examples of separation and correction are presented.

The dynamic errors of the gear grating system and of the master worm grating system are determined by shifting method. The mesh error of the master worm is determined as follows. The profile error of two gears having different number of teeth Z_1 and Z_2 which are ground on two different gear grinders, is measured on a single-flank mesh integrated error measuring instrument using a specific multi-start master worm. Suppose the measuring results of the profile error obtained for gear Z_1 and gear Z_2 can be expressed with time series $S_1(t)$ and $S_2(t)$. Firstly, calculate the cross correlation function $R_{12}(\tau)$ between $S_1(t)$ and $S_2(t)$. Secondly, calculate the cross correlation function $R_{2(12)}(\tau)$ between $S_2(t)$ and $R_{12}(t)$, in which containing all information about the mesh error of the master worm, including the amplitude values and the phase angle for various harmonics, by which the mesh error of the master worm can be obtained.

Author Index

MIX
Papier aus verantwortungsvollen Quellen
Paper from responsible sources
FSC® C105338

If you have any concerns about our products,
you can contact us on
ProductSafety@springernature.com

In case Publisher is established outside the EU,
the EU authorized representative is:
Springer Nature Customer Service Center GmbH
Europaplatz 3, 69115 Heidelberg, Germany

Printed by Libri Plureos GmbH
in Hamburg, Germany